APOPTOSIS

PEZCOLLER FOUNDATION SYMPOSIA

Recent volumes published by Plenum Press:

Volume 4 • CELL ADHESION MOLECULES: Cellular Recognition Mechanisms
Edited by Martin E. Hemler and Enrico Mihich

A Continuation Order Plan is available for this series. A continuation order will bring delivery of each new volume immediately upon publication. Volumes are billed only upon actual shipment. For further information please contact the publisher.

APOPTOSIS

Edited by

Enrico Mihich
Roswell Park Cancer Institute
Buffalo, New York

and

Robert T. Schimke
Stanford University
Stanford, California

PLENUM PRESS • NEW YORK AND LONDON

Library of Congress Cataloging-in-Publication Data

Apoptosis / edited by Enrico Mihich and Robert T. Schimke.
p. cm. -- (Pezcoller Foundation symposia ; 5)
Includes bibliographical references and index.
ISBN 0-306-44733-9
1. Apoptosis--Congresses. I. Mihich, Enrico. II. Schimke, Robert T. III. Series.
QH671.A65 1994
574.87'65--dc20 94-12196
CIP

Proceedings of the Fifth Pezcoller Symposium on Apoptosis, held June 9–11, 1993, in Trento, Italy

ISBN 0-306-44733-9

A Division of Plenum Publishing Corporation
233 Spring Street, New York, N.Y. 10013

Printed in the United States of America

Professor Alessio Pezcoller
(Photo by Dino Panato, Trento, Italy)

THE PEZCOLLER FOUNDATION

The Pezcoller Foundation was created in 1979 by Professor Alessio Pezcoller (1896-1993) who was the chief surgeon of the S. Chiara Hospital in Trento from 1937 to 1966 and who gave a substantial portion of his estate to support its activities; the Foundation also benefits from the cooperation of the Saving Bank Cassa di Risparmio di Trento e Rovereto.

The main goal of this non-profit foundation is to provide and recognize scientific progress on life-threatening diseases, currently focusing on cancer. Towards this goal, the Pezcoller Foundation awards, every two years, the Pezcoller Prize, recognizing highly meritorious contributions to medical research; it also sponsors a series of annual symposia promoting interactions among scientists working at the cutting edge of basic oncological sciences.

The award selection process is managed by the European School of Oncology in Milan, Italy, with the aid of an international committee of experts chaired by Professor U. Veronesi.

The symposia are held in the Trentino Region of Northern Italy and their scientific focus is selected by Enrico Mihich with the collaboration of an international Standing Symposia Committee. A Program Committee determines the content of each symposium.

The first symposium focused on *Drug Resistance: Mechanisms and Reversal* (E. Mihich, Chairman, 1989); the second on *The Therapeutic Implications of the Molecular Biology of Breast Cancer* (M.E. Lippman and E. Mihich, Co-Chairmen, 1990); the third on *Tumor Suppressor Genes* (D.M. Livingston and E. Mihich, Co-Chairmen, 1991), and the fourth on *Cell Adhesion Molecules: Cellular Recognition Mechanisms* (M.E. Hemler and E. Mihich, Co-Chairmen, 1992). The sixth symposium (1994) will be focused on *Normal and Malignant Hematopoiesis: New Advances* (E. Mihich and D. Metcalf, Co-Chairmen).

PREFACE

The fifth Annual Pezcoller Symposium entitled, Apoptosis, was held in Trento, Italy, June 9-11, 1993 and was focused on the specific phenomena leading to Programmed Cell Death (PCD) or Apoptosis, and the mechanisms involved. With presentations at the cutting edge of progress and stimulating discussions, this Symposium addressed the genetics and molecular mechanisms determining PCD and the role of this suicidal process in cancer and the immune system. The functions of p53, c myc and bcl 2 in affecting apoptosis in different cell types and the role of ions and intracellular pH changes and that of intranuclear endonucleases are given particular emphasis as are the effects of anticancer agents, hormone imbalances and growth factors.

The role of p53, a tumor suppressor gene, in inducing PCD is discussed in detail as pertinent to hematological and non-hematological tumors. The requirement of p53 for the induction of apoptosis by ionizing radiation or adenovirus oncoproteins is outlined. Decision points during the cell cycle affecting the cascade of events leading to PCD are discussed as is their role as "switches" under the control of c-myc and bcl-2 proteins or the influence of cycle specific drugs. The concurrent requirement of multiple signals in determining apoptosis is emphasized. The examples of the role of PCD in the regulation of hematopoiesis, and in the generation of antigen-specific immune repertoire are illustrated. The role of calcium ions and ion signaling in determining apoptosis, in part through activation of endonuclease and increased DNA susceptibility to these enzymes consequent to changes in chromatin conformation, are outlined. In conclusion, the multiplicity of mechanisms determining PCD was extensively discussed. The role of genes like wt p53 or c-myc and of accessory factors in directing cells through "switch" decision points towards proliferation or PCD is demonstrated in various cellular systems. The inhibition of PCD, by growth factors or by products of genes like bcl-2 and the induction of PCD by hormones and by growth factors withdrawal, or by DNA damaging and anticancer drugs, are described. The function of PCD in hematopoiesis and in negative and positive selection during the development of the immune repertoire is discussed as is the cascade of events resulting in apoptotic death in each of the models examined and the factors regulating them.

We wish to thank the participants in the Symposium for their substantial contributions and their participation in the spirited discussions which followed. We would also like to thank Drs. J. Michael Bishop, Arnold Levine and David M. Livingston for their essential input as members of the Program Committee, and Ms. A. Toscani for her invaluable assistance. The aid of the Bank Cassa di Risparmio di Trento and Rovereto, and the Municipal, Provincial and Regional Administrations in supporting this Symposium through the Pezcoller Foundation are also acknowledged with deep appreciation.

Finally, we wish to thank the staff of Plenum Publishing Corporation for their efficient cooperation in the production of these Proceedings.

Enrico Mihich
Robert T. Schimke

CONTENTS

GENETIC CONTROL OF PROGRAMMED CELL DEATH IN THE NEMATODE CAENORHABDITIS ELEGANS

H. Robert Horvitz

Howard Hughes Medical Institute
Department of Biology
Massachusetts Institute of Technology
Cambridge, MA 02139 USA

INTRODUCTION

Naturally-occurring or "programmed" cell death appears to be a universal aspect of animal development (e.g., Ref. 1). For example, massive cell death occurs during the development of the mammalian fetus, particularly in the fetal brain; in areas of the developing vertebrate nervous system as many as 85% of the developing neurons die. Similarly, about 95% of developing thymocytes die without ever leaving the thymus. Why such cell deaths occur and how they are regulated are fundamental problems in developmental biology.

In addition, it seems plausible that the mechanisms responsible for programmed cell death might be relevant to many aspects of human disease (e.g., Ref. 2). For example, the clinical manifestations of a variety of disorders are consequences of the occurrence of abnormal cell deaths; Alzheimer's Disease, Huntington's Disease, Amyotrophic Lateral Sclerosis, stroke and myocardial infarction all fall into this category. It is conceivable that the abnormal cell deaths seen in these disorders are "programmed" deaths expressed by the wrong cells.

Furthermore, a decrease in cell death would lead to an increase in cell number. In this way, it is possible that alterations in temporal or spatial patterns of cell death could result in cancerous growth. As noted below, recent findings have established that such a perturbation in the control of programmed cell death indeed seems to be involved in certain human cancers.

For these reasons, we have become interested in the problem of programmed cell death.

RATIONALE

To analyze the molecular genetic control of programmed cell death, we have focused on the nematode *Caenorhabditis elegans*. Although we are studying a worm, a major rationale for our work is the conviction that genes and molecules that control cell death in *C. elegans* will prove to be highly similar to genes and molecules that control cell death in humans. This belief is based upon recent findings that have greatly reinforced the idea that not only general mechanisms but also specific molecules are strikingly conserved from yeast to worms to insects to mammals. One example of such conservation of function derives from studies of a *C. elegans* pathway for signal transduction. The entire pathway consists of molecules closely related to mammalian oncogenes, including a growth factor-like protein, a growth factor receptor, an adapter molecule with three *src*-like domains, a

Apoptosis, Edited by E. Mihich and R.T. Schimke
Plenum Press, New York, 1994

Ras protein, and a Raf protein.[3-7] In short, we believe that worm genes that control cell death will prove to be highly similar to genes with corresponding functions in humans.

C. ELEGANS PROGRAMMED CELL DEATH RESEMBLES APOPTOSIS

The complete cell lineage of *C. elegans* is known.[8-10] In addition to the 959 cells that are generated and found in the adult, another 131 cells are generated only to undergo programmed cell death. Which cells live and which cells die are invariant features of *C. elegans* development. These programmed cell deaths are generally rapid, with cells dying within an hour of the time they are formed and often prior to any morphological signs of differentiation. Ultrastructural studies have revealed that programmed cell death in *C. elegans* shares features with the process of apoptotic cell death seen in vertebrates.[11-12] For example, both types of deaths involve contraction of the cytoplasm, chromatin condensation and removal of the dead cell by phagocytosis.

GENETIC PATHWAY FOR PROGRAMMED CELL DEATH

To understand the mechanisms responsible for programmed cell death, we have been identifying and characterizing genes that function in this process (for review, see Ref. 1). We have defined an 11-gene pathway for programmed cell death (Figure 1). Three genes (*ced-3*, *ced-4* and *ced-9*)(*ced*, cell death abnormal) are involved in killing cells, seven genes (*ced-1*, *ced-2*, *ced-5*, *ced-6*, *ced-7*, *ced-8*, *ced-10*) control the phagocytosis of dying cells by neighboring cells, and one gene (*nuc-1*) (*nuc*, nuclease abnormal) functions in the degradation of the engulfed cell. In addition, three other genes (*ces-1*, *ces-2*, *egl-1*)(*ces*, cell death specification; *egl*, egg-laying abnormal) are involved in specifying which cells should and which cells should not activate the general 11-gene program.

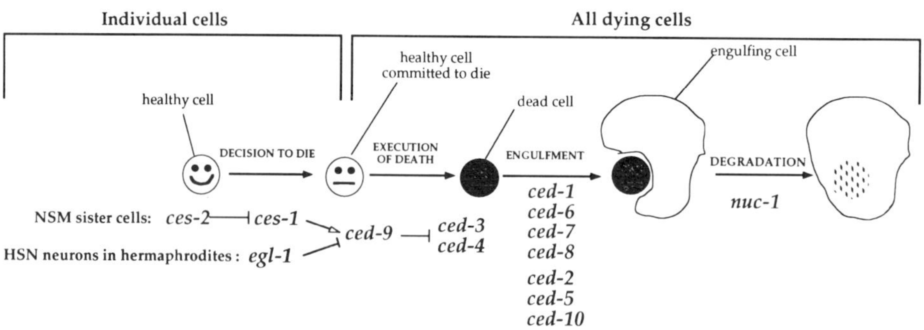

Figure 1. Genetic pathway for programmed cell death in *C. elegans*.

CED-3 AND CED-4 ARE NEEDED FOR CELLS TO UNDERGO PROGRAMMED CELL DEATH

The actions of the genes *ced-3* and *ced-4* are essential for cells to die by programmed cell death.[13] Mutations that inactivate either gene prevent all 131 programmed cell deaths, and the resulting cell-deathless animal has 131 extra surviving cells that in general differentiate into different and recognizable cell types. Remarkably, such an animal matures and is fertile, although it displays a variety of subtle behavioral and developmental defects. Genetic mosaic experiments have indicated that both *ced-3* and *ced-4* act either within cells that die or within their close relatives, suggesting that these genes encode "suicide" functions that cells must express to undergo programmed cell death.[14] Molecular studies of the *ced-4* gene indicate that the 63 kD Ced-4 protein is not similar to any gene in the available databases.[15] Two regions of the inferred Ced-4 amino acid sequence show some similarity to the calcium-binding EF-hand motif, suggesting that this protein and hence programmed cell death in *C. elegans* might be regulated by calcium. However, no biochemical data are available that directly demonstrate that the Ced-4 protein binds calcium. The molecular study of the *ced-3* gene is currently in progress (J. Yuan, S. Shaham, S. Ledoux, H. Ellis and H. R. Horvitz, unpublished results).

CED-9 PROTECTS CELLS FROM PROGRAMMED CELL DEATH

The gene *ced-9* also acts in controlling the life-death decision of cells.[16] However, whereas the activities of *ced-3* and *ced-4* cause cells to die, the activity of *ced-9* causes cells to survive. Two opposite classes of *ced-9* mutations have been identified. Mutations that increase *ced-9* activity cause all 131 cells that should die instead to live, just as do mutations that eliminate *ced-3* or *ced-4* activity; by contrast, mutations that reduce *ced-9* activity cause cells that should live instead to die by programmed cell death (as defined both morphologically and by the requirement of these deaths for the activities of *ced-3* and *ced-4*). These observations have indicated that *ced-9* acts as a developmental switch gene (Figure 2): when *ced-9* is active cells live, and when *ced-9* is inactive cells undergo programmed cell death.

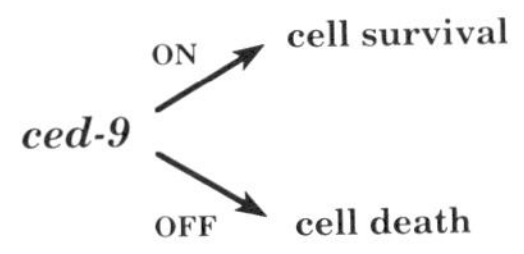

Figure 2. The gene *ced-9* acts as a genetic switch to control the decision between cell survival and programmed cell death.

Interestingly, it seems that most and maybe all cells that survive during the course of *C. elegans* development are protected against dying from programmed cell death by the action of the *ced-9* gene. The DNA sequence of the *ced-9* gene has revealed that this gene encodes a protein that is similar to the protein product of the mammalian proto-oncogene *bcl-2* (M. Hengartner and H. R. Horvitz, unpublished results). The *bcl-2* gene has been proposed to function in mammals just as we believe *ced-9* functions in *C. elegans*: to protect cells against programmed cell death.[17-19] *ced-9* and *bcl-2* are similar not only structurally but also functionally: studies of transgenic *C. elegans* animals have revealed that human *bcl-2* can protect cells from undergoing programmed cell death in *C. elegans* (Ref. 20; M. Hengartner and H. R. Horvitz, unpublished results) and also that human *bcl-2* can substitute for *ced-9* in mutant worms deficient in *ced-9* activity (M. Hengartner and H. R. Horvitz, unpublished results). These observations indicate that the *C. elegans ced-9* gene and the mammalian *bcl-2* gene act similarly.

SEVEN GENES ARE NEEDED FOR THE ENGULFMENT OF DYING CELLS

Seven genes are needed for the efficient phagocytosis of cell corposes generated by programmed cell death.[12,21] Since in mutants defective in these genes this process of phagocytosis is defective and yet in general cells still undergo programmed cell death, phagocytosis is not responsible for causing cells to die. Thus, the engulfment of dying cells by their neighbors provides a mechanism for ridding the organism of cell corpses rather than for generating those corpses. Mutations in any one of the seven genes involved in this process of phagocytosis prevent the engulfment of only a few dying cells; by contrast, double mutants carrying mutations in certain pairs of genes are defective in the engulfment of many more dying cells. This major defect occurs only if one mutation is in *ced-1*, *ced-6*, *ced-7* or *ced-8* and the other mutation is in *ced-2*, *ced-5* or *ced-10*. A plausible interpretation of these genetic data is that these two sets of genes act in two parallel and mostly functionally redundant pathways, so that a mutation in one pathway has little effect because the second pathway remains intact. One possibility is that these two pathways are two distinct systems involving cell-surface ligands and receptors that act in the recognition of a dying cell by its neighbor. Molecular studies of the *ced-5* gene are in progress (Yi-Chun Wu and H. R. Horvitz, unpublished results).

THE NUC-1 NUCLEASE DEGRADES DNA IN DYING CELLS BUT IS NOT NEEDED FOR CELLS TO UNDERGO PROGRAMMED CELL DEATH

The gene *nuc-1* is needed for the degradation of the DNA in dying cells.[21-23] *nuc-1* controls a nuclease, although unlike a nuclease associated with apoptotic deaths in mammalian cells,[24-26] this worm nuclease does not display a dependence on calcium for its activity. Since in *nuc-1* mutants DNA degradation is blocked and yet cells still die, nuclease activity -- at least the activity of the nuclease controlled by the *nuc-1* gene -- is not essential for cells to undergo programmed cell death in *C. elegans*.

SPECIFIC GENES CONTROL THE LIFE-DEATH DECISIONS OF SPECIFIC CELLS

The eleven genes discussed above function in all of the 131 cell deaths that occur during *C. elegans* development. What specifies which of the total 1090 cells generated will live and which will die? So far, we have identified three genes that affect this decision. Each of these genes appears to control only a very few cell deaths. For example, the genes *ces-1* and *ces-2* control the decisions of two cells in the pharynx, the feeding organ of *C. elegans*, to undergo programmed cell death.[27] These genes appear to act in a negative regulatory cascade: *ces-2* prevents *ces-1* from inhibiting the deaths of these two cells, so that in a wild-type animal with normal *ces-1* and *ces-2* function *ces-1* is inactivated allowing these cells to undergo programmed cell death. Similarly, the gene *egl-1* seems to affect whether a single pair of cells, the serotonergic HSN motor neurons, live or die.[13,28] Genomic clones that contain the *ces-1* and *ces-2* genes have been identified, and the molecular characterization of these genes should begin shortly (N. Tsung, M. Hengartner, M. Metzstein and H. R. Horvitz, unpublished results).

CONCLUSION

Genetic studies have identified 14 genes that function to control *C. elegans* programmed cell death, a process that appears morphologically similar to mammalian apoptosis. Molecular studies have revealed that one of these genes, *ced-9*, acts to protect cells against programmed cell death and is structurally and functionally similiar to the mammalian cell death gene *bcl-2*. These observations indicate that *ced-9* and *bcl-2* act in similar ways and suggest that *bcl-2* controls programmed cell death in mammals by interacting with genes similar to those we have identified to control programmed cell death in *C. elegans*.

ACKNOWLEDGMENTS

The studies described in this manuscript performed in the author's laboratory were supported by the United States Public Health Service and the Howard Hughes Medical Institute. The author is an Investigator of the Howard Hughes Medical Institute.

REFERENCES

1. Ellis, R. E., Yuan, J., and Horvitz, H. R., 1991a. Mechanisms and functions of cell death. *Ann. Rev. Cell Biol.* 7: 663-698.
2. Horvitz, H. R. and Chalfie, M., 1991. Implications of nematode neuronal cell death for human neurological disorders. In *Neurodegenerative Disorders: Mechanisms and Prospects for Therapy*, D. Price, H. Thoenen and A. Aguayo, John Wiley & Sons, Chichester.
3. Horvitz, H. R. and Sternberg, P., 1991. Multiple intercellular signalling systems control the development of the *Caenorhabditis elegans* vulva. *Nature* 351: 535-541.
4. Sternberg, P. and Horvitz, H. R., 1991. Signal transduction during *C. elegans* vulval induction. *TIG* 7: 366-371.
5. Clark, S., Stern, M. and Horvitz, H. R., 1992. *C. elegans* cell-signalling gene *sem-5* encodes a protein with SH2 and SH3 domains. *Nature* 356: 340-344.

6. Hill, R. and Sternberg, P., 1992. The gene *lin-3* encodes an inductive signal for vulval development in *C. elegans*. *Nature* 358: 470-476.
7. Han, M., Golden, A., Han, Y. and Sternberg, P., 1993. *C. elegans lin-45 raf* gene participates in *let-60 ras*-stimulated vulval differentiation. *Nature* 363: 133-140.
8. Sulston, J. and Horvitz, H. R., 1977. Post-embryonic cell lineages of the nematode *Caenorhabditis elegans*. *Develop. Biol.* 56: 110-156.
9. Kimble, J. and Hirsh, D., 1979. The postembryonic cell lineages of the hermaphrodite and male gonads in *Caenorhabditis elegans*. *Devel. Biol.* 70: 396-417.
10. Sulston, J., E. Schierenberg, J. White, and Thomson, N., 1983. The embryonic cell lineage of the nematode *Caenorhabditis elegans*. *Develop. Biol.* 100: 64-119.
11. Robertson, A. and Thomson, N., 1982. Morphology of programmed cell death in the ventral nerve cord of *Caenorhabditis elegans* larvae. *J. Embryol. exp. Morph.* 67: 89-100.
12. Ellis, R. E., Jacobson, D. M., and Horvitz, H. R., 1991b. Genes required for the engulfment of cell corpses during programmed cell death in *Caenorhabditis elegans*. *Genetics* 129: 79-94.
13. Ellis, H. and Horvitz, H. R., 1986. Genetic control of programmed cell death in the nematode *C. elegans*. *Cell* 44: 817-829.
14. Yuan, J. and Horvitz, H. R., 1990. The *Caenorhabditis elegans* genes *ced-3* and *ced-4* act cell autonomously to cause programmed cell death. *Develop. Biol.* 138: 33-41.
15. Yuan, J. and Horvitz, H. R., 1992. The *Caenorhabditis elegans* cell death gene *ced-4* encodes a novel protein and is expressed during the period of extensive programmed cell death. *Development* 116: 309-320.
16. Hengartner, M. O., Ellis, R. E., and Horvitz, H. R., 1992. *C. elegans* gene *ced-9* protects cells from programmed cell death. *Nature* 356: 494-499.
17. Vaux, D., Cory, S., Adams, J., 1988. *Bcl-2* gene promotes haemopoietic cell survival and cooperates with *c-myc* to immortalize pre-B cells. *Nature* 335: 440-442.
18. Nunez, G., London, L., Hockenbery, D., Alexander, M., McKearn, J. P., and Korsmeyer, S. J., 1990. Deregulated Bcl-2 gene expression selectively prolongs survival of growth factor-deprived hemopoietic cell lines. *J. Immunol.* 144: 3602-3610.
19. Korsmeyer, S. J., 1992. Bcl-2: an antidote to programmed cell death. *Cancer Surv.* 15: 105-118.
20. Vaux, D., Weissman, I., and Kim, S. K., 1992. Prevention of programmed cell death in *Caenorhabditis elegans* by human *bcl-2*. *Science* 258: 1955-1957.
21. Hedgecock, E., J. Sulston and Thomson, N., 1983. Mutations affecting programmed cell death in the nematode *Caenorhabditis elegans*. *Science* 220: 1277-1279.
22. Sulston, J., 1976. Post-embryonic development in the ventral nerve cord of *Caenorhabditis elegans*. *Philos. Trans. R. Soc. London Ser. B* 275: 287-298.
23. Hevelone, J., and Hartman, P., 1988. An endonuclease from *Caenorhabditis elegans*: partial purification and characterization. *Biochem. Genet.* 26: 447-461.
24. Wyllie, A., 1980. Glucocorticoid-induced thymocyte apoptosis is associated with endogenous endonuclease activation. *Nature* 284: 555-556.
25. Cohen, J. and Duke, R., 1984. Glucocorticoid activation of a calcium-dependent endonuclease in thymocyte nuclei leads to cell death. *J. Immunol.* 132: 38-42.
26. Wyllie, A., Morris, R., Smith, A., and Dunlop, D., 1984. Chromatin cleavage in apoptosis: association with condensed chromatin morphology and dependence on macromolecular synthesis. *J. Pathol.* 142: 67-77.
27. Ellis, R. E. and Horvitz, H. R., 1991. Two *C. elegans* genes control the programmed deaths of specific cells in the pharynx. *Development* 112: 591-603.
28. Trent, C., Tsung, N. and Horvitz, H. R., 1983. Egg-laying defective mutants of the nematode *Caenorhabditis elegans*. *Genetics* 104: 619-647.

Discussion

M. Oren

About the nuclease *nuc-1* especially: What is the mechanism? Is the nuclease induced by the engulfing cell in the engulfed cell? Is *nuc-1* mediating the induction of nuclease or is it encoding the nuclease itself?

R. Horvitz

The basic question is, what do we know about *nuc-1*?, and the answer is, very little!

Firstly, it has not been proved that the *nuc-1* gene directly encodes the nuclease. It is simply known that in the mutant, the activity is down at least two orders of magnitude. Secondly, there have been no mosaic studies of the gene and since the gene has not been cloned and the protein has not been purified, there are no expression studies. So we do not know if the gene is expressed within the dying cell or within the engulfing cell. The guess is that it is within the engulfing cell, only because the corpses do not show DNA degradation if there is no engulfment but, of course, the engulfment could activate an endogenous nuclease. The other thing I should say is that the biochemical characterization of the nuclease activity has indicated that this is not a calcium dependent nuclease, so the nuclease biochemically is distinct from the calcium dependent enzyme that we will hear more about in this meeting and that is associated with apoptosis in mammals.

R. Schimke

Do you have any studies where you have looked at cells undergoing normal developmental programmed death with your probes to see, either qualitatively or quantitatively, whether the *ced* genes are expressed in those particular tissues or specific cells? A second question is: Do you think the expression of these genes might be determining whether cells will or will not die, again within the *ced* family of genes?

R. Horvitz

The first question is, what do we know about the expression patterns of *ced-3* and *ced-4* molecularly as opposed to bygenetic analysis? For both of them we have made fusions with the E. coli *lacZ* gene and we are also raising antibodies using epitope tagging or antibodies against fusion proteins that include the proper protein itself. There are lots of caveats on *lacZ* fusion and even epitope tag studies of this sort, so I think the experiments that are most convincing are the ones that use antibodies raised against the protein *per se*. These experiments are still in early days; we certainly know that cells that are undergoing cell death express these proteins, but what it looks like is that there are cells that are not undergoing cell death that also can express at least one of these proteins. We already know from what I have told you that simply expressing *ced-3* and *ced-4* is not enough to kill, so it is not shocking, again, that we would find these proteins expressed in cells that do not die and the question then comes to *ced-9*. Where is *ced-9* expressed? And there we have no data at this point. We have no *lacZ* fusion or antibody data yet concerning *ced-9*.

G. Guidotti

Two brief questions. One is: Is there any information on the intracellular distribution of the *ced-9* product in *C. elegans*? The second question: You said that *bcl-2* expression protects from cell death in *C. elegans*. Is *ced-9* expression protective in mammals? It should be possible to generate transgenic mice expressing a *ced-9* transgene controlled by the immunoglobulin heavy chain promoter-enhancer and see whether these animals exhibit a lympho-accumulation as do mice transgenic for IgH enhancer-*bcl-2*, a known model for an autoimmune disease resembling human lupus erythematosus.

R. Horvitz

To the first question, we have no antibody to *ced-9*, so for that reason we have no localization study, by any of the various criteria you might imagine. To the second question, I have told you half the story, mainly, that human *bcl-2* in worms will protect and the question is, reciprocally, will worm *ced-9* protect in mammals? We have initiated collaborative experiments with Stan Korsmeyer and he is starting work both in cell lines and in animals but it is really too early to say what the results of those studies are at this point. We are very interested to see what happens there.

P. Krammer

The ALS patients, are they particularly sensitive or are their cells particularly sensitive to TNF?

R. Horvitz

I know of no studies in ALS with TNF, either in cells or patients. I know of nothing that has been done.

D. Delia

I would like to ask you a question concerning *ced-9*. It was recently reported that another gene, *MCL-1,* isolated from myeloid cells in humans if I remember correctly, has some sequence similarities with *bcl-2* particularly in c-terminal regions. Now, the function of this gene is not known and it appears that its regulation is slightly different from *bcl-2*. Have you looked for similarities of this gene *MCL-1* with *ced-9*?

R. Horvitz

Ced-9 is as similar to *MCL-1* as it is to *bcl-2* and as *bcl-2* is to *MCL-1*.

D. Livingston

Is it possible to ask the question whether *ced-9* and *3* or *4*, directly interact? Should it be possible to look for complexes biochemically and/or genetically?

R. Horvitz

So, how direct now is the interaction between *ced-9*, *ced-3* and *ced-4?* There are actually many ways of approaching that problem, both biochemical and genetic, and we have initiated some of those. We are obviously very interested in this question. Unfortunately, I have nothing to say about it whatsoever. In this particular case my suspicion is that for demonstrating direct interactions that the biochemical approach may be more straightforward. We certainly are expressing these proteins to do such studies.

D. Hockenbery

I seem to remember some descriptions of an asymmetric cell division that preceded cell death in which the sister cell that died would receive less cytoplasm and this could be recognized as an early event. Is that really linked to cell death in *C. elegans* and is that reversed by *ced-9* overexpression?

R. Horvitz

Almost all of the cell deaths that occur are asymmetric in size in terms of the division that generates the dying cells with the smaller cell being the one that dies. In a *ced-3* or *ced-4* mutant, it is possible that those divisions could now be symmetric in size. But if so, I think really all that would say is that the decision about cell fate is made prior to the time of division and in fact there is also an ultrastructural sign of that. When you look at a cell that is dividing to make one corpse and one surviving cell, you see at the time before that mitosis is complete, that the engulfment process has already been initiated, so the half of the cell, basically, that will become a corpse is becoming engulfed as that cell division is proceeding. So, in a sense, determination precedes the division and I think that is true for size as well as for other factors

D. Hockenbery

And just to follow that up, are there any phenotypes of mutations in these genes, in the surviving sister cell, if the process is initiated in the mother cell? Does the surviving sister cell have any phenotype due to cell death gene mutant?

R. Horvitz

No, I mean the surviving sisters seem fine. Those are the cells of the animal.

P. Nicotera

I have two questions: The first one concerns the *ced-3* gene. You mentioned that the mutations are on the c-terminal primarily, they do not affect the serine rich region. Can you exclude that this mutation will affect binding into that region? That the mutation that you have in the c-terminal will affect a binding on the other regions although that is not mutated directly, for example, by agents which will phosphorylate that region? The second question is about the final part of your talk, which I found particularly interesting, on the Amyotrophic Lateral Sclerosis and your comments concerning the expression of the SOD protein. Now, this is not a particularly specific reaction because it occurs in other conditions, for example, in TNF induced cell killing; in this case you do also have SOD protein which is supposed to be involved in the mechanism of cell death, and recently there has been a report showing that you have antibodies against L-type calcium channels in the same disease, in Amyotrophic Lateral Sclerosis. So, my question is: Do you think that apoptosis is involved in the killing of these cells and do you think that the SOD is a causative phenomenon, that it is not an epiphenomenon in the disease?

R. Horvitz

I will start with the end. What we know is that mutations, and furthermore, different mutations in this SOD gene, are causally linked to the disease process and the simplest interpretation is that changes in this enzyme's activity, and we notice such changes are there from biochemical studies that have not yet been published, are responsible for the disease process. Exactly what is different physiologically remains to be determined. That is not clear at all. Now, in terms of the L-type calcium channel finding, it is very possible that that is simply an antibody that has been raised in response to the fact that there are antigens released from dying cells. I think that is probably the simplest interpretation of that finding. Having said that, I should tell you that in another series of projects that we are doing in my laboratory, we have cloned the *C. elegans* alpha 2 subunit of the L-type calcium channel equivalent; we have mutations in this chain, and although the animal is quite abnormal and does not behave the way it should at all, there are no signs of perturbation of cell death that we know of. Whether or not that is an extrapolation one would want to make, I do not know. My own guess is that the L-type channel antibodies are just a sign of the disease process rather than its cause. From the genetic studies we can say something fundamentally different with respect to SOD because there we have mutations that lead to the disease. Those mutations, it seems to me, are quite unlikely to be effects of the disease process.

P. Nicotera

The question is: What do we really know about structure-function of *ced-3,* and what do our mutations tell us in terms of possible interactions of different domains.

A. Horvitz

There is nothing I can say about that at this point.

D. Papermaster

Do you know whether there are *ced-9*-like proteins that are cell specific in their expression downstream of this early developmental controlled gene that might indeed induce certain specific cell lines to stay alive that otherwise might not? Do you have those kinds of simple systems in the worm?

R. Horvitz

Are there any cell-type specific genes that are downstream of the generalized program? The answer from the genetic studies is that so far we only see cell-type specificity that is acting upstream of the general program, but it is always possible that there exist genes that we have not identified. But there is no basis for invoking such genes at this point.

T. Taniguchi

I have two brief questions at this point. The first question is whether the *ced-1* and *ced-2* expressions require expression of *ced-3* and *ced-4*? And the second question is whether, under any circumstances, you have tested those *ced-3* and *ced-4* overexpressions in mammalian cells cause cell death.

R. Horvitz

The first question is: What do we know about the expression of the engulfment genes? As I mentioned, there is only one of them that has been cloned at this point: it is very recent and we do not have enough data to say really anything meaningful about expression there. We have not yet put *C. elegans ced-3* and *ced-4* into mammalian cells to see if they will kill mammalian cells and the reason is very simple. There are lots of things that kill cells. I mean, if you want to kill mammalian cells, I am sure all of us know ways of doing that. And what we wanted to do before we tried experiments like that was to learn enough about how these genes kill in *C. elegans* to better design the experiments and the control experiments for mammalian cells and for mammals in general. So that for example, if we had done the naive experiment of simply putting these genes into mammalian cells it might well not have worked, just as it seems that a comparable experiment in *C. elegans* works better if *ced-9* is inactivated. Now we have a more sophisticated way to try the mammalian experiment. But there are still other questions. We do not know whether both of these genes must function to kill or whether one of them is sufficient to kill: one may just act as a positive regulator of the expression of the other and may or may not be needed if the latter is overexpressed. We are doing all of those kinds of experiments in worms to really define the genetics, if you will, of the pathway to tell us what to do in mammals.

M. Kastan

In the ALS patients, when the SOD is mutated is there a reason that the motor neurons are particularly susceptible to that, or are there other tissues in their body that seem to be sensible to the increased numbers of radicals?

R. Horvitz

You have touched upon one of the most intriguing features of ALS and, in fact, many neural degenerative disorders which is the cell-type specificity. In fact motor neurons are quite specifically affected in ALS: the SOD gene is expressed very broadly so its expression pattern is not sufficient to account for that specificity and we do not know at this point why there is specificity. The fact that there is specificity is not terribly shocking, and the reason for that is that motor neurons are different from other cells. In fact, every cell type is different in some way from other cell types and one could imagine that there is some aspect of that difference that leads to a differential sensitivity in any of a variety of cases. If you wanted me to, I could speculate wildly and idly about what is different about motor neurons, but the bottom line is I have no idea what aspect of their difference is really related to this phenomenon. The observation is that the disease is quite specific for these cells. I am hoping that further studies of the gene and its function and dysfunction in the motor neurons of ALS patients will help answer your question.

J. Ashwell

In view of your observation that the dying cells are being engulfed even during mitosis, do you know from your mosaic studies whether the engulfment genes are expressed by the dying cell, or are, for example, recognition elements put out by the cell that is being targeted for death. Or are the genes in the phagocytic cell encoding a receptor, for example.

R. Horvitz

Right, so the question is where do the engulfing genes work? If you remember the cartoon I showed I basically had two pathways, each of which had some functions in the engulfing cell and some functions in the cell corpse, as if the cell corpse is putting up little molecular flags saying I am a corpse, come and get me", and there are receptors for these ligands that are present on the engulfing cell. That is our model. Now, we have not yet done molecular expression studies of these genes so we have no information from such studies. We are just beginning to do mosaic analyses to resolve questions in that way. However, what we do know, and you have now gotten into a bit of genetic nitty gritty, is that there is a difference amongst these genes with respect to whether they act maternally or zygotically. And without going through all the data and all the arguments, basically what it looks like is that certain genes may function in any cell type and other genes may be turned on in specific cell types. Our hypothesis is that the genes that are turned on in specific cell types may be turned on in the corpses, and the genes that can function anywhere are functioning in the engulfing cells, because we think that any cell in the animal may well be capable of engulfment if it is in the vicinity of a cell corpse. I could go through the details of those experiments but they're indirect and therefore not compelling. Molecular and genetic mosaic data would be much more direct.

T. McDonnell

Can you tell us anything about the *ces-1* structure and also, more specifically, the relationship of *ces-1* mutations with *ced-9* mutations.

R. Horvitz

OK, so *ces-1* is a negative regulator of *ced-9*. We have cloned *ces-1* by germline rescue. What that means is we have a cosmid that if injected, making a transgenic animal, rescues *ced-9* functions. We have not obtained any sequence information for *ces-1* as of yet though, so I do not know what the gene looks like. We should know that very soon.

J. Isaacs

Would you clarify something for me? As your *C. elegans* becomes an adult, what about cell turnover? I assume that the *ced-3* and *ced-4* mutants reach adulthood, is that correct?

R. Horvitz

Yes.

J. Isaacs

OK, after they have reached adulthood, although the cell number does not change over time, do the cells proliferate? Is there proliferation or turnover of the adult cells?

R. Horvitz

No, the only tissue that continues to proliferate in an adult is the germline and that is it. There is no further cell division in the adult whatsoever. There is no death really that occurs except when the organism dies. So there is no turnover, there is no homeostasis.

J. Isaacs

How does the organism die?

R. Horvitz

Well, how does any organism die? What really is meant by organismic death we could talk about, but with *C. elegans* what I can tell you is that *C. elegans* ages, which means that the probability of death increases with age; that it undergoes a variety of behavioral and morphological changes so that it acquires the dark pigmentation that is characteristic of age, its muscles become flaccid, it is much less energetic, it moves less and so on and so forth. So you can see that the organism dies eventually when various biological processes have stopped. We have done experiments to ask the question "are the cell death genes that I have told you about involved in the aging process and, in particular, in longevity?" The answer is that they are not. The longevity of mutants defective in *ced-3*, *ced-4*, *ced-9*, you know, in *ced 1,2,3,4,5,* all the way through, all the things I have told you about, there is no effect on aging *per se*. So, our conclusion from this is that what causes aging in *C. elegans* and what causes organismic death are mechanistically distinct from what causes programmed cell death during development.

D. Livingston

Regarding inherited ALS, what is the typical penetrance in families? Is it high?

R. Horvitz

ALS is said to be between 5% and 10% familial. Which means that you see clear cut autonomal dominant inheritance in families in approximately that percentage of cases. When you look at ALS, at the statistics of age of onset and of duration, what one finds is that these are the same overall for the sporadic cases and for the familial cases. The average age of onset is 50. The average duration from time of diagnosis to death is about three years. When you ask then, how sharp is the peak of age of onset, it turns out it is very broad. Individuals can get ALS at a much younger age, or at a much older age. And that statement is true even within families where you have the same gene defect responsible for the disease. So, within that context, when you look at penetrance, the penetrance is very high as long as people live long enough to get the disease. But there are examples, for instance, we have a family that just in the last two months has the unfortunate circumstance where a daughter who has had the disease for a period of time, is the daughter of a mother that we knew from the pedigree studies to be carrying the disease gene and the mother has now just come down with symptoms of the disease, obviously many years later than the daughter showed it comparably.

M. D'Incalci

Can you show an allelic link, for example, to a time of onset? Are there certain alleles that you have identified, where you get really early onset and some with really late onset?

R. Horvitz

If you look at given families and given alleles, there are families that seem to have a much longer duration and families that have a much shorter duration. There do not seem to be families that have an earlier onset or a later onset. One of the intriguing observations now is why there is alleles specificity for duration but not for onset.

M. D'Incalci

What do you know about the effect of chemicals and drugs on the expression of *ced-3* and *4*?

R. Horvitz

We really have not done very much in terms of pharmacologically attempting to perturb worm cell death. We have played around a little bit with a variety of things, but I do not think I have anything meaningful to say about that. I think actually that is a very interesting line of experimentation. One of our goals is to set up *C. elegans* in a way that it could be used for screening and evaluating drugs. We already have evidence that we can use *C. elegans* in this way for other kinds of things such as serotonin uptake. *C. elegans* has been used to analyze serotonin uptake and in fact to identify novel classes of serotonin uptake inhibitors. Recent studies have indicated that *C. elegans* can be used in comparable ways to study single transduction and, in particular, Ras protein function. So we hope we will be able to do it with cell death as well.

D. Green

We know at this meeting and we shall hear about lots of ways to introduce apoptotic death and many of those are drug related. I think a question that is often pressing us is the relationship between the process of apoptosis *per se* and the genetic regulation of programmed cell death. If you were to add Actinomycin D or etoposide or various drugs which have been shown to induce apoptosis, basically just kill the worm with them, would there be any protective effect of *ced-9* especially or perhaps even of *ced-3* or *ced-4* mutants? We know *bcl-2* can protect in these settings.

R. Horvitz

Right, and we have not done those experiments *per se*. Let me say that pharmacological experiments with the worm are sometimes a little difficult to interpret because many things do not get in. Remember, these are whole animals we are talking about and many drugs do not get into the animal. So we have not actually tried experiments with RNA or protein synthesis inhibitors. My guess is we would have to come up with some tricks to actually get those to work.

P. Amati

My question is: In the worm, how far back in the pedigree of a cell that will go to apoptosis can you trace the definition of that apoptosis? In other words, is it possible to look back and define the moment when a given cell has been programmed to die?

R. Horvitz

In the worm, when does the cell death process really begin? I think the best we can say from what we have seen is that that process is certainly underway by the time of the mitosis that generates the dying cell. Beyond that I do not think I can say anything.

APOPTOSIS IN INHERITED RETINAL DEGENERATIONS

David S. Papermaster and Izhak Nir

Department of Pathology
University of Texas Health Science Center
San Antonio, Texas, 78284

INTRODUCTION

οἵη περ φύλλων γενεή, τοίη δὲ καὶ ἀνδρῶν
As is the generations of leaves, so is that of men
-Homer, *The Iliad*- vi:146

The molecular biological descriptions of genetic defects in several forms of inherited blindness in humans, mice, and dogs leave us bewildered about the causes of photoreceptor death. Mutations of many different proteins interfere with normal development of photoreceptors or initiate a process that leads to premature photoreceptor death for obscure reasons. These mutations include the rhodopsin gene at many different sites (reviewed by Berson[1]), the β subunit of cyclic-GMP phosphodiesterase in *rd* mice and *rcd1* Irish setters[2,3], the *rds/peripherin* gene in mice and man[4,5,6,7] and the A component of geranylgeranyl transferase in human choroideremia[8,9]. Two features are common to some of these retinal degenerations: the loss of the polarized distribution of rhodopsin and the absence of an inflammatory response to the dying cells.

The polarized distribution of the rod photoreceptor's major protein, rhodopsin, the light sensitive visual pigment, is not sustained in many degenerating retinas. Normal rods sequester rhodopsin to a special light-sensitive organelle, the outer segment, which is composed of stacks of rhodopsin-laden disk-shaped membranes enveloped by a plasma membrane that has bilayer continuity with adjacent cellular domains. Despite this continuity, rhodopsin is confined at high density to the disks and the plasma membrane of the outer segment in normal cells. The adjacent inner segment, axon and synaptic terminal contain little rhodopsin. Once degeneration begins, however, rhodopsin is readily detected throughout the plasma membrane of the degenerating photoreceptors, including the synapse. The normal sequestration of rhodopsin is not a consequence of the incapacity of the other plasma membrane domains to tolerate the molecule since mechanical injury leading to fusion of the outer and inner segments rapidly generates a widespread diffusion of rhodopsin to all compartments of the plasma membrane. Na,K-ATPase, on the other hand, remains

Apoptosis, Edited by E. Mihich and R.T. Schimke
Plenum Press, New York, 1994

restricted in its localization to the lateral plasma membrane of the inner segment[10], perhaps as a consequence of tethering of some of its molecules to the underlying cytoskeleton[11]. It is not clear if this phenotypic modification of the distribution of rhodopsin is part of the pathogenesis of rod cell injury or a reflection of it. This change is observed in each species in which it has been studied; mice with the *rd* and *rds* mutations, RCS rats with the *rdy* mutation[12,13,14] and humans with autosomal dominant retinitis pigmentosa caused by a Thr-17-Met mutation in the rhodopsin molecule[15].

Of greatest relevance to this conference, is the remarkable histology of the dying retinas. In each case, these retinas have little or no invasion by leucocytes or macrophages. In some cases, e.g., the *rd* mouse, the onset of death is sudden at the 12th post-natal day and is completed within a few days, but no evidence of participation of extra-retinal cells in cleaning up the debris is found. In other forms, the duration of the degeneration is more prolonged and photoreceptor cells die in a few weeks or over months or years. Eventually, all these degenerations result in a final state of loss of photoreceptors, collapse of the remaining retina against the adjacent pigment epithelium which may also be altered, narrowing of the retinal vasculature and finally a useless remnant of neuronal tissue. In humans and some animal models, the entire retina is not homogeneously involved.

Where do the dying cells go--how is the debris handled? Will the study of these questions provide any insights into the mechanism of cell death? Could such insights contribute to an approach which could delay the onset of photoreceptor cell death and thereby prolong useful vision or prevent the loss of vision entirely? In retrospect, these obvious questions should have provoked those of us studying retinal pathology to test them within the current intellectual ferment of the studies of cell death in developing tissues and pathologic conditions occasioned by the seminal work on apoptosis by Kerr et al.[16]. Surprisingly, little attention to this approach has appeared until some recent studies. Once asked, however, the answers support an important role of apoptotic cell death in the degenerating retina. Because these studies have been of short duration, the data is, accordingly, still quite preliminary. We will describe some of our own recent studies of the *rds* mouse and RCS rat retinas and review some issues of relevance to human inherited retinal dystrophies and degenerations.

To introduce our studies of this topic, we provide an extremely abbreviated review of the normal development of the retina and a summary of prior studies of some of the retinal degenerations. During embryonic life, numerous retinoblasts differentiate and some cells die by apoptosis prior to the creation of the multi-layered adult structure[17, 18]. Shortly after birth (the exact date and rate varies among mammalian species) the last divisions of primitive retinoblasts are completed and the outermost layer begins to differentiate into elongated cells that will eventually become the light-sensitive photoreceptor layer composed of rods and cones. These cells are specialized sensory neurons which project their outer segments into the interphotoreceptor space until they abut the adjacent pigment epithelium on the back surface of the eye. The membranes of the outer segment are under a dynamic rhythm of renewal. The tips of rod outer segments are shed each morning and destroyed by phagocytosis and phagolysosomal degradation in the adjacent pigment epithelium (about 10% of the length of the rod outer segments is shed in mammals[19]). New disks containing a high density of rhodopsin and other proteins of the phototransduction cascade are assembled each day in rods and displaced apically by new membrane assembly at the base[20,21]. This morphogenetic activity is occurring at the tip of the immotile (9+0) cilium which joins the outer segment to the inner segment. All protein synthesis, sorting, and transport to the base of the cilium for final assembly into the outer segments is completed in the inner segment[20,22]. To a much lesser extent, the cell also renews its other domains, the lateral cell membrane which houses the Na,K-ATPase that generates the membrane potential required for the light-sensitive dark current, and the synaptic terminal where contacts are made with the next layer of interneurons, the bipolar and horizontal cells and amacrine cells. They in turn, transmit the information of photon capture to the innermost layer, the ganglion cells, which send axons to the center of the brain for eventual distribution of visual information to the remainder of the central nervous system.

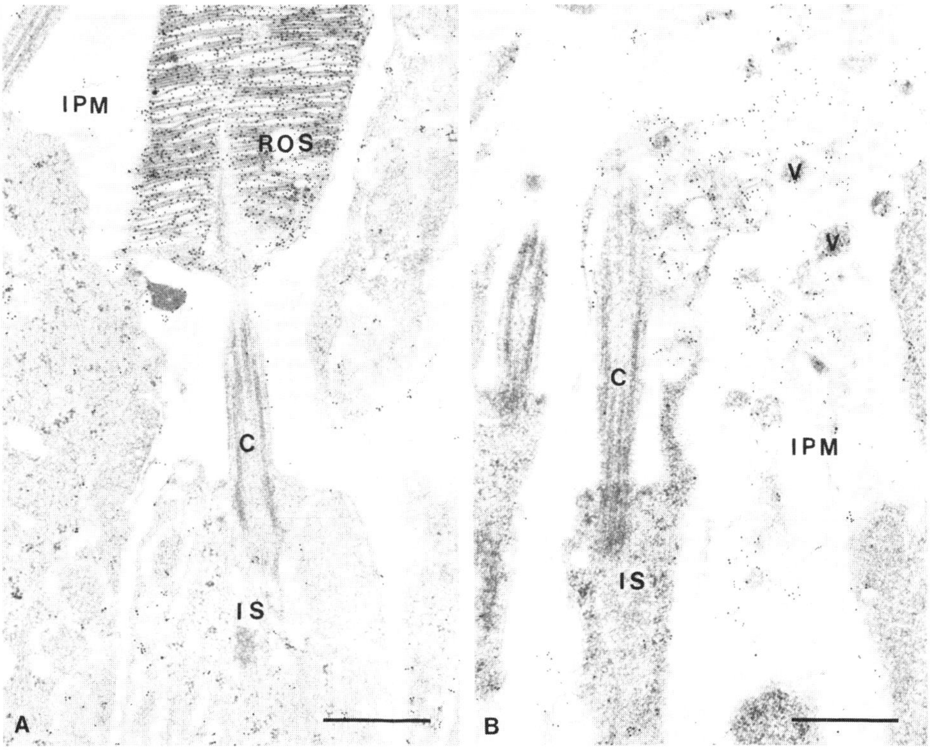

Figure 1. Electron micrographs of a normal BALB/c and an *rds* retina at the junction of the inner and outer segments. The thin sections have been labeled with an anti-rhodopsin antibody that is detected with gold conjugates. **A.** In the normal mouse, rhodopsin is assembled into thin disks of the rod outer segment (ROS) that is enveloped by a plasma membrane. Both of these ROS membranes have abundant rhodopsin detectable immunocytochemically. The connecting cilium (C) is slightly labeled on the distal half. The inner segment (IS) plasma membrane is barely labeled. **B.** The *rds* retina is unable to form an outer segment as a consequence of a mutation of its *rds*/peripherin gene. This protein is normally localized to the margins of the disks. As a consequence, newly synthesized opsin is transported to the tip of the cilium (C). There, vesicles (V) containing abundant rhodopsin form at the ciliary tip and fall off into the interphotoreceptor space (IPM). (Adapted from Agarwal et al.[23] with permission). Bar .5 μm.

Evidence for apoptotic photoreceptor cell death in retinas of *rds* mice and RCS (*rdy*) rats

Although several convenient models of inherited retinal degeneration exist for study in the laboratory, the *rds* mouse model is very interesting because this degeneration is slow, its exact molecular genetic defect has been identified, and a similar disease exists in humans. These mice cannot form an outer segment because the gene for one of the outer segment proteins, peripherin, which is confined to the disk margin[24] and apparently contributes to disk morphogenesis, is interrupted by an insertion so that a truncated protein is encoded and probably not translated[7,25]. Homozygous mutant *rds* retinas develop normally for the first 10 days postnatally, but then fail to generate the outer segment upon completion of the migration of the basal body to the apex of the inner segment[26,27,28]. The cilium projects normally into the interphotoreceptor space and rhodopsin and other proteins are synthesized and transported to the ciliary tip, as they would be in the developing retina[23,29]. But in the affected mice, rhodopsin-containing vesicles and malformed membranes form at the ciliary tip and fall off into the interphotoreceptor space (Figure 1). There, they are taken up by the adjacent pigment epithelial cells by phagocytosis[29]. Accordingly, the *rds* rod cells have only a trivial amount of rhodopsin on their cell surfaces and have a very low-amplitude electrical response to stimulation by light[30].

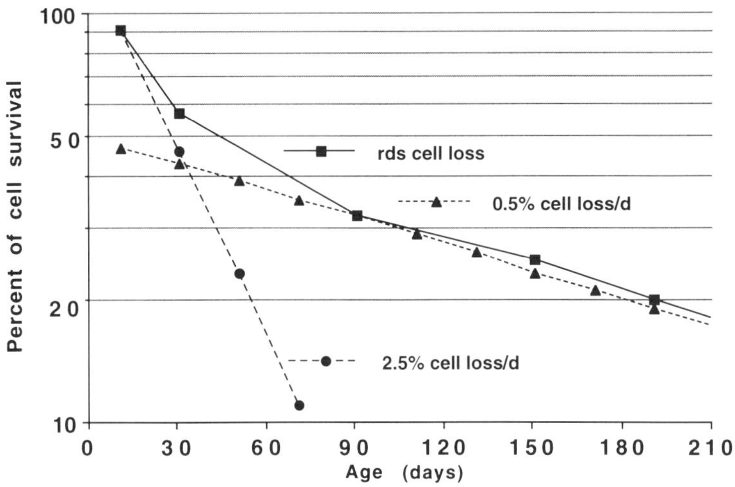

Figure 2. Survival of *rds* photoreceptors. The decline in photoreceptor cell nuclei as the *rds* mice age (semi-logarthmic plot). In the early months, cells are dying at about five times the rate of cell death after 2 months (adulthood). The loss of photoreceptor nuclei (open squares) is compared to theoretical curves of cell loss at rates of 2.5%/d (circles) and 0.5%/d (triangles).

From the beginning of retinal life, this dystrophy leads to premature cell death[26,31]. By the 11th post-natal day, these retinas have fewer photoreceptor neurons than normal (about 90% of controls). Cell death progresses at a pace indicative of at least two rates, a rapid phase in the first month and a slower phase thereafter (Figure 2). The morphology of the degenerating retinas reveals two features. The inner segments and connecting cilia and synapses appear normal but the absence of the outer segment layer is complete. The photoreceptor nuclear layer thins with the passage of time. Finally, this

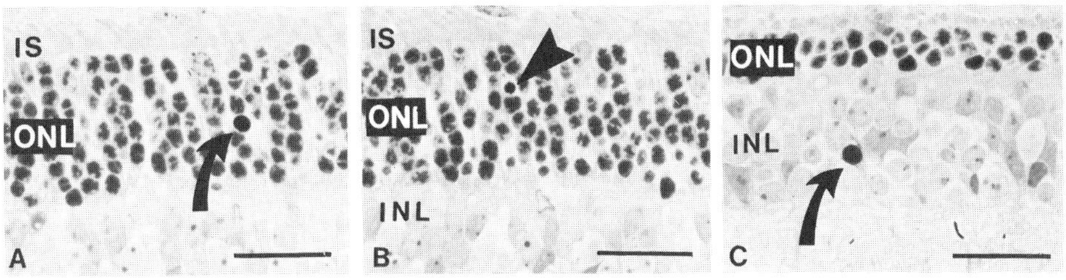

Figure 3. Light micrographs of *rds* retinas at 17d. (**A, B**) and 6 mo. of age (**C**). Pyknotic cells (arrows) are seen throughout the photoreceptor outer nuclear layer (ONL) and inner nuclear layer (INL) at low frequency. As the mice age, the photoreceptor outer nuclear layer thins. Nuclear fragments (apoptotic bodies?) are occasionally detected. (B. arrow-head). Macrophages are not apparent. 1μm thick sections stained with toluidine blue. Bar 15 μm.

layer, which starts off 8 to 10 cells thick becomes a monolayer with gaps by 1 year of age (Figure 3).

Heterozygotes are also affected but to a lesser extent. They form abnormal outer segments and the cells die at a slower rate[27]. Humans with a mutation of this gene also have a slow degeneration as heterozygotes[4,5]. Transgenic *rds* mice provided with normal copies of the *rds/peripherin* gene are rescued and form normal retinas[25].

Because several genes are up- or down-regulated in dying cells, we first determined if the expression of some of the genes involved in the phototransduction cascade and in vitamin A transport were altered. We found that both rhodopsin and interphotoreceptor retinol binding protein (IRBP) were synthesized normally and that rhodopsin was transported to the tip of the cilium where it was lost by vesicular shedding into the interphotoreceptor space[12,31-33]. By contrast, expression of arrestin, a 48kDa protein with high affinity binding to phosphorylated opsin, was remarkably altered. Arrestin normally varies two-fold in its expression in a diurnal fashion with highest levels of the mRNA and protein in the late afternoon. In *rds* mice, however, its expression is elevated both in light and dark at levels comparable to those in light-exposed normal mice, as if their retinas were in constant light[34]. A molecule implicated in cell death, *c-fos*, also varies diurnally in retinas with its greatest expression in darkness after at least 2 hours of light exposure. Only a brief, low-level rise in expression occurs at light onset[35]. In preliminary experiments, *rds* mice appear to have a normal cycle of *c-fos* expression indicating the persistence of diurnal periodicity in some retinal systems despite the loss of the rhythmicity of arrestin expression (Nir and Agarwal, unpublished observations). Clusterin expression is also elevated in *rds* mice[36] and in human retinitis pigmentosa[37]. This combination of unaltered and altered protein turnover does not readily provide evidence for a mechanism for cell death.

Light microscopic examination of *rds* retinas reveals several features which are suggestive of apoptotic cell death (Figure 3). Dense round pyknotic nuclei are scattered at low density throughout the degenerating retina at all ages. Crenation of nuclei and small nuclear fragments are occasionally seen. In some instances, the pyknotic nuclei are thrust out of the photoreceptor layer and lie on the apical surface in the interphoto-receptor space where they might be phagocytosed by pigment epithelial cells, but this is rare. Electron microscopy provides further evidence for apoptosis. Nuclear and cytoplasmic condensation are prominent features. Nuclear fragmentation and preservation of organelles within apoptotic bodies can be found (Figure 4). No evidence for cell swelling, dilation of endoplasmic

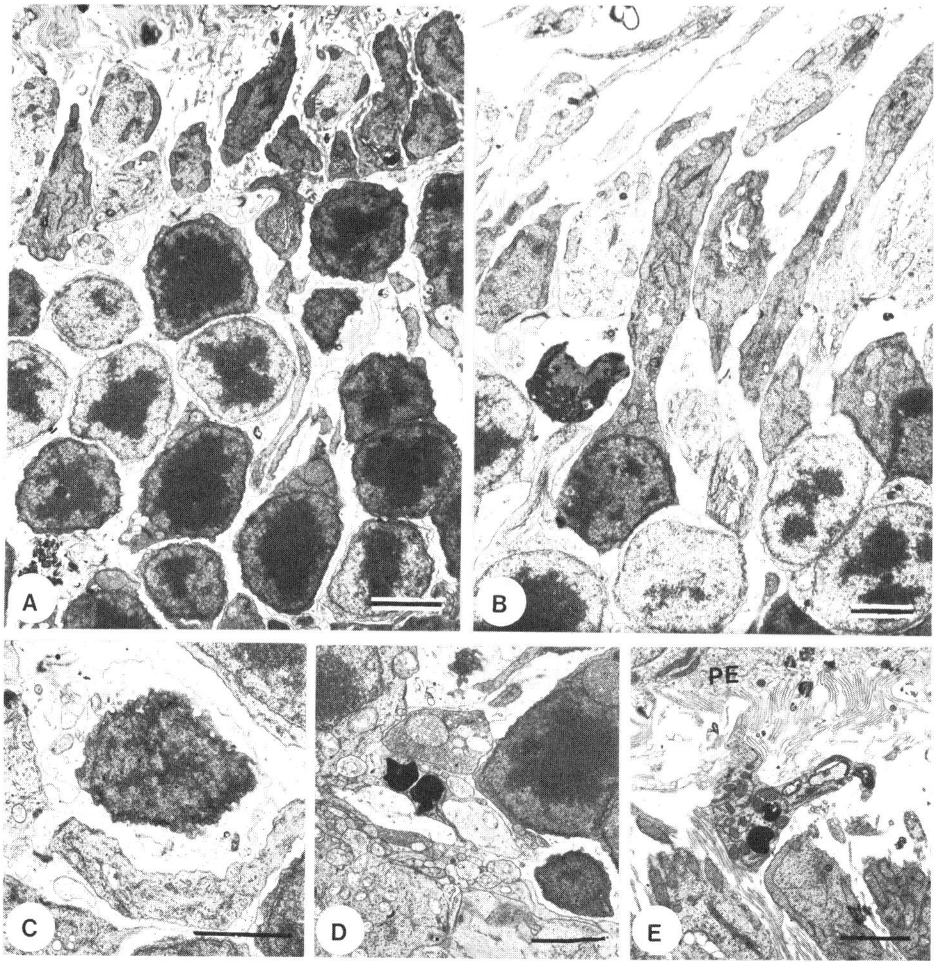

Figure 4. Electron micrographs of *rds* retinas at various ages (**A-E**). Outer segments are not formed. Dying cells condense their cytoplasm (**E**). Intracellular organelles are also condensed--vesiculated mitochondria and endoplasmic reticulum are not apparent. Nuclei are hyperchromatic and the nuclear membrane collapses around the nuclear remnant occasionally in a crenated pattern (**C**). Nuclear fragments comparable to apop-totic bodies of other cell types are present within the outer nuclear layer (**A,B**), near the plexiform (synaptic) layer (**D**) and (**E**) at the interface of the outer nuclear layer and microvilli of the retinal pigment epithelium (PE). Bar 2 μm.

reticulum or Golgi cisternae is revealed and neither mitochondria or phagosomes become dilated. However, phagocytosis by adjacent photoreceptors and Mueller cells is not readily demonstrated.

Curiously, the adjacent interneuronal cell layer seems remarkably unaffected by this catastrophic destruction of the outer nuclear layer. Only in the latest stages of the disease does the thickness of the inner nuclear layer decrease. Rare pyknotic nuclei are revealed there (Figure 3C).

The rates of cell death can be approximated by fitting the curve of cell loss to tangents which indicate, most simply, that the rate of cell death in the first few months is about 2.5%/day. After three months, death of photoreceptors proceeds at a slower rate of about 0.5%/day until the age of 7 months when 80% of the photoreceptor cells are lost (Figure 2). We therefore sought a sensitive indicator of endonuclease action to study these rare events *in situ* and turned to the assay reported recently by Gavrieli et al.[38] which end-labels nicked DNA by sequential action of terminal de-oxyribonucleotidyl transferase, biotinyl-dUTP and streptavidin-peroxidase. We modified the technique slightly by using diaminobenzidine to detect the bound peroxidase. Stained nuclei and nuclear fragments are revealed throughout the affected retina of 28 day old *rds mice* (Figure 5). The numbers of stained nuclei and nuclear fragments varies from none to 5% of the nuclei in the microscopic field. These extremes were rare, most areas had 1-2% of nuclei stained, a density which matches the expected loss of cells at that age. Control mice had 1-2 stained nuclei in an entire retinal section. The speed with which these cells are destroyed and their debris is removed is unknown. Obviously, that would affect the results of this assay but the close correlation of the rates of cell death and the numbers of stained nuclei indicates that they probably do not persist for more than a day. Dr. Fulton Wong (Duke University, personal communication) has seen typical nucleosome sized DNA fragment ladders in agarose electrophoretic gels of *rds* retinas. Thus, several lines of evidence support the interpretation that cell death in *rds* retinas is apoptotic.

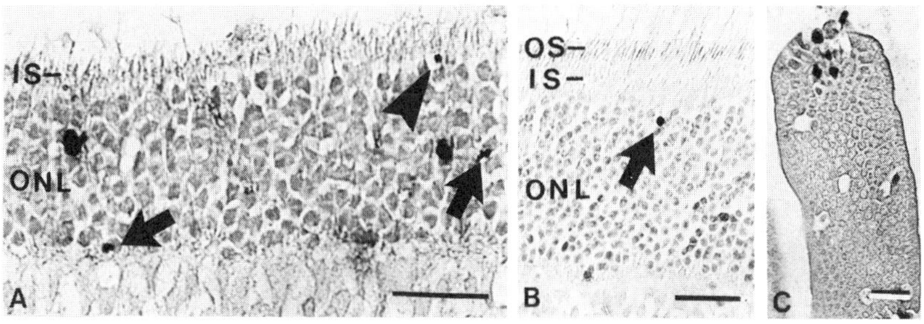

Figure 5. Light micrographs of a 28 d. old *rds* mouse retina, BALB/c retina and small intestinal villus of the *rds* mouse stained by the procedure of Gavrieli et al.[38]to detect apoptosis. **A.** Rare scattered stained nuclei and nuclear fragments (arrows) are distributed at low density throughout the *rds* retina outer nuclear layer (ONL) One apoptotic body (arrowhead) lies among inner segments (IS). **B.** In BALB/c controls only 1-2 stained nuclei (arrow) are seen in the entire retinal section. Outer segments (OS) are formed in this normal retina. **C.** Epithelial cells at the tip of the villus are stained. Nicked DNA is revealed in apoptotic nuclei by enzymatic end-labeling with biotinyl dUTP. Bound streptavidin-peroxidase is detected by diaminobenzidine. Deparaffinized section counterstained by 1% aqueous methyl green and photographed through a Wratten 25A filter. Bar 10 μm.

RCS rats are affected by an extraordinary defect in the pigment epithelial (RPE) cells: their RPE cells are unable to phagocytose and destroy rods tips. The exact molecular defect is not yet defined. As noted above, mammalian rods shed and replace about 10% of their outer segment discs each day. At first, RCS rat rods appear entirely normal. And indeed they are, for in chimeric rats with a mosaic of normal and affected RPE, the fate of the underlying rods is determined by the parentage of the RPE cells[39]. Gradually the debris of undigested disks builds up for about 28 days of post-natal life (Figure 6). Then the rod cells become overwhelmed for obscure reasons. Death is rapid and most of the cells die in the next month. Opsin polarity is lost[12] and gradually the capacity to synthesize opsin mRNA and rhodopsin declines to trivial levels as the photoreceptors die. Until they die, there appears to be no detectable decrease in biosynthetic activity when data are correlated with surviving photoreceptors[33]. This catastrophe is accompanied by numerous shrunken cells and pyknotic nuclei (Figure 6). End-labeling of the retinal DNA in a 28 day old RCS rat retina at an early stage of the degeneration reveals the extent of endonuclease activity. At least

10% of the nuclei are labeled and numerous, small, densely stained nuclear fragments are scattered not only throughout the outer nuclear layer but also among the inner segments and outer segment debris as if some of the dying nuclei had been thrust out of their domain (Figure 7). Control rats of the same age are entirely unlabeled.

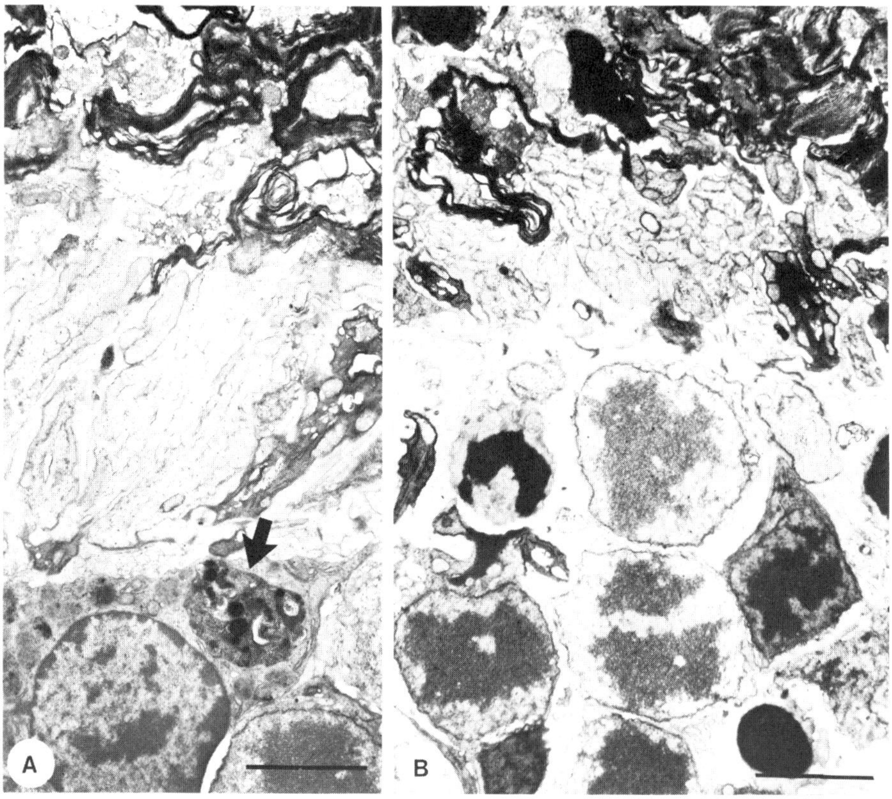

Figure 6. Electron micrographs of 45 day old RCS rat retinas. Note the lamellar debris accumulating at the interface between the tips of the outer segments and the apical surfaces of the retinal pigment epithelial (RPE) cells as a consequence of the phagocytic defect of the RPE. **A.** A photoreceptor cell (? cone) has ingested a large apoptotic body (arrow). **B.** Inner segment cytoplasm of some of the dying cells is condensed, mitochondria are contracted and nuclei are shrunken, pyknotic and some are fragmented. Bar 3 μm.

LaVail's group has demonstrated partial preservation of these retinas by injection of 1μg of basic fibroblast growth factor (bFGF) into the vitreous or into the interphotoreceptor space[39,40]. Recent reports, however, indicate that the RCS rat expresses normal levels of bFGF prior to the onset of degeneration[41]. Thus, the response to exogenous bFGF may either reflect a vulnerable system which is responsive to elevated levels of bFGF or a problem in release and/or uptake. Faktorovich et al.[42], point out, however, that bFGF induces macrophage migration into the retina. Shahinfar et al.[43] have decreased the extent of photoreceptor cell loss by inhibition of protein synthesis, also a characteristic feature of apoptotic cell death. Ischemia is thought to kill cells by a necrotic, not an apoptotic mechanism. Yet protection against ischemic neuronal cell death is achieved with flunarizine, an intracellular calcium overload blocker[44]. The same drug also reduced

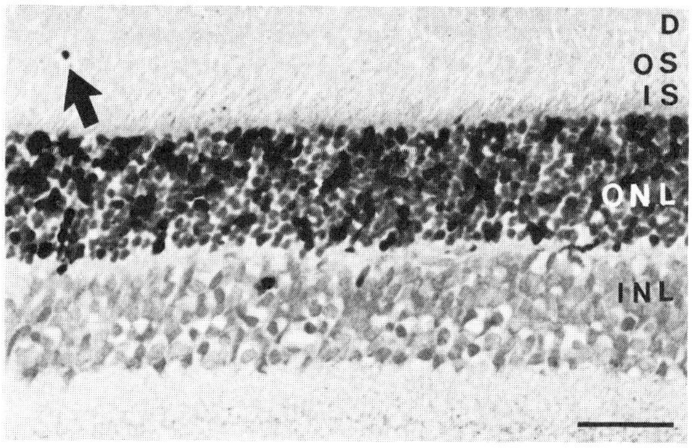

Figure 7. Light micrographs of a 28 day old RCS rat retina stained by the procedure of Gavrieli et al.[38] to detect apoptosis. Numerous stained nuclei and nuclear fragments are distributed at high density throughout the outer nuclear layer. About 10% of the photo-receptor nuclei are stained and small stained fragments are scattered throughout the outer nuclear layer (ONL) and between inner segments (IS) and outer segment (OS) debris (D) (arrow). Only a rare stained nucleus or fragment is seen in the inner nuclear layer (INL). In normal rat controls no stained nuclei are seen in the entire retinal section (data not shown). Counterstained by 1% aqueous methyl green. Bar 20 μm.

photoreceptor cell death in light induced retinal degeneration[45]. Buchi[46] has observed morphologic changes in retinas injured by ischemia-reperfusion that resemble apoptosis. Moreover, Oberhammer et al.[47], have emphasized that bFGF induces apoptosis of hepatocytes. Thus, the opportunity to alter photoreceptor death pharmacologically appears to be valuable but generalization about mechanisms of photoreceptor death is still difficult.

Faktorovich et al.[40] made the important observation that photoreceptor preservation was greater along the needle track and suspected other cytokines and growth factors might also be effective. Using an alternate model of constant light exposure for 7 days which induces a rapid rod cell death, they tested numerous agents and found several which could replace bFGF, including aFGF and IL-2[39]. We can anticipate their further studies of these agents on inherited retinal degenerations. Their results support another aspect of apoptotic cell death in nervous tissues and endocrine sensitive organs, i.e., the protection from apoptosis of various cells by additions of growth factors, cytokines and hormones[48,49]. The studies of this aspect of retinal degeneration are obviously at an early stage.

These results provide some of the morphological and biochemical correlates to fulfill the test of the hypothesis that apoptotic cell death is the mechanism of cell loss in these models of inherited retinal degeneration. Obviously we need to evaluate more mice and rats at more ages, and more varieties of retinal degeneration to extend these studies before any generalizations would be appropriate. But at the very least, they finally provide a rational starting point for the study of the dying photoreceptor.

The lack of concordance of the molecular defect and the rate and course of retinal degeneration is also a feature in humans affected by autosomal dominant retinitis pigmentosa and some cases of macular dystrophy caused by mutations of the rhodopsin and the peripherin gene. One important kindred highlights the entire dilemma. The heterozygous, middle-aged propositus carries a missense mutation of rhodopsin at proline 23 (Pro-23-His) and is completely blind as a result of degeneration of both her rods and cones[1]. Why should her cones die when they don't express the rhodopsin gene? Her heterozygous older sister, who carries the same mutation, drives a truck at night and yet has virtually no rod function. She functions well because her cones have survived and artificial lighting is sufficient for her to see. Clearly, if we can learn why her cones are protected, we would have information of great value for that family, especially if the protective agent could be isolated and provided to the affected offspring.

We are just initiating studies of human retinas by the assays described above. Hopefully we will be able to ascertain more about the mechanisms of cell death and design studies of their mouse homologues that can be fruitfully tested for anti-apoptotic therapies to ameliorate or eliminate the premature cell death of these important neurons.

Acknowledgments: The authors are indebted to Ms. Nancy Ransom, Shellye Lampkin and Dena Edwards for their excellent technical support and Ms. Carol Masch for preparing the manuscript. We benefitted greatly by helpful discussions and methods of procedure contributed by Dr. Tim McDonnell, M.D. Anderson Hospital, Houston and the kind gift of RCS rat eyes by Dr. Matthew LaVail, UCSF, San Francisco.

REFERENCES

1. E.L. Berson, Retinitis pigmentosa, The Friedenwald Lecture. *Invest. Ophthamol. Vis. Sci.* 34:1659 (1993).
2. C. Bowes, T. Li, M. Danciger, L.C. Baxter, M.L. Applebury, and D.B. Farber, Retinal degeneration in the *rd* mouse is caused by a defect in the beta subunit of rod cGMP-phosphodiesterase, *Nature* 347:677 (1990).
3. M.L. Suber, S.J. Pittler, N. Qin, G.C. Wright, V. Holcombe, R.H. Lee, C.M. Craft, R.N. Lolley, W. Baehr, and R.L. Hurwitz, Irish Setter dogs affected with rod/cone dysplasia contain a nonsense mutation in the rod cGMP phosphodiesterase beta subunit gene, *Invest. Ophthalmol. Vis. Sci.* 34:769 (1993).
4. G.J. Farrar, P. Kenna, S.A. Jordan, R. Kumar-Singh, M.M. Humphries, E.M. Sharp, D.M. Sheils, and P. Humphries, A three-base-pair deletion in the peripherin-*RDS* gene in one form of retinitis pigmentosa, *Nature* 354:478 (1991).
5. K. Kajiwara, L.B. Hahn, S. Mukai, G.H. Travis, E.L. Berson, and T.P. Dryja, Mutations in the human retinal degeneration slow gene in autosomal dominant retinitis pigmentosa, *Nature* 354: 480 (1991).
6. G.H. Travis, M.B. Brennan, P.E. Danielson, C.S. Kozak, and J.G. Sutcliffe, Identification of a photoreceptor-specific mRNA encoded by the gene responsible for retinal degeneration slow (*rds*), *Nature* 338:70 (1989).
7. G. Connell, R. Bascom, L. Molday, D. Reid, R.R. McInnes, and R.S. Molday, Photoreceptor peripherin is the normal product of the gene responsible for retinal degeneration in the *rds* mouse, *Proc. Natl. Acad. Sci. USA* 88:723 (1991).
8. M. Seabra, J. Goldstein, T. Sudhof, and M. Brown, Rab geranylgeranyl transferase - A multi-subunit enzyme that prenylates GTP-binding proteins terminating in Cys-X-Cys or Cys-Cy, *J. Biol. Chem.* 267: 14497 (1992).
9. M. Seabra, M. Brown, and J. Goldstein, Retinal degeneration in choroideremia: Deficiency of Rab geranylgeranyl transferase, *Science* 259: 377 (1993).
10. B.G. Schneider, Na^+, K^+-ATPase isoforms in the retina, *Int. Rev. Cytol.* 133:151 (1991).
11. M. Spencer, P.B. Detwiler, and A.H. Bunt-Milam, Distribution of membrane proteins in mechanically dissociated retinal rods, *Invest. Ophthalmol. Vis. Sci.* 29: 1012 (1988).
12. I. Nir, N. Agarwal, G. Sagie, and D.S. Papermaster, Opsin distribution and synthesis in degenerating photoreceptors of *rd* mutant mice, *Exp. Eye Res.* 49: 403 (1989).
13. I. Nir and D.S. Papermaster, Immunocytochemical localization of opsin in the inner segment and ciliary plasma membrane of photoreceptors in retinas of *rds* mutant mice, *Invest. Ophthalmol. Vis. Sci.* 27:836 (1986).
14. I. Nir and D.S. Papermaster, Immunocytochemical localization of opsin in degenerating photoreceptors of RCS rats and *rds* mice, *in:* "Inherited and

Environmentally Induced Retinal Degenerations," M. LaVail, J. Hollyfield and R.E. Anderson, eds, Alan R. Liss, Inc, New York, p 251 (1989).
15. Z.Y. Li, S.G. Jacobson, and A.H. Milam, Histopathology of the retinas from a patient with autosomal dominant retinitis pigmentosa, Thr-17-Met rhodopsin mutation, *Invest. Ophthalmol. Vis. Sci.* 34:1459 (1993).
16. J.F.R. Kerr, A.H. Wyllie, and A.R.Curie, Apoptosis: A basic biological phenomenon with ranging implications in tissue kinetics, *Br. J. Cancer* 26:239 (1972).
17. P.L. Penfold and J.M. Provis, Cell death in the development of the human retina: Phagocytosis of pyknotic and apoptotic bodies by retinal cells, *Graefe's Arch. Clin. Exp. Opthalmol.* 224:549 (1986).
18. R.O.L. Wong and A. Hughes, Role of cell death in the topogenesis of neuronal distributions in the developing cat retinal ganglion cell layer, *J. Comp. Neurol.* 262:496 (1987).
19. M.M. LaVail, Circadian nature of rod outer segment disc shedding in the rat, *Invest. Ophthalmol. Vis. Sci.* 19:407 (1980).
20. R.W. Young, Visual cells and the concept of renewal, *Invest. Ophthalmol. Vis. Sci.* 15:700 (1976).
21. D.S. Papermaster and B.G. Schneider, Biosynthesis and morphogenesis of outer segment membranes in vertebrate photoreceptor cells *in:* "Cell Biology of the Eye," David S. McDevitt, ed. Academic Press, NY, p. 475 (1982).
22. D.S. Papermaster, B.G. Schneider, D. DeFoe, and J.C. Besharse, Biosynthesis and vesicular transport of opsin on vesicles in retinal rod photoreceptors, *J. Histochem. Cytochem.* 34:5 (1986).
23. N. Agarwal, I. Nir, and D.S. Papermaster, Opsin synthesis and mRNA levels in dystrophic retinas devoid of outer segments in retinal degeneration slow (*rds*) mice, *J. Neuroscience* 10:3275 (1990).
24. R.S. Molday, D.D. Hicks, and L. Molday, Peripherin, a rim-specific membrane protein of rod outer segment discs, *Invest. Ophthalmol. Vis. Sci.* 28:50 (1987).
25. G.H. Travis, K.R. Groshan, M. Lloyd, and D. Bok, Complete rescue of photo-receptor dysplasia and degeneration in transgenic retinal degeneration slow (*rds*) mice, *Neuron* 9:113 (1992).
26. S. Sanyal, A. de Ruiter, and R.K. Hawkins, Development and degeneration of retina in *rds* mutant mice: light microscopy, *J. Comp. Neurol.* 194:193 (1980).
27. R.H. Hawkins, H.G. Jansen, and S. Sanyal, Development and degeneration of retina in *rds* mutant mice: Photoreceptor abnormalities in the heterozygous, *Exp. Eye Res.* 41:701 (1985).
28. I. Nir and N. Agarwal, Arrestin mRNA expression, biosynthesis and localization in degenerating photoreceptors of mutant *rds* mice retinas, *J. Comp. Neurol.* 308:1 (1991).
29. H.G. Jansen and S. Sanyal, Development and degeneration of retina in *rds* mutant mice: Electron microscopy, *J. Comp. Neurol.* 224:71 (1984).
30. J.H. Reuter and S. Sanyal, Development and degeneration of retina in *rds* mutant mice: the electroretinogram, *Neurosci. Lett.* 48:231 (1984).
31. I. Nir, N. Agarwal, and D.S. Papermaster, Opsin gene expression during early and late phases of retinal degeneration in *rds* mice, *Exp. Eye Res.* 51:257 (1990).
32. I. Nir, N. Agarwal, and D.S. Papermaster, IRBP gene expression in the dystrophic retina of the mutant *rds* mouse *in:* Progress in Clinical and Biological Research."Inherited and Environmentally Induced Retinal Degenerations", RE Anderson, JG Hollyfield and MM LaVail, eds. Alan Liss, Inc. p. 489 (1991).
33. N. Agarwal, I. Nir, and D.S. Papermaster, Expression of opsin and IRBP genes in mutant RCS rats, *Exp. Eye Res.* 54:545 (1992a).
34. N. Agarwal, I. Nir, and D.S. Papermaster, A lack of diurnal cycle of S-Ag in *rds* mutant mice retina may be deleterious to photoreceptors, *Invest. Opthalmol. Vis. Sci.* 33:Suppl. 1064 (1992b).
35. I. Nir and N. Agarwal, Diurnal expression of *c-fos* in the mouse retina. *Molec. Brain Res.* in press (1993).
36. K. O'Rourke, C. Jomery, S.E. Jones, and N. Agarwal, Clusterin mRNA is overexpressed in *rds* mutant mouse retina, *Invest. Opthalmol. Vis. Sci.* 34:768 Suppl. (1993).
37. S.E. Jones, J.M.A. Meerabux, D.A. Yeats, and M.J. Neal, Analysis of differentially expressed genes in retinitis pigmentosa retinas, *FEBS Lett.* 300:279 (1992).

38. Y. Gavrieli, Y. Sherman, and S. Ben-Sasson, Identification of programmed cell death *in situ* via specific labeling of nuclear DNA fragmentation, *J. Cell Biol.* 119:493 (1992).
39. M.M. LaVail, K. Unoki, D. Yasumura, M.T. Matthes, G.D. Yancopoulos, and R.H. Steinberg, Multiple growth factors, cytokines, and neurotrophins rescue photoreceptors from the damaging effects of constant light, *Proc. Natl. Acad. Sci. USA.* 89:11249 (1992).
40. E.G. Faktorovich, R.H. Steinberg, D. Yasumura, M.T. Matthes, and M.M. LaVail, Basic fibroblast growth factor and local injury protect photoreceptors from light damage in the rat, *J. Neurosci.* 12(9):3554 (1992).
41. H. Naka, A. Hayashi, H. Kuriyama, and Y. Tano, Difference in the post-natal expression of bFGF between RCS and Long Evans rats, *Invest. Ophthalmol. Vis. Sci.* 34:1077 Suppl. (1993).
42. E.G. Faktorovich, R.H. Steinberg, D. Yasumura, M.T. Matthes, and M.M. LaVail, Photoreceptor degeneration in inherited retinal dystrophy delayed by basic fibroblast growth factor, *Nature* 347:83 (1990).
43. S. Shahinfar, D.P. Edward, and M. Tso, A pathologic study of photoreceptor cell death in retinal photic injury, *Curr. Eye Res.* 10:47 (1991).
44. P.J. Pauwels, J.E. Leysen, and P.A.J. Janssen, Minireview - Ca^{++} and Na^{+} channels involved in neuronal cell death, Protection by Flunarizine, *Life Sciences* 48:1881 (1991).
45. D.P. Edward, T.T. Lam, S. Shahinfar, J. Li, and M. Tso, Amelioration of light-induced retinal degeneration by a calcium overload blocker, *Arch. Ophthalmol.* 105:554 (1991).
46. E.R. Buchi, Cell death in rat retina after pressure-induced ischaemia-reperfusion insult: Electron microscopic study II. Outer nuclear layer. *Jpn. J. Ophthalmol.* 36:62 (1992).
47. F.A. Oberhammer, M. Pavelka, S. Sharma, R. Tiefenbacher, A.F. Purchio, W. Bursch, and R. Schulte-Hermann, Induction of apoptosis in cultured hepatocytes and in regressing liver by transforming growth factor beta 1, *Proc. Natl. Acad. Sci. USA* 89:5408 (1992).
48. R.W. Oppenheimer, Cell death during development of the nervous system, *Am. Rev. Neuro.* 14:453 (1991).
49. W. Bursch, F. Oberhammer, R. Schulte-Herman, Cell death by apoptosis and its protective role against disease, *Trends Pharmacol. Sci.* 13:245 (1992).

Discussion

S. Schlossman

When you shed rhodopsin, do you shed the whole membrane or do you cleave a portion of the molecule on the membrane?

D. Papermaster

What appears to happen in the frog is that the plasma membrane of distal 10% of the outer segment begins to invaginate. The plasma membrane seems to undergo almost a cytokinesis event that separates the distal from the proximal end of the outer segment. The pigment epithelial processes follow that invagination inward. As the shed tip separates it is phagocytosed and fuses with a phagolysosome of the pigment epithelial cell. An alternative model is that the disk that lies between the distal end that is going to be shed and the proximal end that is going to be saved is split or vesiculates so that the disk membrane becomes part of the distal plasma membrane creating a new plasma membrane which is the gap. Both of these models have evidence to support them.

S. Schlossman

Are other proteins transported with rhodopsin?

D. Papermaster

Yes, at least a fraction of the transduction cascade is shed in the tip and resynthesized like rhodopsin. Some of the proteins appear to be co-transported at the same time with rhodopsin as we showed recently [Deretic and Papermaster, J. Cell Biol. 113:1281 (1991)]. These proteins appear to be transported at least in part on vesicles that are carrying rhodopsin from the Golgi apparatus out to the site of the insertion.

D. Hockenbery

In the *rds* mouse model, do alterations in the light-dark cycle influence the timing of cell death? Are there any experiments which maintain photoreceptor cell viability?

D. Papermaster

We have not yet looked at any of the factors affecting survival of *rds* mouse retinas. The only studies that have gone on, that I know of, in the case of retinal degenerations and factors, are the studies of RCS rats I have cited from La Vail's laboratory by Faktorovich, *et al.*, where they showed that basic FGF also inhibits cell death. There is another model which is not an inherited disorder. If you expose mice or rats to bright light for 7 days, their outer segments, and retinas, will degenerate. This is a faster way to do the study so LaVail's group has been using that model of late. Blasting the eye with light and giving the various factors prevents them from dying. Our problem in applying their approach to *rds* mice, of course, would be that, unlike the light induced model where all the cells die within a week, we would have to watch over a prolonged period.

G. Lenaz

What is known about the structure of peripherin and is it a membrane protein or is it known how it is involved in the sorting of rhodopsin and stabilization in the outer segment membrane?

D. Papermaster

The peripherin molecule has, in its sequence, four putative transmembrane domains. It is a glycoprotein. In the disk, the glycoprotein domains of both rhodopsin and peripherin would be inside the disk. There is a ctyoplasmic domain and it is that domain which is thought, possibly, to bind also to some kind of cytoskeleton, to link the disk also to some plasma membrane component so that these disks do not just go spinning around in the cytoplasmic space of the outer segment. They do remain ordered along the length of the outer segment. In the case of the peripherin *rds* mutations, there are several now that have been discovered in humans and one in the *rds* mouse. In the mouse, a retroviral insertion interrupts the sequence and yields a truncation mutation. The protein may not be synthesized; the messenger may be unstable. We do not know if the protein is synthesized in the RDS human retina because the tissues are not yet available for study. Comparable mutations occur in the related protein which is called ROM-1. They are missense mutations. What is peripherin's role in this morphogenesis? We do not know, except that it is in the right place.

G. Evan

Can you tell me anything about retinopathies associated with diabetes? The reason I ask is that IGF's, and obviously insulin, are very potent promiscuous survival factors.

D. Papermaster

We do not really understand the nature of diabetic retinopathy. What we can say is what we see phenomenologically. The disease does not, *a priori* appear to be terribly different from diabetic glomerulopathy. It appears to be primarily a disorder induced by vascular injury, and secondarily a disorder in the retina in the sense that we see the same kinds of little berry aneurysms in the vessels of the retina as one sees in the Kimmeisteil Wilson changes of the glomerulae of the diabetic kidney. Excess extracellular forms matrix forms around the pericytes and the endothelial cells of the diabetic vessel. The retina becomes focally ischemic as a consequence of the closure of these vessels and then a vast proliferation of vessels occurs into the ischemically injured area. That is also why the current therapy for this disease is to "staple" the retina down with many laser burns in a matrix all over the retina, hoping to forestall a massive bleed to preserve retinal life. It is a crude and yet necessary clinical response until we can learn more about vascular proliferation.

R. Schimke

I am not a phagocyte biologist by any means but I was fascinated by several of your comments including the RDS rat where the retinal epithelia appears to be acting as phagocytic cells. I wonder if you could comment, in general, about how you activate adjacent cells that are not necessarily wandering macrophages to become phagocytic, and what "signals" they might recognize in the cell to be engulfed?

D. Papermaster

With respect to activation of phagocytosis, RPE phagocytosis is phenotypically delayed in normal retinas until the outer segment reaches the right size. The diurnal shedding event is then triggered. How this cross-talk is maintained, so far, is not at all clear. We do know, for example, that the retinal pigment epithelium of an RCS rat, is perfectly capable of phagocytosing bacteria or latex particles without difficulty at any age. So they do not have a phagocytosis defect *per se*. They have a defect in the specific phagocytosis of the shed outer segment debris. The defect is on the PE side not on the photoreceptor side, for the reasons I cited in the chimeric rats. With regard to the role of these pigment epithelial cells in phagocytosis in the apoptotic RCS rat retina, I have no information about that yet because I have really only begun to look at these retinas recently. What I can tell you is that there is an enormous layer of outer segment debris which is nearly the thickness of a normal outer segment layer lying between the pigment epithelium and the dying cells. As I searched within these retinas, I did not see the kind of intimate relationship of the outer segment surface of dying retinal cells with the apical surface of the PE cells. In contrast, I was at least able to see that phenomenon in the *rds* mouse. Because there is so much cell death throughout the entire layer of nuclei of the RCS rat photoreceptors and the death is occurring at all levels, my first impressions are that the dying photoreceptors are being taken up by the adjacent cells and not predominantly by the PE cells.

M. Fried

You said the FGF and 7 other compounds stop cells from dying. Do they have a similar mechanism? Are they inducing other cytokines or growth factors?

D. Papermaster

No, these are not my experiments, they are done by Faktorovich in Matthew LaVail's laboratory and by Matt LaVail in his laboratory, and these studies have been published and were cited in my paper. Basically what they have shown is; if I can remember the list of active ones, IL-1b, acidic FGF, CNTF, and BDNF were especially active as I recall. The point was that there was no apparent specificity of type as far as I know.

M. Fried

But they are all growth factors or cytokines?

D. Papermaster

That is correct.

WILD TYPE p53 ACTIVITY CONTRIBUTES TO DEPENDENCE ON HEMATOPOIETIC SURVIVAL FACTORS

Eyal Gottlieb[1], Rebecca Haffner[1], Elisheva Yonish-Rouach[1,2] Thomas von Ruden[3], Erwin Wagner[3], and Moshe Oren[1]*

[1]Department of Chemical Immunology, The Weizmann Institute of Science, Rehovot 76100, Israel
[2]Institut de Recherches Scientifiques sur le Cancer, 94891 Villejuif Cedex, France
[3]Research Institute of Molecular Pathology (IMP), A-1030, Vienna, Austria

INTRODUCTION

The p53 protein is the product of a tumor suppressor gene, whose inactivation is most probably involved in the development of many types of cancer (for recent reviews on p53 see 1-6). In most cases, tumor cells harbor point mutations in the p53 gene, resulting in the overproduction of mutant forms of the protein. Whereas the wild-type (wt) form of p53 can exhibit a variety of anti-proliferative and tumor-inhibitory activities, mutants of the types found in cancer cells are typically devoid of such activities. Hence, the principal outcome of such mutations is probably the inactivation of the tumor suppressor function of wt p53. In addition, at least certain mutations may also confer a distinct gain of function[7], thereby possibly contributing a novel oncogenic activity to the affected cell.

The biochemical basis for the tumor suppressor properties of wt p53 is still under investigation. There is mounting evidence that the underlying mechanism relies on the capability of p53 to act as a sequence-specific transcription factor (see 1-6 for reviews). The wt p53 protein can bind tightly to specific DNA elements, and activate transcription from promoters residing in the vicinity of such p53 binding sites. On the other hand, typical tumor-derived p53 mutants fail to interact with these p53 response elements, and consequently do not transactivate target promoters. In addition, wt p53 can repress transcription from a wide array of promoters, probably through interacting with components of the basal transcription machinery[8-12]. It therefore seems plausible that the biological effects of wt p53 may be exerted through its ability to modulate the expression of a battery of target genes. Further insight into this issue will require the identification of these putative genetic targets of p53.

The rate of growth of a tumor is dictated by the balance between two opposing processes: cell proliferation and cell death[13]. In principle, a tumor suppressor gene may affect any of the two processes. Thus wt p53 could suppress tumor development either

Apoptosis, Edited by E. Mihich and R.T. Schimke
Plenum Press, New York, 1994

through inhibiting the preferential proliferation of neoplastic cells, or through accelerating their death. Early studies provided compelling evidence for an anti-proliferative capacity of wt p53[1-6], believed to account for an important part of its *in vivo* effects. More recent findings have raised the possibility that, under physiological conditions, the principal signal responsible for triggering the anti-proliferative action of wt p53 is DNA damage[14-16]. In addition, however, it is becoming clear that wt p53 may also interfere with tumorigenic processes by contributing to the induction of cell death.

The first clues to a possible role of wt p53 in mediating cell death came from experiments with the M1 mouse myeloid leukemic cell line[17]. These cells are devoid of any detectable p53 protein, and actually fail to produce the corresponding mRNA. In order to study the consequences of wt p53 activation in M1 cells, such cells were stably transfected with a temperature sensitive (ts) mutant of murine p53, p53val135. The protein encoded by the ts mutant has features similar to those of typical oncogenic p53 mutants when the cells are maintained at 37.5°C. When such cells are shifted down to 32°C, the protein regains properties very much like those of authentic wt p53[18]. M1 cells harboring the ts p53 could be propagated efficiently at 37.5°C, where they were practically indistinguishable from parental M1. However, upon down-shift to 32°C and the subsequent induction of wt-like p53 activity, extensive cell death was induced very readily[17]. The dying cells exhibited typical apoptotic features, including distinct morphological alterations and internucleosomal DNA fragmentation, giving rise to a diagnostic DNA "ladder"[17]. Hence, reconstitution of wt p53 activity in a p53-deficient leukemic cell line induced apoptosis.

Subsequent studies have confirmed the ability of wt p53 to elicit apoptosis in a variety of tumor cell types, of both hematopoietic[19,20] and non-hematopoietic[21] origin. Moreover, the expression of functional wt p53 has recently been shown to be required for the efficient induction of apoptosis by ionizing radiation[22,23] and by the adenovirus E1A oncoprotein [24,25]. All these findings argue in favor of a physiological role for wt p53 in mediating signal-dependent cell death in a variety of systems.

RESULTS

Relationship of p53-mediated apoptosis to cell cycle progression

Perhaps the best documented biological effect of excess wt p53 activity is the induction of a viable growth arrest, typically occurring in the G1 phase of the cell cycle[1-6]. The apoptotic activity of wt p53 could therefore either constitute a secondary effect of a p53-dependent growth arrest, or else could represent a distinct facet of wt p53 function. In an attempt to distinguish between these two possibilities, we investigated the relationship between cell cycle progression and p53-mediated apoptosis in M1 cells expressing the ts p53. No evidence for any type of growth arrest could be found in such cells maintained at 32°C[26]. In fact, cells appeared to move uninterruptedly through the cell cycle until they eventually died. Thus, the ability of wt p53 to elicit apoptosis in M1 cells did not require a prior establishment of a block to cell proliferation.

There is, however, a clear connection between cell cycle progression and the apoptotic action of wt p53. Fractionation of proliferating ts p53-containing M1 cells by centrifugal elutriation, followed by a shift-down to 32°C, revealed that cells which possess a G1 DNA content at the time of wt p53 activation are the first to undergo apoptosis[26]. This suggested the existence of a particular point in G1, where cells can become committed to apoptosis in the presence of active wt p53. A very similar situation was also encountered in murine erythroleukemia cells subjected to the wt-like activity of p53val135 at 32°C[19]. It therefore appears likely that p53 acts preferentially at a decision point in late G1, perhaps owing to the availability of appropriate molecular targets during that particular window. The consequences of such action will depend on the specific cellular context. Whereas in some cell types p53 will merely block transition into S phase and establish a reversible growth arrest, in others it will set off an apoptotic process, resulting in cell death.

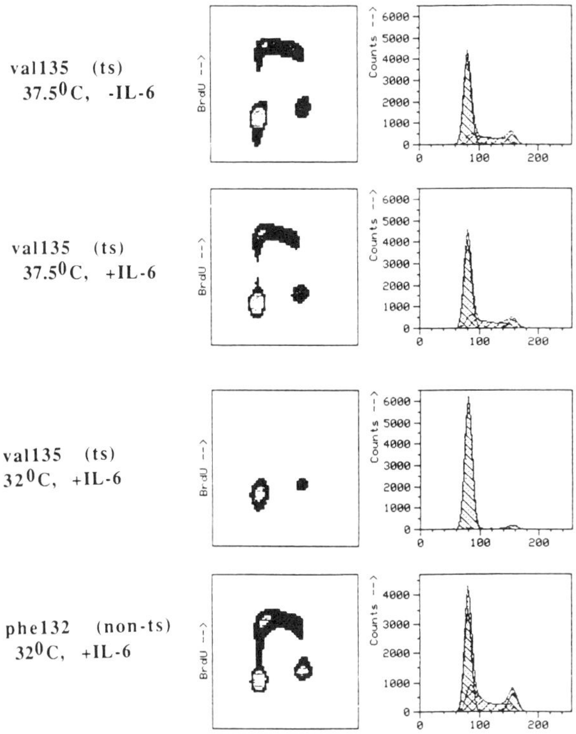

Figure 1. Induction of an efficient G_0-like growth arrest by the joint action of wt p53 and IL6. M1 cells transfected with p53val135 were maintained at either 37.5°C or 32°C, in the absence or presence of IL6. M1 cells expressing comparable levels of a non-ts mutant, p53phe132, were used as controls for temperature effects. Cell cycle distribution was assayed by FACS analysis of propidium iodide-stained permeabilized cells (right), whereas DNA synthesis was assayed by BrdU incorporation. See (26) for further experimental details.

Induction of apoptosis by wt p53 is inhibited by IL6

M1 cells are highly responsive to interleukin 6 (IL6). The effect of IL6 on p53-mediated M1 apoptosis was therefore tested. It was found that IL6 could exert a strong protective effect on cells exposed to wt-like p53 activity at 32°C[17]. Interestingly, the combined action of wt p53 and IL6 resulted in a pronounced alteration in cell cycle distribution. Whereas exposure to wt p53 alone had no effect at all on the cell cycle distribution of M1 cells[26], and treatment with IL6 alone (at 37.5°C) resulted only in a partial accumulation of the population in G1 (Fig. 1), cells exposed to a combination of wt p53 and IL6 exhibited an almost complete G1 arrest (Fig. 1). In the latter case, the cells were also smaller and contained a reduced amount of RNA (data not shown). These properties suggest that IL6 and wt p53 actually synergize to induce a G_0-like state, which is probably similar to that found under physiological conditions in normal non-proliferating myeloid cells. Such notion is in line with the idea that apoptosis may be brought about by a combination of concurrent conflicting signals which can not be reconciled[27,28]. M1 cells may be continuously exposed to a constitutive growth promoting signal, which keeps them moving incessantly through the cell cycle. Activation of wt p53 probably generates a contrasting inhibitory signal, which can not be reconciled with the constitutive growth promoting one; such unresolvable conflict may turn on the apoptotic pathway. In contrast, when IL6 is added, the balance is shifted strongly towards the growth inhibitory end. This could either result from a direct ability of IL6 to shut off the otherwise constitutive positive signal, or from the synergistic effect of two inhibitory stimuli (wt p53 and IL6), which in concert give rise to a negative signal which is much more prominent than the constitutive positive one. In either event, the combination of IL6 and wt p53 may now allow the cells to undergo a more efficient growth arrest, and thereby be spared from apoptosis. If there is indeed a requirement for transit through a particular window in G1 in order to enter the apoptotic pathway, it seems plausible that cells exposed to wt p53 plus IL6 may arrest prior to reaching that window, and thus be excluded from the apoptotic pathway.

IL6 has pleiotropic effects on M1 cells, including the induction of differentiation, inhibition of cell proliferation and modulation of the expression of numerous genes[29,30]. It could therefore be argued that IL6 also acts as a general inhibitor of cell death in M1 cells, and that there is no specific relationship between its action and the presence of wt p53. This, however, is clearly not the case. In fact, IL6 has absolutely no protective effect when M1 apoptosis is induced by serum starvation[26].

Another possible explanation for the protective effect of IL6 may have to do with its ability to trigger M1 cell differentiation. If differentiated cells are intrinsically immune to the apoptotic effects of wt p53, IL6 could simply be acting through the induction of a differentiated state. In this case, however, the presence of IL6 should become non-essential once cells have undergone efficient differentiation. Experimental evidence has clearly established that this is not the case. In fact, cells which have undergone extensive IL6-triggered differentiation in the presence of wt p53 are still at least as dependent on the continuous presence of IL6 for survival as are their untreated counterparts[26]. All the above observations argue that IL6 can selectively interfere with p53-mediated cell death.

A most attractive way of interpreting the ability of IL6 to protect cells from p53-dependent apoptosis is by assuming that activation of wt p53 renders M1 cells survival factor-dependent. Normal myeloid progenitors, from which the M1 line was derived, probably depend on survival factors for maintaining their viability within the bone marrow. When such factors become limiting, excess cells will undergo apoptosis. Any genetic alteration which can reduce the extent of dependence on survival factors is expected to confer a selective survival advantage in the very competitive bone marrow environment. One could propose that the ongoing expression of wt p53 plays a role in maintaining the need for survival factors under physiological conditions. In that case, the loss of wt p53 expression may cause myeloid progenitor cells to become less factor dependent; such mechanism may have contributed to the emergence of the M1 leukemia *in vivo* or to its establishment as a stable leukemic cell line *in vitro*. According to this model, reconstitution of wt p53 expression may regenerate dependence on the appropriate survival factors. As long as the factors are not present, such p53-reconstituted cells will be triggered to undergo apoptosis. However, addition of the original limiting survival factor, or of another factor which can efficiently substitute for it in generating the correct signals, will satisfy the requirement and ensure cell viability. In support of this conjecture, p53-mediated cell death can be inhibited by either IL6 or IL3[26], both of which can serve as survival factors for normal myeloid

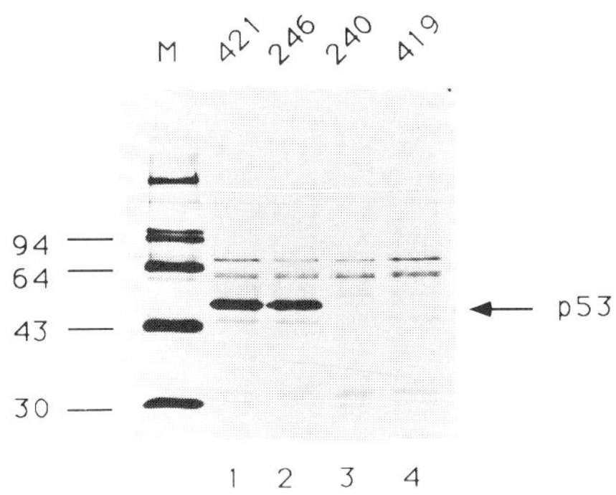

Figure 2. Analysis of biosynthetically labeled p53 from 32D cells. Extracts were immunoprecipitated with the p53-specific monoclonal antibodies PAb421 (lane 1), PAb246 (lane 2), PAb240 (lane 3), or with a control monoclonal antibody (PAb419, directed against the SV40 large T antigen; lane 4). M, molecular weight markers.

progenitors under a variety of experimental conditions. In the case of IL3, protection from p53-mediated apoptosis requires the prior induction of the synthesis of IL3 receptors[26].

Inactivation of wt p53 interferes with apoptosis in IL3-dependent cells deprived of IL3

The ability of IL6 to inhibit p53-mediated M1 cell death may be due to the ability of wt p53 to reinstate dependence on hematopoietic survival factors. However, this particular experimental system is very artefactual, being based on the vast overexpression of transfected wt-like p53 in cells which have probably been selected to proliferate in the absolute absence of p53. It was therefore of importance to determine whether p53 played a role in more physiological situations, in which apoptosis was induced by survival factor deprivation. To that end, we sought to monitor the effects of interfering with p53 function in factor-dependent hematopoietic cells which express endogenous wt p53. Two cell lines were chosen for those studies: 32D and DA-1. 32D is an IL3-dependent mouse myeloid cell line, which has previously been shown to undergo apoptosis in response to IL3 withdrawal[27]. The p53 protein of 32D is reactive with the p53-specific monoclonal antibody PAb246, recognizing the wild type (wt) conformation of the protein, but not with PAb240, reactive with mutant forms of p53 (Fig. 2). In addition, the p53 protein in 32D has a short half-life, and is capable of sequence-specific DNA binding to p53 targets (data not shown). All these properties suggest that 32D cells express functional wt p53.

Antisense p53 DNA and DNA encoding the ts mutant p53val135 were introduced into plasmid pLXSN. The rationale for using the ts mutant is based on the observation that at 37.5^{o}C, when it possesses a typical mutant conformation, it can act as an effective negative dominant inhibitor of co-expressed wt p53 in cells harboring endogenous wt p53[31]. On the other hand, at 32°C p53val135 does not inhibit wt p53, but rather contributes its own wt-like p53 activity. Expression plasmids for the ts p53 and for antisense p53 were stably transfected, by electroporation, into 32D cells. This gave rise to clone 32Dv7, expressing p53val135 (Fig. 3, lane 4), and to clones 32Da2 and 32Da6, transfected with antisense DNA; 32Da2 displayed markedly reduced p53 levels (Fig. 3, lane 3). In 32Dv7 maintained at 37.5^{o}C, the mutant p53 protein was relatively stable, and reached high steady state levels (data not shown).

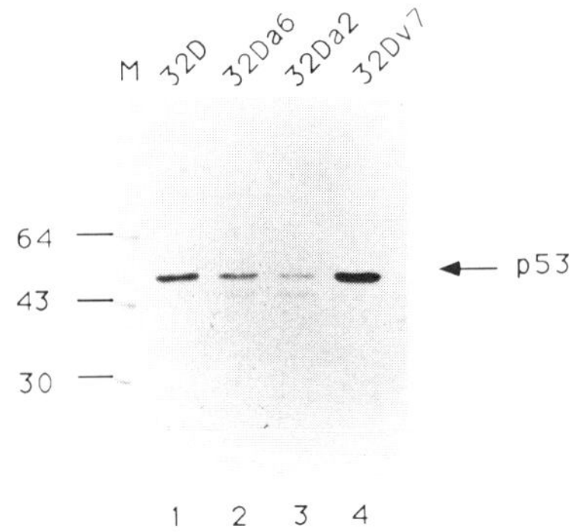

Figure 3. Analysis of biosynthetically labeled p53 from 32D-derived clones. Extracts were immunoprecipitated with PAb421. M, molecular weight markers.

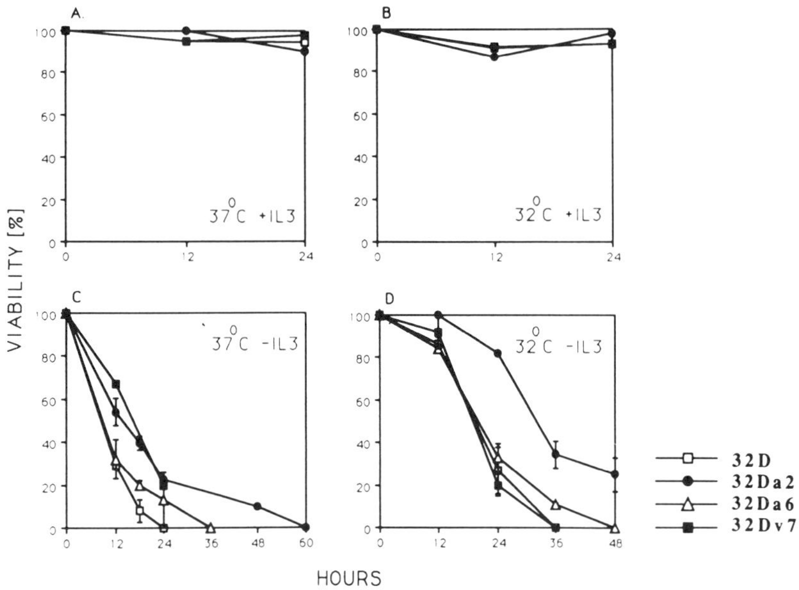

Figure 4. Kinetics of 32D cell death in the absence of IL3. 32D and their derivatives were maintained routinely at 37.5^{0}C in the presence of IL3. For factor withdrawal experiments, cells were washed twice in IL3-free medium, and resuspended in IL3-free medium at a cell density of 2x10^5/ml. Where appropriate, cells were transferred to 32^{0}C immediately after being washed and resuspended. Viability was determined at various time points by the ability to exclude trypan blue. Each determination was made in duplicate; error bars indicate standard deviation. See text for details of individual clones.

When parental 32D were deprived of IL3 at 37.5ºC, rapid cell death ensued. Both 32Dv7 cells, expressing ts p53, and 32Da2 cells, producing very little p53, died more slowly (Fig. 4C). In antisense clone 32Da6, retaining more p53, death was only marginally slower than in parental 32D. Hence abrogation of wt p53 activity, either by a negative dominant mutant or via antisense RNA, delayed 32D cell apoptosis.

A somewhat different picture emerged at 32ºC (Fig. 4D). Under those conditions, the p53val135 in 32Dv7 should regain wt-like activity, whereas 32Da2, expressing antisense RNA, will remain deficient in p53. Consistent with this notion, 32Dv7 now died as rapidly as parental 32D upon IL3 deprivation, whereas 32Da2 maintained extended viability.

A similar approach was taken with DA-1 cells. To achieve more efficient gene transfer and override clonal variation, we generated a high-titer recombinant retrovirus encoding p53val135. DA-1 were infected by co-cultivation with virus-producing cells at 37.5°C, giving rise to a G418-resistant population (DIV). When the viability of such p53val135-expressing cells was analysed following IL3 withdrawal at 37.5°C, it was found to be increased relative to that of DA-1 cells infected with a control retrovirus (DIN), conferring only G418 resistance (Fig.5). As with 32D, the protective effect of p53val135 on DA-1 was seen only at 37.5°C, where it acts as a dominant negative mutant. At 32°C, where the ts p53 exhibits wt-like p53 activity, its expression even slightly accelerated cell death. Importantly, excess wt p53 did not induce any loss of viability as long as IL3 was present.

VIABILITY OF INFECTED DA1 CELLS

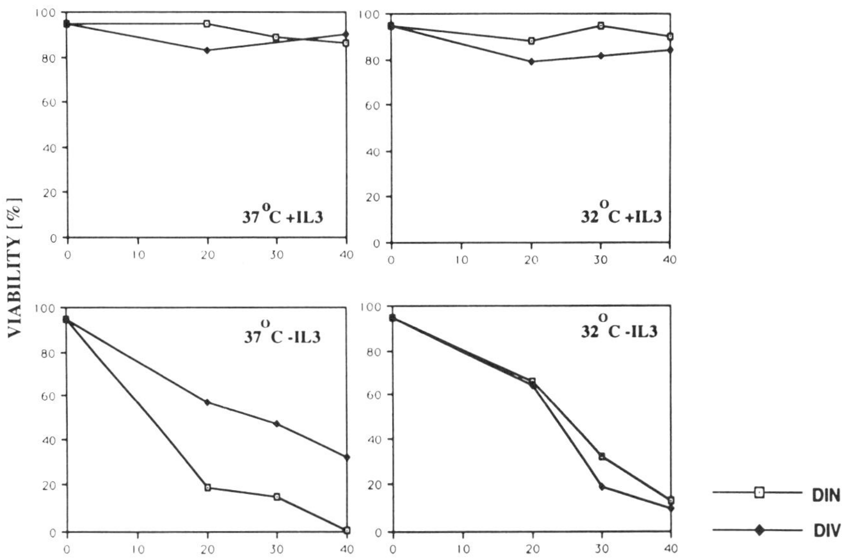

Figure 5. Kinetics of DA-1 cell death in the absence of IL3. DA-1 cells were infected with recombinant retroviruses encoding either p53val135 plus neo (DIV) or neo alone (DIN). Analysis of cell viability at different time points after IL3 withdrawal was performed essentially as in Fig. 4.

To obtain a more efficient negative-dominant inhibition of endogenous wt p53 in DA-1 cells, a recombinant retrovirus was generated which encodes a miniprotein (DD) containing the C-terminal 89 residues of the p53 protein[31]. The DD protein is an extremely effective negative dominant blocker of wt p53 function, apparently acting by forming functionally-defective oligomers with the wt protein[31]. DA-1 cells were infected with the DD retrovirus, and a G418-resistant polyclonal population of infected cells was obtained. The viability of these cells was determined following IL3 withdrawal; viability was monitored by staining non-permeabilized cells with propidium iodide, followed by FACS analysis. Only dead cells take up the dye under those conditions, and consequently they are far more fluorescent than viable cells. The results of such analysis are shown in Fig. 6. It is clear that whereas DA-1 cells infected only with the *neo* virus die rapidly upon factor withdrawal, the expression of the negative dominant DD miniprotein renders these cells far more refractory to apoptosis. Hence in DA-1, too, reduced wt p53 activity interferes with the induction of apoptosis in response to survival factor deprivation.

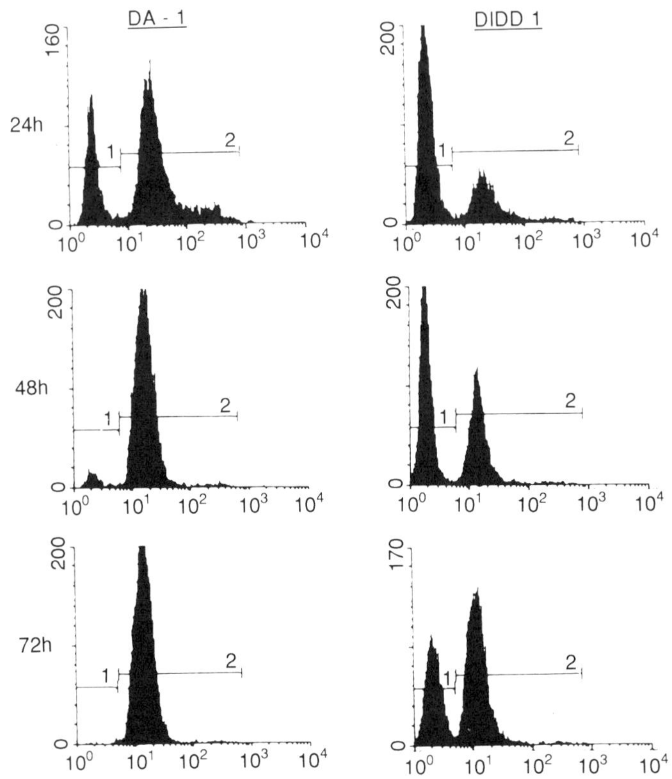

Figure 6. Inhibition of DA-1 cell death by overexpression of the DD miniprotein. Cell viability was assayed at various time points after IL3 withdrawal. DA1 and DIDD1 cells were incubated with propidium iodide without prior permeabilization, and subjected to FACS analysis. The high-intensity peak (fraction 2) represents dead cells, whereas the low-intensity peak (fraction 1) represents viable cells capable of excluding the dye.

DISCUSSION

The findings reported above suggest that one of the normal functions of wt p53 may be to maintain dependence on survival factors. In the absence of such factors, factor-dependent cells will be triggered to undergo apoptosis. The available data indicate that this function of wt p53 is not exerted uniformly throughout the cell cycle. Rather, wt p53 seems to drive the cell into the apoptotic pathway only within a defined window ("decision point") in the G1 phase. Presumably, wt p53 acts to establish a requirement for a survival factor-generated signal at this particular decision point (Fig. 7). When wt p53-expressing cells reach this point in the cycle, they will traverse it successfully only when such signal is present. In the absence of a survival factor-generated signal, the cell will be triggered to undergo apoptosis. However, in cells lacking functional wt p53, the decision point can be traversed even in the absence of the survival factor.

The nature of the signal emanating from the occupied survival factor receptor still remains to be defined. Conceivably, different receptors belonging to different families may use non-identical signals. Alternatively, all pathways may converge onto a common effector, which can overcome the apoptotic effect of activated wt p53. It is also possible that the physiological signal generated by the occupied survival factor receptor can be replaced by other stimuli, which impinge on the same end target. It will be of importance to identify the precise nature of the signal that is induced by the survival factor and can override p53-dependent apoptosis. An obvious candidate is the activation of bcl2, which is known to inhibit many types of apoptosis. It is possible that p53-mediated apoptosis will also be repressed by bcl2. However, at least in the case of M1 cells, bcl2 may not be a player in the prevention of p53-mediated apoptosis. First of all, M1 cells constitutively express relatively high levels of bcl2[32]. Moreover, these levels are not further increased following exposure to IL6 (data not shown). Alternative mechanisms are therefore most probably responsible for the protective effect of IL6 in this system.

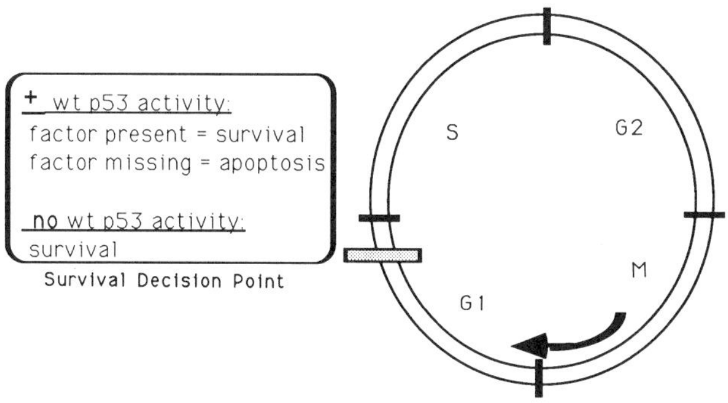

Figure 7. Model for the proposed involvement of p53 in the apoptosis decision point.

The suggestion that wt p53 mediates survival factor dependence can account, at least in part, for the contribution of p53 inactivation to the emergence of hematopoietic malignancies. It is of note that the most common type of cancer observed in p53 knock-out mice is malignant lymphoma[33]. The relatively long latency of such lymphomas indicates that, in addition to the inherent absence of p53 function, additional genetic lesions must be involved in the development of the lymphoma. In this sense, the increased survival advantage provided by the lack of p53 could serve as an initial step towards establishing a clonal population of malignant cells, capable of uninterrupted proliferation in an otherwise restrictive environment.

Finally, while so far our data address only the role of p53 in mediating survival factor dependence in hematopoietic cells, this relationship may not be restricted to cells of this type. It will be of interest to determine the contribution of wt p53 to apoptosis in other systems in which viability is conditional upon the availability of appropriate survival factors.

ACKNOWLEDGEMENTS

This work was supported by grants from the Minerva Foundation (Munich), the US-Israel Binational Science Foundation and the Forchheimer Center for Molecular Genetics.

REFERENCES

1. Levine, A. J., J. Momand, and C.A. Finlay. The p53 tumour suppressor gene. Nature 351: 453-456 (1991).
2. Hollstein, M., D. Sidransky, B. Vogelstein and C.C. Harris. p53 mutations in human cancers. Science 253: 49-53 (1991).
3. Oren, M. p53 - the ultimate tumor suppressor gene? FASEB J. 6, 3169-3176 (1992).
4. Vogelstein, B.., and Kinzler, K.W. (1992). p53 function and dysfunction. Cell 70, 523-526.
5. Mercer, W.E. Cell cycle regulation and the p53 tumor suppressor protein. Critical Rev. Eukaryotic Gene Expression, 2, 251-263 (1992).
6. Prives, C. and J.J. Manfredi. The p53 tumor suppressor protein: meeting review. Genes Dev. 7, 529-534 (1993).
7. Michalovitz, D., O. Halevy, and M. Oren. p53 mutations: gains or losses? J. Cell. Biochem. 45:22-29 (1991).
8. Seto, E., Usheva, A., Zambetti, G.P., Momand, J., Horikoshi, N., Weinmann, R., Levine, A.J., and Shenk, T. Wild-type p53 binds to the TATA-binding protein and represses transcription. Proc. Natl. Acad. Sci. USA, 849, 12028-12032 (1992).
9. Ragimov, N., Krauskopf, A., Oren, M., and Aloni, Y. Wild type p53 can repress transcription initiation in vitro. Oncogene 8, 1183-1193 (1993).
10. Mack, D.H., J. Vartikar, J.M. Pipas, and L.A. Laimins. Specific repression of TATA-meidated but not initiator-mediated transcription by wt p53. Nature 363, 281-283 (1993).
11. Agoff, S.N., Hou, J., Linzer, D.I.H., and Wu, B. Regulation of the human hsp70 promoter by p53. Science, 259, 84-87 (1993).
12. Truant, R., Xiao, H., Ingles, C.J., and Greenblatt, J. Direct interaction between the transcriptional activation domain of human p53 and the TATA box-binding protein. J. Biol. Chem. 268, 2284-2287 (1993).
13. Tomey, J.D., and F.O. Cope (eds). Apoptosis: the molecular basis of cell death. Cold Spring Harbor Laboratory Press, Plainview, NY (1991).
14. Kuerbitz, S.J., B.S. Plunkett, V.W. Walsh, and M.B. Kastan. Wild type p53 is a cell cycle checkpoint determinant following irradiation. Proc. Natl. Acad. Sci. USA **89**: 7491-7495 (1992).
15. Lane, D.P. p53, guardian of the genome. Nature 358: 15-16 (1992).
16. Kastan, M.B., Zhan, Q., El-Deiry, W.S., Carrier, F., Jacks, T., Walsh, W.V., Plunkett, B.S., Vogelstein, B., and Fornace, A.J. Cell 71, 587-597 (1992).
17. Yonish-Rouach, E., D. Resnitzky, J. Lotem, L. Sachs, A. Kimchi, and M. Oren. Wild-type p53 induces apoptosis of myeloid leukaemic cells that is inhibited by interleukin-6. Nature 352:345-347 (1991).
18. Michalovitz, D., O. Halevy, and M. Oren. Conditional inhibition of transformation and of cell proliferation by a temperature-sensitive mutant of p53. Cell 62:671-680 (1990).
19. Ryan, J.J., R. Danish, C.A. Gottlieb, and M.F. Clarke. Cell cycle analysis of p53-induced cell death in murine erythroleukemia cells. Mol. Cell. Biol. 13, 711-719 (1993).
20. Johnson, P., S. Chung, and S. Benchimol. Growth suppression of Friend virus-transformed erythroleukemia cells by p53 protein is accompanied by hemoglobin production and is sensitive to erythropoietin. Mol. Cell. Biol. 13, 1456-1463 (1993).
21. Shaw, P., R. Bovey, S. Tardy, R. Sahli, B. Sordat, B., and J. Costa. Induction of apoptosis by wild type p53 in a human colon tumor-derived cell line. Proc. Natl. Acad. Sci. USA 89:4495-4499 (1992).
22. Lowe, S.W., E.M. Schmitt, S.W. Smith, B.A. Osborne, and T. Jacks. p53 is required for radiation-induced apoptosis in mouse thymocytes. Nature 362, 847-849 (1993).
23. Clarke, A.R., C.A. Purdie, D.J. Harrison, R.G. Morris, C.C. Bird, M.L. Hooper, and A.H. Wyllie. Thymocyte apoptosis induced by p53-dependent and independent pathways. Nature 362, 849-852 (1993).
24. Debbas, M., and E. White. Wild type p53 mediates apoptosis by E1A, which is inhibited by E1B. Genes Dev. 7, 546-554 (1993).
25. Lowe, S.W., and H.E. Ruley. Stabilization of the p53 tumor suppressor is induced by adenovirus 5 E1A and accompanies apoptosis. Genes Dev. 7, 535-545 (1993).
26. Yonish-Rouach, E., Grunwald, D., Wilder, S., Kimchi, A., May, E., Lawrence, J.J., May, P., and Oren, M. p53-mediated cell death: relationship to cell cycle control. Mol. Cell. Biol. 13, 1415-1423 (1993).
27. Askew D.S., R.A. Ashmun, B.C. Simmons, and J.L. Cleveland. Constitutive c-myc expression in an IL-3-dependent myeloid cell line suppresses cell cycle arrest and accelerates apoptosis. Oncogene 6:1915-1922 (1991).

28. Evan, G.E., A.H. Wyllie, C.S. Gilbert, T.D. Littlewood, H. Land, M. Brooks, C.M. Waters, L.Z. Penn, and D.C. Hancock. Induction of apoptosis in fibroblasts by c-myc protein. Cell 69, 119-128 (1992).
29. Sachs, L. The control of growth and differentiation in normal and leukemic blood cells. Cancer 65: 2196-2206 (1990).
30. Resnitzky D., N. Tiefenbrun, H. Berissi, and A. Kimchi. Interferons and interleukin 6 suppress phosphoryrlation of the retinoblastoma protein in growth-sensitive hematopoietic cells. Proc. Natl. Acad. Sci. USA 89: 402-406 (1992).
31. Shaulian, E., Zauberman, A., Ginsberg, D., and Oren, M. Identification of a minimal transforming domain of p53: negative dominance through abrogation of sequence-specific DNA binding. Mol. Cell. Biol. 12, 5581-5592 (1992).
32. Lotem, J., ans L. Sachs. Regulation by bcl-2, c-myc and p53 of susceptibility to induction of apoptosis by heat shock and cancer chemotherapy compounds in differentiation-competent and -defective myeloid leukemic cells. Cell Growth Diff. 4, 41-47 (1993).
33. Donehower, L.A., Harvey, M., Slagle, B.L., McArthur, M.J., Montgomery, C.A., Butel, J.S., and Bradley, A. p53-deficient mice are developmentally normal but susceptible to spontaneous tumours. Nature 356, 215-221 (1992).

Discussion

D. Hockenbery

Does IL-6 modify any aspect of p53 function that you have found, for example, a reporter gene construct?

M. Oren

So far we have looked at two things. We looked at conformational changes in p53, because one simple possibility would be that IL-6 causes the protein to revert into the mutant conformation again, overcoming the temperature effect; the answer is no. We looked at whether IL-6 may keep the protein at the cytoplasm rather than allowing it to go to the nucleus; the answer is no. We looked at DNA binding. DNA binding still goes on. The answer to your question is that this experiment still has to be done. We have pretty much gone off the IL-6 and M1 paradigm because I think it is probably the least physiological of all the systems we can think of and the next step we want to move into is really to work with primary myeloid cells rather than with leukemic cells which are totally abnormal.

D. Hockenbery

Is there any data indicating that in another system in which p53 is present, IL-6 modulates p53 when it differentiates?

M. Oren

Not directly. There are some data which you can interpret this way but it is not evidence in any serious way.

D. Green

In your discussion at the end, you were pointing out the possibility of secondary effects, redundancy in these systems, things like that, and one of the things I want to point out and, we have some experience with it as well, is that when p53 is mutated or absent from a cell, there is an increased frequency of gene conversion, gene amplification and other facts. Of course, that is one reason why you went to primary transfection, to see what the immediate effect would be.

M. Oren

Infection.

D. Green

But suppose you exploit that fact and use these p53 mutants to now derive cells that are completely resistant to the effects of growth factor withdrawal or to some of your other factors. Do you see that and do you have a much greater frequency of mutations?

M. Oren

The answer is yes. This is very preliminary. As I noted, those cells, if you take IL-3 away, are fairly viable, but they do not proliferate. If you keep those cells long enough without IL-3 you can select, relatively easily, a population which now proliferates in the absence of IL-3 so it becomes factor independent both for proliferation and for survival. The most obvious possibility is that, in fact, you have now incurred a second hit in another gene which is involved in allowing proliferation without IL-3. This is what we know so far.

D. Green

That may turn out to be a very useful approach towards tracing pathways of resistance to apoptotic death, which may even be independent of p53. But because of the increased frequency of mutation you will see those much more frequently.

M. Oren

You are right.

J. Cleveland

I was wondering if you had looked at whether the levels of wild-type p53 protein or RNA are actually regulated by ligands such as IL-3? Or whether IL-3 affects half-life of wild-type p53?

M. Oren

Unfortunately, we have very little data on that. We have done preliminary experiments looking at p53 stability upon IL-3 withdrawal and for the time being, more work needs to be done. What we are doing now is timing it at very short intervals. We believe there is an effect on stabilizing p53. It is probably transient and probably peaks and goes away and we really have to time it at short intervals. I prefer to leave it open until we have better data.

J. Cleveland

I have looked on Northerns and p53 RNA is continuously expressed in the absence of ligands.

M. Oren

In Northern, in many of the systems in which p53 is regulated by environmental cues, nothing happens at the level of RNA. The p53 gene seems to be rather a dull gene and almost everything that is altered, does so at the level of protein. Very few exceptions exist to this rule.

J. Cleveland

And you also mentioned in passing that gain of function mutations in mutant p53 actually can suppress apoptosis and that this was Leo Sachs' work. I was wondering if you could expand on that a little bit.

M. Oren

This is the paper that Lotem and Sachs published in *Cell Growth and Differentiation*, in which they looked at M1 cells transfected with mutant p53. If you induce apoptosis in those cells by a variety of treatments, the mutant has no effect. However, if you take those cells and super-infect them with a myc retrovirus so that now they have constitutive myc there is an enhanced apoptotic response to a variety of stimuli. Yet, cells which have the mutant p53 are less responsive to apoptosis than cells which overexpress myc in the absence of mutant p53.

J. Cleveland

The last question has to do with the knockouts and ES cells that are homozygous. Do they undergo rapid cell death when you remove LIF or do they fail to differentiate in the blood islands, or do you know?

M. Oren

Those experiments have to be done. We do not have access to those cells.

M. Kastan

I may have missed it, but what happened when you put in the anti-sense vectors and removed IL-3?

M. Oren

You get more extended survival compared to the parental cells. So the anti-sense and the ts mutant at 37° have a similar effect: cells die, but less efficiently.

M. Kastan

And the second question was: 32D's are becoming a very good model of haematopoietic cell differentiation. If you remove the IL-3 and put in GCSF, they differentiate along the granulocytic lineage. Have you looked at the effects of putting in your mutant p53 on that type of differentiation?

M. Oren

No, we have not. We prefer to address this issue in cells that are a little bit more normal. 32D have been in culture for a long time and you start worrying about their having undergone a lot of changes.

M. Kastan

Well, in fact, we looked at 32D's about a year ago and we did not see any p53 expression in the cells we were working with, so that clearly there are different lineages.

M. Oren

I noticed it in your paper and was very surprised. Our 32D, in fact, originate in Cleveland's lab. Different people may derive different sublines by single-cell cloning. We have observed different patterns of p53 expression in various clones of M1; 32D may be another example.

R. Schimke

Do you have any information about whether the progression of these cells through the cell cycle is not uniform. Do you know where in the cell cycle these cells might be dying? These questions come out of issues Steve Sherwood is going to talk about on Friday where the rate of progression through the cell cycle can be rather critical on whether these cells die or not.

M. Oren

Let me give you a partial answer. In the M1 system, Ellisheva Yonish-Rouach made two observations which relate to your question. First of all, if you activate p53 in those cells, the cell cycle goes on as usual. You do not see any evidence of growth arrest and yet the cells go into apoptosis. So, you do not need to arrest the cells in order to induce apoptosis by p53. In addition, if you elutriate those cells and turn on p53 at different times, and ask at what stage in the cell cycle are the cells most responsive, you can clearly see that cells that are in G1 go into apoptosis much faster than cells which are in other stages of the cell cycle. If you take different time windows, the results suggest that cells have to go first to G1, have p53 there functioning, and only then go into apoptosis. I believe that Michael Clark, in a comparable system using Friend cells with a ts p53, has results which are very much in line with ours. This means that in haematopoetic cells G1 is the exit point into apoptosis, driven by p53.

R. Schimke

You say that the cells are in G1. Is that by sorting definition or, in fact, are they in s-phase? So the issue is, are they synthesizing any DNA because a G1 sort will not differentiate whether they have initiated S-phase?

M. Oren

I do not think I can give you a clear-cut answer. The way the experiment was done, the cells had been sorted at 37° when p53 is not active. Only after sorting they were shifted to 32°, activating p53. If that is taken as a criterion, they probably were in G1 to begin with. If you take cells which are sorted at S-phase, these are the ones that take longest to go into apoptosis, so probably that would be consistent with the fact that they have to go through the entire cell cycle and hit the point in G1. Is it possible that it is very early S and we call it G1? You may be right there.

T. Taniguchi

Some cytokines are known to induce *bcl*-2 expression, and have you looked at *bcl*-2 expression in IL-6-induced M1 cells?

M. Oren

That was an obvious experiment. IL-6 does not increase *bcl*-2 expression in M1. Now what is interesting, in fact, M1 have a relatively high level of *bcl*-2, constitutively more than other haematopoetic cell lines. In fact, M1 without p53 are relatively resistant to a variety of anti-cancer drugs and other apoptotic treatments. That is the work of Leo Sachs. So *bcl*-2 does not seem to be a player here, even though overexpression of *bcl*-2 in other p53 induced apoptotic responses can inhibit cell death. In M1, apparently, this does not seem to be the case.

C. Riccardi

Do you know if IL-6 modifies the cell cycle of your cell lines?

M. Oren

Yes, in many ways.

C. Riccardi

And is this related to the protective effect of IL-6?

M. Oren

That is a very important point. Just to reiterate the question: Does IL-6 have any effect on the cell cycle of p53-expressing M1 cells and is that related to its protective effect on apoptosis? We believe the answer is yes for the second and we know the answer is yes for the first. What is the evidence? IL-6 by itself, if you apply it to M1 cells, would cause a partial G1 arrest. You get like 70% of the cells accumulating in G1, and other cells kind of drag slowly through the cell cycle and eventually arrest throughout the cell cycle. But formally, M1 is responding by a growth arrest to IL-6. If you now take the cells which have the ts p53 at 32° and add IL-6, the cells are protected from apoptosis. If you look at the cell cycle you get a perfect G1 arrest which means that better than 97% of the cells are in G1. And, in fact, if you look at the nuclear density and the cell size and the nucleus size, they fall into the definition of G0 more than G1. One way of interpreting that, in line with the work of Cleveland and Evan, perhaps when we overexpress wild type p53 in M1 cells, you generate a conflict between positive signals which are delivered by something which is not myc and p53. One is signaling positively, one is signaling negatively, and the end result is that the cells die. If we add IL-6 in the presence of excess p53, what perhaps happens is a dominant growth inhibitory signal which now installs a proper growth arrest. This overrides the positive signal and the cells feel happy and stay alive. This right now is our working hypothesis.

D. Delia

The 32D cells, when treated with p53 anti-sense oligos, how susceptible are they to gamma radiation?

M. Oren

We have not done the experiment.

REGULATION OF APOPTOSIS BY THE TRANSFORMING GENE PRODUCTS OF ADENOVIRUS

Eileen White, Lakshmi Rao, Shiun-Kwei Chiou, Ching-Chun Tseng, Peter Sabbatini, Michelle Gonzalez and Philippe Verwaerde

Center for Advanced Biotechnology and Medicine
Rutgers University
Department of Biological Sciences
679 Hoes Lane
Piscataway, NJ 08854

INTRODUCTION

The DNA tumor virus adenovirus infects human cells, recruits them into a proliferative state, and borrows elements of the host cell transcription, translation, and DNA replication machinery to reproduce viral proteins and DNA. In rodent cells which are semipermissive for adenovirus infection, cell growth is deregulated but virus replication is ineffective. As the viral infection does not progress to completion, the deregulation of cell growth control produces transformation. The viral genes required for oncogenic transformation are the E1A and E1B oncogenes.

Regulation of programmed cell death (apoptosis) by E1A and E1B is important for sustaining a productive infection in human cells and is an integral part of the transformation process in rodent cells. The E1A proteins are responsible for initiating a proliferative response and an indirect consequence of this required function of E1A is the induction of apoptosis. The E1B gene encodes overlapping, redundant functions to suppress apoptosis, the 19K and 55K proteins. Thus, E1A expression and subsequent growth deregulation can occur unimpeded by cell death in the presence of E1B expression. Without inhibition of apoptosis by E1B, transformation of rodent cells is rare and premature death of the host cell impairs virus yield in productively infected human cells.

There is substantial evidence that the product of the p53 tumor suppressor gene mediates the induction of apoptosis by E1A and that the E1B gene products function by disabling p53. First, E1A expression causes p53 accumulation. Second, dominant-interfering mutant forms of p53 will block apoptosis by E1A and produce transformation with high frequency. Third, returning p53 to the wild-type conformation produces rapid and complete apoptosis. Fourth, the E1B 55K protein which binds to p53 and directly inhibits its activity will suppress the induction of apoptosis by E1A. Fifth, expression of the E1B 19K gene will prevent p53-dependent apoptosis and is likely to indirectly modify p53 function.

The human *bcl-2* protooncogene, which is known to inhibit apoptosis in other systems, will also suppress apoptosis induced by E1A expression and cooperate with E1A to transform primary rodent cells. This functional substitution of the *bcl-2* gene for the adenovirus E1B gene suggested that *bcl-2* may also interfere with p53 function. *bcl-2* expression was found to be sufficient to block p53-dependent apoptosis. *bcl-2* may be the

Apoptosis, Edited by E. Mihich and R.T. Schimke
Plenum Press, New York, 1994

cellular equivalent of the E1B 19K gene, with its oncogenic and anti-apoptotic activity, in all or in part, attributed to bypassing the function of p53.

RESULTS

Regulation Of Apoptosis In Adenovirus Infected Cells

Identification of the E1B 19K gene product as an inhibitor of apoptosis: Once the E1A and E1B genes were identified as the oncogenes of adenovirus, numerous investigations were undertaken to determine their biochemical mode of action. The initial approach taken by several laboratories to investigate the function of the E1B 19K gene was to study the consequence of E1B 19K gene mutations on virus replication and transformation. Analogous studies of the other viral early genes had turned up prominent defects in viral gene transcription, protein synthesis and DNA replication, and had led to the identification of viral gene products required for regulation of these essential processes. In contrast, inactivation of E1B 19K gene function through mutation produced novel and peculiar phenotypes; the degradation of DNA (*deg* phenotype), and enhanced cytopathic effect or cytocidal (*cyt*) phenotype[1-3]. There is no apparent specificity to the DNA fragmentation, as both cellular and viral DNA become degraded, although the degradation of viral genomes is surely detrimental to virus production[4]. Among host DNA sequences, there is also no differential sensitivity between transcribed and untranscribed genes, suggesting that the DNA fragmentation is non-specific and complete[4]. The prevailing consequence of the *cyt* and *deg* phenotypes is the premature death of the host cell which impairs virus yield[1,5]. Thus the E1B 19K protein was required to maintain the viability of the host cell thereby maximizing virus production[4,6].

The ability of the E1B 19K protein to prevent DNA fragmentation, an indicator of apoptosis, suggested that it was acting as an inhibitor of apoptosis[6]. This was confirmed by demonstration that E1B 19K expression blocks cell death by the known inducers of apoptosis, TNF-α and Fas antigen[7,8,9]. Therefore, the *cyt* and *deg* phenotypes originally used to describe the obvious physiological manifestations of lack of E1B 19K gene function are actually the apoptotic response. Inhibition of apoptosis by the E1B 19K protein may serve the added purpose of permitting adenovirus to escape anti-viral immune surveillance mechanisms implemented by TNF-α and Fas.

Induction of apoptosis by the adenovirus E1A proteins: Since uninfected cells do not normally undergo apoptosis, the induction of apoptosis by E1B 19K gene mutant viruses indicated that the product of another viral gene was responsible for triggering the apoptotic response. Apoptosis (*cyt* and *deg*) is a viral early gene function (E1, E2, or E4), occurring prior to the onset of viral DNA synthesis or when DNA synthesis is blocked by hydroxyurea[3]. Genetic analysis of a series of adenovirus mutants demonstrated that apoptosis was a result of expression of the E1A gene[10]. Specific mutational analysis of E1A indicated that apoptosis was inseparable from the ability of the E1A proteins to induce DNA synthesis[6]. Apoptosis appears to be the indirect consequence of a required function of E1A. It is very likely that some aspect of the deregulation of cell growth by E1A is interpreted by the cell as a signal to undergo apoptosis. Viral intervention in host cell growth control pathways followed by the cell actively committing suicide may be common, and necessitated the evolution of functions to suppress apoptosis and promote survival or latency of the infected cell (reviewed in: 11,12).

The E1A proteins consist of three conserved, functional regions, CR1, CR2, and CR3 (reviewed in: 13,14). CR3 is predominantly required for transactivating transcription of the viral early genes and is largely dispensable for transformation (reviewed in: 15). CR1 and CR2 both independently encode transforming activity and are collectively required for the transformation process when coexpressed with E1B or activated *ras* genes. However, expression of either CR1 or CR2 is sufficient to induce cellular DNA synthesis. CR1 and CR2 serve as binding sites for cellular proteins and insight into how cell growth control is deregulated came from the determination of the identity and function of these cellular-E1A binding proteins (reviewed in: 16).

The first cellular E1A binding protein to be identified was the product of retinoblastoma susceptibility gene or Rb[17]. Rb is a gene whose loss of function is the cause of hereditary and sporadic human retinoblastoma and is thereby categorized as a tumor suppressor gene[18]. Restoration of Rb function suppresses cell growth by halting progression through the cell cycle. In late G1 Rb is hypophosphorylated and found in a complex with the transcription factor E2F (reviewed in: 19). Near the G1/S boundary Rb is hyperphosphorylated, E2F is released to activate the transcription of a variety of cellular genes, and DNA synthesis is initiated. By binding to Rb, E1A causes the release of E2F and deregulates cell growth control. E1A also binds cyclin A, associated kinases, and a 300K phosphoprotein, which also contribute to deregulation of cell growth (reviewed in: 13).

Induction of apoptosis by E1A maps within CR1 and CR2[6]. The only E1A mutant defective at induction of apoptosis was also defective at induction of DNA synthesis and transformation[6]. This E1A mutation deletes residues 22 to 107 and probably effects both CR1 and CR2. In summary, the E1A proteins contribute to the transformation process by binding to and interfering with the function of cellular proteins with key regulatory roles in cell cycle control. It is this aberrant perturbation of cell growth that, is so far, genetically inseparable from, and is the probable cause of, apoptosis.

Regulation Of Apoptosis In Transformation

Transformation of primary cells is a multistep process that requires independent activities encoded by the E1A and the E1B genes. The roles of the E1A and E1B proteins in transformation are very much the same as during productive infection of human cells, except that without the infection progressing to completion, the deregulation of cell growth leads to the emergence of transformed cells. E1A expression is sufficient to drive quiescent cells into a proliferative state, but in the process, apoptosis results [9,20]. Any net accumulation of cell number through cell division is counteracted by elimination through apoptosis. Only rare transformants emerge that express E1A and are immortal in culture[9]. These transformants have undergone a selection process to overcome apoptosis and have sustained mutations in one and perhaps more cellular genes (see below). Thus E1A deregulates growth and promotes apoptosis during the transformation process in a manner that is indistinguishable from that which occurs in productively infected human cells.

The E1B gene has no measurable capacity to stimulate cell proliferation on its own[21]. When cotransfected with E1A, E1B does not outwardly change the growth characteristics of the cells but markedly suppresses all indications of apoptosis[20]. This ability to suppress apoptosis and cooperate with E1A to transform primary rodent cells can be provided by either the E1B 19K or the 55K proteins, although the 19K protein is apparently more active[20].

If induction of apoptosis by E1A was indeed a significant obstacle to transformation then the *bcl-2* gene should cooperate with E1A in a transformation assay. *bcl-2* is the most common translocation in human B cell follicular lymphoma[22-24]. *bcl-2* overepression has been shown to suppress apoptosis upon growth factor withdrawal in cell lines[25-28]. In animal models *bcl-2* overepression results in extended B cell memory and monoclonal lymphomas[29-31]. Thus *bcl-2* has been identified as an inhibitor of apoptosis in independent situations and should act as a valid test for the significance of apoptosis in transformation by E1A.

bcl-2 was found to suppress apoptosis by E1A and cooperate with E1A to transform primary rodent cells[20]. All rodent cell transformants derived from E1A plus *bcl-2* transfections overexpressed the human Bcl-2 protein[20]. Inhibition of apoptosis and transformation with E1A can be accomplished by either of the E1B proteins (19K or 55K) or overexpression of the human Bcl-2 protein.

E1A-Induced Apoptosis Is Mediated By Wild-Type p53

The positive and negative regulation of apoptosis by E1A and E1B provided an opportunity to get at the cellular apoptotic machinery and its regulation. The cell is apparently responding to aberrant or untimely induction of DNA synthesis by E1A, but what senses this, and how its activity is perverted by E1B and *bcl-2*, was not known. One clue was provided by the E1B 55K protein which is known to bind to and directly inhibit the function of p53[32,33]. Furthermore, although ectopic p53 expression most commonly produces cell cycle arrest[34,35], its expression in a myeloid leukemic line produces apoptosis which suggested that p53 may function as a tumor suppressor by triggering apoptosis[36]. To test the involvement of p53 in E1A-associated apoptosis directly, dominant, interfering mutant p53 proteins were utilized.

E1A and mutant alleles of p53 efficiently cooperate to transform primary rodent cells demonstrating that direct interference with p53 function will block apoptosis in cells expressing E1A[37] (Fig. 1). The cooperation between E1A and mutant forms of p53 was the most efficient transforming combination that we have observed in this assay, surpassing E1A plus E1B 19K transfections[37] (Fig. 1). Interfering with p53 function by transfection of the murine MDM-2 gene whose product has been shown to bind to and inhibit p53[38-40], will also suppress apoptosis and permit transformation, but at a reduced frequency (Fig. 1).

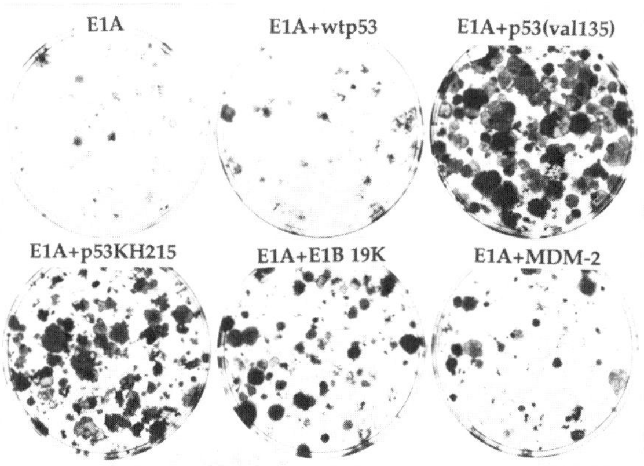

Fig. 1. Focus formation between E1A and viral and cellular oncogenes. Primary BRK cells were transfected with the oncogenes as indicated. Plates were stained after 34 days in culture to reveal foci.

By utilizing a temperature sensitive p53 mutant p53(val135) we have determined that p53 must be maintained in the mutant conformation for cell viability and the transformed state to be maintained. Returning p53 to the wild-type conformation by shift of E1A plus p53(val135) transformants from the restrictive to permissive temperature results in rapid and complete apoptosis[37]. Therefore, induction of apoptosis by E1A requires the presence of wild-type p53 function.

One caveat to placing p53 in a pivotal role in directing apoptosis following E1A expression was the result that apoptosis was induced in HeLa cells upon E1B mutant adenovirus infection. HeLa cells are derived from a human cervical carcinoma that have very likely been transformed by a human papilloma virus (HPV), as HeLa cells express the HPV E6 and E7 transforming genes. While the E7 gene functionally and structurally resembles E1A, the E6 product binds to and targets p53 to degradation via the ubiquitin pathway[41]. As a result, the p53 gene in HeLa cells is wild-type, the mRNA is made, but the p53 protein is not detectable. The only way apoptosis could be p53-dependent in HeLa cells would be if E1A expression induced p53 accumulation beyond the capacity of E6 to facilitate its degradation.

E1A expression has been shown to induce p53 accumulation in other cell lines[42] and is dramatically induced in HeLa cells following E1A expression (Chiou, Rao and White, in preparation). E1A expression by viral infection, or by transient expression with E1A plasmid expression vectors in HeLa cells produces a dramatic accumulation of p53 protein as determined by indirect immunofluorescence and Western blotting. Thus, the cellular response to E1A expression is the production of p53 protein accumulation. The induction of p53 accumulation and apoptosis by E1A in HeLa cells is consistent with a role for p53 in directing apoptosis in response to E1A.

The E1B 19K and Bcl-2 Proteins Inhibit Wild-Type p53-Dependent Apoptosis

The dependence of E1A-initiated apoptosis in p53, and the ability of the E1B 19K and Bcl-2 proteins to block apoptosis by E1A, suggested that E1B 19K and Bcl-2 may function by inhibition of p53. This was tested directly by introduction of either an E1B 19K gene or a human *bcl-2* gene into E1A plus p53(val135) transformed BRK lines that undergo apoptosis upon conversion of p53 from the mutant to the wild-type form[37]. Unlike the parental and control cell lines, E1B 19K or Bcl-2 expressing cell lines were completely resistant to the induction of apoptosis by wild-type p53[37,43]. Some of the lines not only remained viable, but continued to grow at the permissive temperature[37,43]. E1B 19K or Bcl-2 can, if expressed at sufficient levels, overcome both the anti-proliferative and apoptotic effects of p53.

Functional Substitution Of *bcl-2* For The E1B 19K Gene

The similar ability of the E1B 19K and Bcl-2 protein to block p53-dependent apoptosis suggested that they may function by the same mechanism. Furthermore, both possess the same intracellular distribution, and are found in association with cytoplasmic membranes and the nuclear envelope[44-46]. The E1B 19K protein has the capacity to block apoptosis in response to TNF-α and anti-Fas antibodies[7-9], and during adenovirus infection of human cells[6]. These E1B 19K gene functions cosegregate with transforming activity suggesting that all are related to a solitary function of inhibition of apoptosis[9,45]. Bcl-2 was similarly examined for these activities.

HeLa cells express the Bcl-2 protein at extraordinarily low levels and rapidly undergo apoptosis following infection with E1B 19K gene mutant viruses[45]. The human *bcl-2* gene under the control of a strong viral promoter[47] was introduced into HeLa cells and clones that expressed high levels of Bcl-2 were identified[45]. Infection of parental or control HeLa cells with an E1B 19K gene deletion mutant adenovirus produces the *cyt* and *deg* phenotypes within 48 hours[45]. In contrast, HeLa lines overexpressing Bcl-2 did not display the *cyt* phenotype and showed a marked reduction in the extent of DNA fragmentation (*deg* phenotype). The extent of complementation of Bcl-2 for the E1B 19K protein was directly proportional to the amount of Bcl-2 expressed in the different HeLa lines[45]. Thus, Bcl-2 has a significant capacity to replace E1B 19K gene function in adenovirus infected human cells.

There have been reports of the ability of Bcl-2 to suppress apoptosis by TNF-α and Fas antigen[48-50]. Assays for measuring the ability of the E1B 19K protein to block apoptosis following TNF-α and anti-Fas antibody treatment have been developed in our laboratory[9,45] and HeLa cell lines that stably express the E1B 19K protein are available as positive controls for resistance to apoptosis[51]. When examined for resistance to TNF-α and Fas antigen-induced apoptosis, Bcl-2 expressing HeLa cells were as resistant as cells expressing the E1B 19K protein[45]. In conclusion, the Bcl-2 protein was functionally indistinguishable from the E1B 19K proteins in all assays currently available.

The apparent functional equivalency of the E1B 19K and human Bcl-2 proteins suggested a similar if not identical mode of action and prompted a comparison of the amino acid sequences of the two proteins. The E1B 19K protein sequence is available, not only from the Ad2 and 5 serotypes described here, but from 10 other serotypes of adenovirus. The original *cyt* mutants of adenovirus were isolated from the highly oncogenic Ad12 serotype[52] indicating that inhibition of apoptosis by the E1B 19K protein is a conserved function of the protein. When the amino acid sequence for all 12 19K proteins is compared, three regions with distinct levels of homology are apparent. Most striking is the central region of the protein between residues 81 and 116 which is highly conserved with

58% identity across all serotypes[9,45]. The amino terminal residues 1 to 80 are moderately conserved (44%) and the carboxy-terminus between residues 117 to 175 is poorly conserved (22%)[9,45]. Introduction of missense mutations in the highly conserved central regions without exception impairs function, where mutations in the carboxy-terminus have no measurable consequences[9,45]. Some mutations in the amino terminus result in wild-type activity where others are defective[9,45]. In conclusion, the region between residues 44 and 113 is most critical for maintaining 19K protein structure and/or functional activity.

The amino acid sequence of Bcl-2 is available from human, mouse and chicken[53-56]. Bcl-2 is very highly conserved throughout and mutational analysis is limited. Three other gene products have been identified with limited homology to *bcl-2*, the murine Mcl-1[57], Epstein-Barr virus BHFR1[58], and *C. elegans ced9* genes[59]. Although the function of Mcl-1 and BHFR1 is not known, *ced9* functions as an inhibitor of programmed cell death in *C. elegans* [59]. Alignments with Bcl-2 and related sequences with the E1B 19K protein sequence show a modest degree of similarity (Verwaerde and White, unpublished observations). Most of the homology lies within the highly conserved central region and not at the nonconserved carboxy-terminus[45]. Domain swapping experiments between E1B 19K and Bcl-2, and the other related proteins, should determine if any regions of homology represent conserved functional domains, and if these proteins represent members of a family of proteins with a similar anti-apoptosis function.

DISCUSSION

Apoptosis May Be The Result Of Conflicting Or Incompatible Growth Signals

It is interesting to note that apoptosis is often associated with situations where cell proliferation is forced or when an important growth factor is withdrawn. The conflicting or incompatible growth signals that arise may be uninterpretable and the signal for the cell to commit suicide (Fig. 2). E1A expression potently stimulates proliferation[60,61], but also induces p53 accumulation[42], which is the cellular response to DNA damage[62] (Fig. 2). There is compelling evidence that one function of p53 is to induce growth arrest in response to DNA damage[63-65]. The simultaneous induction of cell proliferation by E1A and growth arrest by p53 may be the conflicting growth signals responsible for apoptosis in this case[37]. These conflicting growth signals could produce cell cycle alterations that would be grossly aberrant and fatal. However, in some cell types p53 function is required for induction of apoptosis in response to DNA damage[66,67]. Therefore, p53 must be considered as a potential regulator of growth arrest and apoptosis, and perhaps both involve intervention in cell cycle control by p53 with different outcomes depending on the physiological state of the cell.

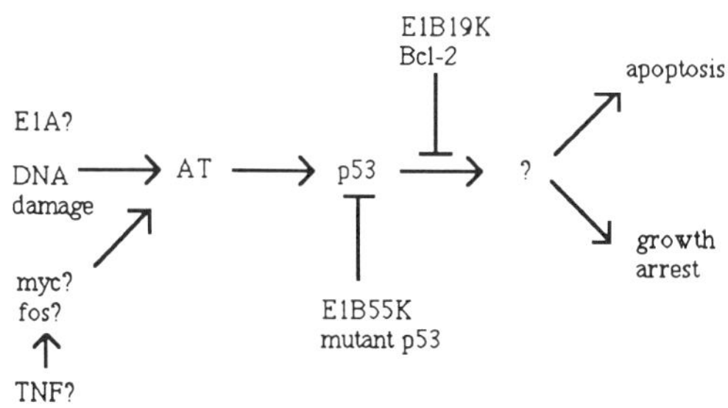

Fig. 2. Regulation of apoptosis (see text for explanation).

c-myc amplification and *c-fos* expression, which are associated with growth stimulatory activities similar to those of E1A, also produce apoptosis[68-72] (Fig. 2). As in the case of E1A, *bcl-2* will suppress apoptosis by *c-myc* and cooperate with *myc* to transform cells[28,73,74]. Whether *c-myc* or *c-fos*-induced apoptosis is p53-dependent is not yet known.

TNF-α induces *c-myc* and *c-fos* expression[75], raising the possibility that TNF-α may induce apoptosis through the *myc* or *fos* pathways (Fig. 2).

The absence of specific growth factors is also known to be a potent inducer of apoptosis in many circumstances. Apoptosis may be the default response to inappropriate cell proliferation, as would occur in the absence of specific growth factors, and may be one mechanism for regulation of apoptosis in normal development (reviewed in: 76). Insufficient positive growth stimulation caused by a growth factor imbalance could produce apoptosis. *bcl-2* will often rescue apoptosis upon factor depravation, suggesting a role for *bcl-2* in regulating apoptosis in development[25-27]. This is consistent with the expression pattern of Bcl-2[77] and with the proposed role for *bcl-2* in regulating appropriate lymphoid cell survival in the immune system[31,78,79].

Transformation Requires Induction Of Cell Proliferation To Be Coupled To Suppression Of Apoptosis

Multiple genetic and functional events are required to progress from a normal to a malignant state. Identification of these events, their biological consequences, and order of occurrence, is fundamental to understanding the cause of cancer, and for identifying points of intervention. Based on the E1A/E1B paradigm, stimulation of cell proliferation and suppression of apoptosis must be coupled to sustain malignant growth. Oncogenic events may possibly be categorized to fall into at least these two functional categories with E1A and *c-myc* representing the former, and E1B, *bcl-2*, and mutant p53 representing the later. As activated *ras* will also cooperate with E1A or *c-myc*[80,81], it will be interesting to determine the ability of *ras* to suppress p53-dependent apoptosis.

Multiple Pathways For Regulation Of Apoptosis

With the limited information currently available it is already apparent that there will be multiple pathways for induction and inhibition of apoptosis (Fig. 2). p53 is required for apoptosis induced by DNA damage and E1A expression, but not by glucocorticoids, or that which occurs in mouse development[37,66,67,82]. Bcl-2 will block apoptosis in response to DNA damage, glucocorticoids, and growth factor withdrawal, but not negative selection in thymocytes[25,27,78,79,83]. As neither Bcl-2 nor E1B 19K proteins will block induction of p53 levels[43], and Bcl-2 will block both p53-dependent and independent pathways, both may function downstream of p53[43] (Fig. 2). The ataxia telangiectasia (AT) gene is required for induction of p53 levels in response to DNA damage and has been classified as a potential upstream regulator of p53[64] (Fig. 2). Whether induction of p53 levels by E1A also requires AT function will be of great interest.

Mechanism of Regulation of Apoptosis

There are many issues that remain to be addressed before the process of apoptosis can be clearly understood. First, the specific biochemical events perpetrated by E1A that induce p53 accumulation and apoptosis need to be established. The elegant genetic and biochemical information available on E1A function accumulated by many laboratories will make this possible. Second, the mechanism by which p53 induces apoptosis needs to be established. There is evidence that p53 may function by regulating transcription (reviewed in: 84). If p53-dependent apoptosis requires new protein synthesis then p53 may activate the expression of genes which in turn kill the cell. If apoptosis does not require protein synthesis, then p53 may have a more direct role in causing cell death, perhaps by direct effects on the cell cycle. Third, the means by which Bcl-2 and the E1B 19K protein prevent p53-dependent apoptosis remains to be determined. The E1B 19K protein does not detectably bind to p53 in vivo or in vitro, nor is p53 stability or localization affected (Verwaerde and White, in preparation), but it may modulate the transcriptional activity of p53 and/or its phosphorylation state. Identification of cellular proteins which physically interact with the E1B 19K and Bcl-2 proteins is also essential to defining the mechanism of inhibition of apoptosis.

FUTURE PROSPECTS

The most interesting new frontier will be to identify new gene products which function in the positive and negative regulation of apoptosis. These genes would be expected to be required for critical aspects of development, immune regulation, tumor suppressor function, and may regulate viral pathogenesis. Genes that regulate apoptosis can be identified following selection for resistance to apoptosis. Selection for rare survivors that have overcome apoptosis induced by E1A expression results in mutations in p53 with high frequency (Rao and White, unpublished observations). Identification of other mutational events in this and other systems should establish the pathways and regulators of apoptosis.

Another source of regulators of apoptosis is viral gene products. It is now apparent that apoptosis is an important aspect of the pathology of adenovirus and other viruses (reviewed in: 11, 12). Reports of induction of apoptosis following virus infection are ever increasing, as are the identification of viral inhibitors of apoptosis. Understanding the means by which viruses elicit apoptosis and how inhibitory strategies have evolved will be invaluable.

Acknowledgments

This work was supported by grants from the National Cancer Institute, the New Jersey Commission on Science and Technology, and a tranining grant in biotechnology from the National Institutes of Health.

REFERENCES

1. S. Pilder, J. Logan, and T. Shenk, Deletion of the gene encoding the adenovirus 5 early region 1B - 21,000-molecular weight polypeptide leads to degradation of viral and cellular DNA, *J. Virol.*, 52:664 (1984).
2. N. Takemori, C. Cladaras, B. Bhat, A.J. Conley, and W.S.M. Wold, cyt gene of adenovirus 2 and 5 is an oncogene for transforming function in early region E1B and encodes the E1B 19,000-molecular-weight polypeptide, *J. Virol.*, 52:793 (1984).
3. E. White, T. Grodzicker, and B.W. Stillman, Mutations in the gene encoding the adenovirus E1B 19K tumor antigen cause degradation of chromosomal DNA, *J. Virol.*, 52:410 (1984).
4. E. White, S.H. Blose, and B. Stillman, Nuclear envelope localization of an adenovirus tumor antigen maintains the integrity of cellular DNA, *Mol. Cell. Biol.*, 4:2865 (1984).
5. T. Subramanian, M. Kuppuswamy, J. Gysbers, S. Mak, and G. Chinnadurai, 19-kDa tumor antigen coded by early region E1b of adenovirus 2 is required for efficient synthesis and for protection of viral DNA, *J. Biol. Chem.*, 259:11777 (1984).
6. E. White, R. Cipriani, P. Sabbatini, and A. Denton, The adenovirus E1B 19-Kilodalton protein overcomes the cytotoxicity of E1A proteins, *J. Virol.*, 65:2968 (1991).
7. L.R. Gooding, L. Aquino, P.J. Duerksen-Hughes, D. Day, T.M. Horton, S. Yei, and W.S.M. Wold, The E1B-19K protein of group C adenoviruses prevents cytolysis by tumor necrosis factor of human cells but not mouse cells, *J. Virol*, 65:3083 (1991).
8. S. Hashimoto, A. Ishii, and S. Yonehara, The E1B oncogene of adenovirus confers cellular resistance to cytotoxicity of tumor necrosis factor and monoclonal anti-Fas antibody, *Int. Immunol*, 3:343 (1991).
9. E. White, P. Sabbatini, M. Debbas, W.S.M. Wold, D.I. Kusher, and L. Gooding, The 19-kilodalton adenovirus E1B transforming protein inhibits programmed cell death and prevents cytolysis by tumor necrosis factor α, *Mol. Cell. Biol.*, 12:2570 (1992).
10. E. White and B. Stillman, Expression of the adenovirus E1B mutant phenotypes is dependent on the host cell and on synthesis of E1A proteins, *J. Virol.*, 61:426 (1987).
11. E. White and L.R. Gooding, Regulation of apoptosis by human adenoviruses, *in* Apoptosis: The Molecular Basis for Cell Death II, (in press) (1993).
12. E. White, Regulation of apoptosis by the transforming genes of the DNA tumor virus adenovirus, *Proc. Soc. Exp. Biol. Med.*, (in press) (1993).
13. E. Moran, E1A/T antigen/E7 and the cell cycle, *Cur. Opin. Gen. Dev.*, 3:63 (1993).
14. T. Shenk and J. Flint, Transcriptional and transforming activities of the adenovirus E1A proteins, *Adv. Cancer Res.*, 57:47 (1991).
15. A.J. Berk, Adenovirus promoters and E1A transactivation, *Annu. Rev. Genet.*, 20:5 (1986).

16. N. Dyson and E. Harlow, Adenovirus E1A targets key regulators of cell proliferation, *Cancer Surv.*, 12:161 (1992).
17. P. Whyte, K. Buchkovich, J.M. Horowitz, S.H. Friend, M. Raybuck, R.A. Weinbert, and E. Harlow, Associtation between an oncogene and an anti-oncogene: the adenovirus E1A proteins bind to the retinoblastoma gene product, *Nature (London)*, 334:124 (1988).
18. R.A. Weinberg, Tumor suppressor genes, *Science*, 254:1138 (1992).
19. J.R. Nevins, E2F: A link between the Rb tumor suppressor protein and viral oncoproteins, *Science*, 258:424 (1992).
20. L. Rao, M. Debbas, P. Sabbatini, D. Hockenberry, S. Korsmeyer, and E. White, The adenovirus E1A proteins induce apoptosis which is inhibited by the E1B 19K and Bcl-2 proteins, *Proc. Natl. Acad. Sci. USA*, 89:7742 (1992).
21. E. White and R. Cipriani, Role of adenovirus E1B proteins in transformation: altered organization of intermediate filaments in transformed cells that express the 19-kilodalton protein, *Mol. Cell. Biol.*, 10:120 (1990).
22. Y. Tsujimoto, J. Gorham, J. Cossman, E. Jaffe, and C.M. Croce, The t(14;18) chromosome translocations involved in B cell neoplasms result from mistakes in VDJ joining, *Science*, 229:1390 (1985).
23. A. Bakhshi J.P. Jensen, P. Goldman, J.J. Wright, O.W. McBride, A.L. Epstein, and S.J. Korsmeyer., Cloning the chromosomal break-point of the t(14:18) human lymphomas: cloustering around JH on Chromosome 14 and near a transcriptional unit on 18, *Cell*, 41:889 (1985).
24. M.L. Cleary and J. Sklar, Nucleotide sequence of a t(14;18) chromosomal breakpoint in follicular lymphoma and demonstration of a breakpoint cluster region near a transcriptionally active loucus on chromosome 18, *Proc. Natl. Acad. Sci. USA*, 82:7439 (1985).
25. T.E. Allsopp, S. Wyatt, H.F. Paterson, and A.M. Davies, The proto-oncogene bcl-2 can selectively rescue neurotrophic factor-dependent neurons from apoptosis, *Cell*, 73:295 (1993).
26. I. Garcia, I. Martinou, Y. Tsujimoto, and J.-C. Martinou, Prevention of programmed cell death of sympathetic neurons by the bcl-2 proto-oncogene, *Science*, 258:302 (1992).
27. G. Nuñez, L. London, D. Hockenbery, J.P. McKearn, and S.J. Korsmeyer, Deregulated Bcl-2 gene expression selectively prolongs survival of growth factor-deprived hematopoietic cell lines, *J. Immunol.*, 144:3602 (1990).
28. D.L. Vaux, S. Cory, and T.M. Adams, Bcl-2 promotes the survival of haemopoietic cells and cooperates with *c-myc* to immortalize pre-b cells, *Nature (London)*, 335:440 (1988).
29. T.J. McDonnell and S.J. Korsmeyer, Progression from lymphoid hyperplasia to high-grade malignant lymphoma in mice transgenic for the t(14:18), *Nature (London)*, 349:254 (1991).
30. T.J. McDonnell, N. Deane, F.M. Platt, G. Nunez, U. Jaeger, J.P. McKearn, and S.J. Korsmeyer, bcl-2-immunoglobulin transgenic mice demonstrate extended B cell survival and follicular lymphoproliferation, *Cell*, 57:79 (1989).
31. G. Nuñez, D. Hockenbery, T.J. McDonnell, C.M. Sorensen, and S.J. Korsmeyer, Bcl-2 maintains B cell memory, *Nature (London)*, 353:71 (1991).
32. P. Sarnow, Y.S. Ho, J. Williams, and A.J. Levine, Adenovirus E1b-58 kd tumor antigen and SV40 large tumor antigen are physically associated with the same 54 kd cellular protein in transformed cells, *Cell*, 28:387 (1982).
33. P.R. Yew and A.J. Berk, Inhibition of p53 transactivation required for transformation by adenovirus early 1B protein, *Nature (London)*, 357:82 (1992).
34. J. Martinez, I. Georgoff, and J.M.A.J. Levine, Cellular localization and cell cycle regulation by a temperature-sensitive p53 protein, *Genes & Dev.*, 5:151 (1991).
35. L. Diller, J. Kassel, C.E. Nelson, M.A. Gryka, G. Litwak, M. Geghardt, and B. Bressac, p53 functions as a cell cycle control protein in osteosarcomas, *Mol. Cell. Biol.*, 10:5772 (1990).
36. E. Yonish-Rouach, D. Resnitzky, J. Lotem, L. Sachs, A. Kimchi, and M. Oren, Wild-type p53 induces apoptosis of myeloid leukaemic cells that is inhibited by interleukin-6, *Nature (London)*, 352:345 (1991).
37. M. Debbas and E. White, Wild-type p53 mediates apoptosis by E1A which is inhibited by E1B, *Genes Dev.*, 7:546 (1993).
38. J. Momand, G.P. Zambetti, D.C. Olson, D. George, and A.J. Levine, The mdm-2 oncogene product forms a complex with the p53 protein and inhibits p53-mediated transactivation, *Cell*, 69:1237 (1992).
39. J.D. Oliner, K.W. Kinzler, P.S. Meltzer, D.L. George, and B. Vogelstein, Amplification of a gene encoding a p53-associated protein in human sarcomas, *Nature (London)*, 358:80 (1992).
40. J.D. Oliner, J. Pietenpol, S. Thiagalingam, J. Gyuris, K.W. Kinzler, and B. Vogelstein, Oncoprotein MDM2 conceals the activation domain of tumor suppressor p53, *Nature (London)*, 362:857 (1993).
41. M. Scheffner, B.A. Werness, J.M. Hulbregtse, A.J. Levine, and P.M. Howley, The E6 oncopretein encoded by human papillomavirus types 16 and 18 promotes the degradation of p53, *Cell*, 63:1129 (1990).
42. S. Lowe and H.E. Ruley, Stabilization of the p53 tumor suppressor is induced by adenovirus-5 E1A and accompanies apoptosis, *Genes Dev.*, 7:535 (1993).

43. S.-K. Chiou, L. Rao, and E. White, Bcl-2 blocks p53-dependent apoptosis, (submitted).
44. M.D. Jacobson, J.F. Burne, M.P. King, T. Miyashita, J.C. Reed, and M.C. Raff, Bcl-2 blocks apoptosis in cells lacking mitochondrial DNA, *Nature (London)*, 361:365 (1993).
45. S.K. Chiou, C.C. Tseng, L. Rao, P. Verwaerde, and E. White, Funtional complementation of the *bcl-2* protooncogene for the adenovirus E1B 19K gene, (in preparation).
46. P. Monagan, D. Robertson, T.A. Amos, M.J. Dyer, D.Y. Mason, and M.F. Greaves, Ultrastructural localization of Bcl-2 protein, *J. Histochem. Cytochem.*, 40:1819 (1992).
47. D. Hockenbery, G. Nuñez, C. Milliman, R.D. Schreiber, and S. Korsmeyer, Bcl-2 is an inner mitochondrial membrane protein that blocks programmed cell death, *Nature (London)*, 348:334 (1990).
48. T. Hennet, G. Bertoni, C. Richter, and E. Peterhans, Expression of Bcl-2 protein enhances the survival of mouse fibrosarcoid cells in tumor necrosis factor-mediated cytotoxicity, *Can. Res.*, 53:1456 (1993).
49. M.Y. Mapara, R. Bargou, C. Zugek, H. Dohner, F. Ustaoglu, R. Jonker, P.H. Krammer, and B. Dorken, APO-1 mediated apoptosis or proliferation in human chronic B lymphocytic leukemia: correlation with bcl-2 oncogene expression, *Eur. J. Immunol.*, 23:702 (1993).
50. B. Vanhaesebroeck, J.C. Reed, D. DeValck, J. Grooten, T. Miyashita, S. Tanaka, R. Beyaert, F. VanRoy, and W. Fiers, Effect of bcl-2 proto-oncogene expression on cellular sensitivity to tumor necrosis factor-mediated cytotoxicity, *Oncogene*, 8:1075 (1993).
51. C.H. Herrmann and M.B. Mathews, The Adenovirus E1B 19-Kilodalton Protein Stimulates Gene Expression by Increasing DNA Levels, *Mol. Cell. Biol.*, 9:5412 (1989).
52. N. Takemori, J.L. Riggs, and C. Aldrich, Genetic studies with tumorigenic adenoviruses. I. Isolation of cytocidal (cyt) mutants of adenovirus type 12, *Virology*, 36:575 (1968).
53. Y. Tsujimoto and C.M. Croce, Analysis of the structure, transcripts, and protein products of bcl-2, the gene involved in human follicular lymphoma, *Proc. Natl. Acad. Sci. USA*, 83:5214 (1986).
54. M.L. Cleary, S.D. Smith, and J. Sklar, Cloning and structural analysis of cDNAs for bcl-2 and a hybrid bcl-2/immunoglobulin transcript resulting from the t(14;18) translocation, *Cell*, 47:19 (1986).
55. M. Negrini, F. Silinui, C. Kozak, Y. Tsujimoto, and C.M. Croce, Molecular analysis of mbcl-2: structure and expression of the murine gene homologous to the human gene involved in follicular lymphoma, *Cell*, 49:455 (1987).
56. D.L. Cazals-Hatem, D. Louie, S. Tanaka, and J.C. Reed, Molecular cloning and DNA sequence analysis of cDNA encoding chicken homologue of the Bcl-2 oncoprotein, *Biochem. Biophys. Acta*, 1132:109 (1992).
57. K.M. Kozopas, T. Yang, H.L. Buchan, P. Zhou, and R.W. Craig, MCL1, a gene expressed in programmed myeloid cell differentiation, has sequence similarity to BCL-2, *Proc. Natl. Acad. Sci. USA*, 90:3516 (1993).
58. G.R. Pearson, J. Luka, L. Petti, J. Sample, M. Birkenbach, D. Braun, and E. Kieff, Identification of an Epstein-Barr virus early gene encoding a second component of the restricted early antigen complex, *Virology*, 160:151 (1987).
59. M.O. Hengartner, R.E. Ellis, and H.R. Hovitz, *Caenorhabditis elegans* gene *ced-9* protects cells from programmed cell death, *Nature (London)*, 356:494 (1992).
60. L. Kaczmarek, B. Ferguson, M. Rosenberg, and R. Baserga, Induction of cellular DNA synthesis by purified adenovirus E1A proteins, *Virology*, 152:1 (1986).
61. S. Stabel, P. Argos, and L. Philipson, The release of growth arrest by microinjection of adenovirus E1A DNA, *EMBO J.*, 4:2329 (1985).
62. W. Maltzman and L. Czyzyk, UV irradiation stimulates levels of p53 cellular tumor antigen in nontransformed mouse cells, *Mol. Cell. Biol.*, 4:1689 (1984).
63. M.B. Kastan, O. Onyekwere, D. Sidransky, B. Vogelstein, and R.W. Craig, Participation of p53 protein in the cellular response to DNA damage, *Cancer Res.*, 51:6304 (1991).
64. M.B. Kastan, Q. Zhan, W.S. El-Deiry, F. Carrier, T. Jacks, W.V. Walsh, B.S. Plunkett, B. Vogelstein, and A.J. Fornace, A mammalian cell cycle checkpoint pathway utilizing p53 and GADD45 is defective in ataxia-telangiectasia, *Cell*, 13:587 (1992).
65. S.J. Kuerbitz, B.S. Plunkett, W.V. Walsh, and M.B. Kastan, Wild-type p53 is a cell cycle checkpoint determinant following irradiation, *Proc. Natl. Acad. Sci. USA*, 89:7491 (1992).
66. A.R. Clarke, C.A. Purdie, D.J. Harrison, R.G. Morris, C.C. Bird, M.L. Hooper, and A.H. Wyllie, Thymocyte apoptosis induced by p53-dependent and independent pathways, *Nature (London)*, 362:849 (1993).
67. S.W. Lowe, E.M. Schmitt, S.W. Smith, B.A. Osborne, and T. Jacks, p53 is required for radiation-induced apoptosis in mouse thymocytes, *Nature (London)*, 362:847 (1993).
68. A.H. Wyllie, K.A. Rose, R.C. Morris, C.M. Steel, E. Foster, and D.A. Spandidos, Rodent fibroblast tumours expressing human myc and ras genes: growth, matastasis and endogenous oncogene expression, *Brit. J. Cancer*, 56:251 (1987).
69. D.S. Askew, R.A. Ashmun, B.C. Simmons, and J.L. Cleveland, Constitutive c-*myc* expression in an IL-3-dependent myeloid cell line suppresses cell cycle arrest and accelerates apoptosis, *Oncogene*, 6:1915 (1991).

70. G.I. Evan, A.H. Wyllie, C.S. Gilbert, T.D. Littlewood, H. Land, M. Brooks, C.M. Waters, L.Z. Penn, and D.C. Hancock, Induction of apoptosis in fibroblasts by c-*myc* protein, *Cell*, 69:119 (1992).
71. Y. Shi, J.M. Glynn, L. Guilbert, T.G. Cotter, R.P. Bissonnette, and G.D. R., Role for c-myc in activation-induced apoptotic cell death in T cell hybridomas, *Science*, 257:212 (1992).
72. R.J. Smeyne, M. Vendrell, M. Hayward, S.J. Baker, G.G. Maio, K. Schilling, L.M. Robertson, T. Curran, and J.I. Morgan, Continuous c-fos expression prededes programmed cell death in vivo, *Nature (London)*, 363:166 (1993).
73. R.P. Bissonnette, F. Echeverri, A. Mahboubi, and D. Green, Apoptotoc cell death induced by c-myc is inhibited by bcl-2, *Nature (London)*, 359:552 (1992).
74. A. Fanidi, E.A. Harrington, and G. Evan, Cooperative interaction between c-myc and bcl-2 proto-oncogenes, *Nature (London)*, 359:554 (1992).
75. J.-X. Lin and J. Vilcek, Tumor necrosis factor aand interleukin-1 cause a rapid and transient stimulation of c-fos and c-myc mRNA levels in human fibroblasts, *J. Biol. Chem.*, 262:11908 (1987).
76. M.C. Raff, Social controls on cell survival and cell death, *Nature (London)*, 356:398 (1992).
77. D.M. Hockenbery, M. Zutter, W. Hickey, M. Nahm, and S.J. Korsmeyer, Bcl-2 protein is topographically restricted in tissues characterized by apoptotic death, *Proc. Natl. Acad. Sci. USA*, 88:6961 (1991).
78. C.L. Sentman, J.R. Shutter, D. Hockenbery, O. Kanagawa, and S.J. Korsmeyer, bcl-2 inhibits multiple forms of apoptosis but not negative selection in thymocytes, *Cell*, 67:879 (1991).
79. A. Strasser, A.W. Harris, and S. Cory, bcl-2 transgene inhibits T cell death and perturbs thymic self-censorship, *Cell*, 67:889 (1991).
80. H. Land, L.F. Parada, and R.A. Weinberg, Tumorigenic conversion of primary embryo fibroblasts requires at least two cooperating oncogenes, *Nature (London)*, 304:596 (1983).
81. H.E. Ruley, Adenovirus early region 1A enables viral and cellular transforming genes to transform primary cells in culture, *Nature (London)*, 304:602 (1983)
82. L.A. Donehower, M. Harvey, B.L. Slagle, M.J. McArthur, J. Montgomery C. A., J.S. Butel, and A. Bradley, Mice deficient for p53 are developmentally normal but suceptible to spontaneous tumors, *Nature (London)*, 356:215 (1992).
83. E.S. Alnemri, T.F. Fernandes, S. Haldar, C.M. Croce, and G. Litwack, Involvement of BCL-2 in glucocorticoid-induced apoptosis of human pre-B-leukemias, *Cancer Res.*, 52:491 (1992).
84. G.P. Zambetti and A.J. Levine, A comparison of the biological activities of wild-type and mutant p53, *FASEB J.*, (in press) (1993).

Discussion

M. Fried

You showed that MDM-2 plus E1A induce a lot of transformation. That combination should have efficiently blocked p53.

E. White

Well, MDM-2 has to be overexpressed because it is usually amplified in human tumors or murine tumors. And we are only transfecting in a cosmid clone, so I think that is why it was inefficient.

M. Fried

Do the homologues of *bcl*-2 you showed, also act like *bcl*-2 in blocking apoptosis?

E. White

I think that we may hear from Gerard Evan about one of them. Maybe you should ask him.

M. Oren

Will E1A induce apoptosis in cells which do not have wild type p53?

E. White

Yes, Tyler Jacks and Scott Lowe at M.I.T. have put E1A into primary cells nullizygous for p53 and they get transformation without apoptosis, that is E1A will transform them by itself.

M. Oren

Does ElA affect the functional state of p53?

E. White

That we have not done. We have looked at a number of things about p53 function and the ability of E1B 19K to affect p53. We know the levels of p53 were unaltered when we shifted to the permissive temperature, they are unchanged whether E1B 19K is there or not. We have looked at the localization of p53, it goes to the nucleus just fine with or without E1B 19K. We have also looked at the phosphorylation state and the phosphorylation state is radically different. We have begun to look at DNA binding but we have no data. We have looked at CAT assays, transactivation of the of the MCK sequence, E1B 19K will greatly accelerate the ability of wild type p53 to transactivate the MCK sequence. But it also obliterates repression by p53, so right now our results are consistent with believing that repression is something that is associated with apoptosis by p53.

P. Krammer

Do you really inhibit with *bcl*-2 or do you shift kinetics?

E. White

We really inhibit it because we actually left the APO-1 antibody in there for a very long period of time. I do not know why that was. I expect a graduate student forgot about the experiment, so I think it was in there for at least a week before we evaluated the cells. We actually removed the antibody and looked to see if the cells would grow afterwards and they did. So they can recover from it as well.

P. Krammer

It is unclear to me at the moment what is going on. There are cases where the level of *bcl*-2 message or even the level of *bcl*-2 protein are not correlated at all with anti APO-1 sensitivity or resistance. So there must be something else in the cell in addition to *bcl*-2, maybe interacting with *bcl*-2, which determines the resistant state.

E. White

I agree there are probably other factors, but I am also concerned when you are dealing with transformed cell lines you do not know what the genetic makeup of that cell is. And I think that that can have a great influence. If inhibition of programmed cell death in various forms can contribute to the transformation process, then every cell line in some way or another is going to be affected in programmed cell death.

J. Cleveland

I was not sure if you had done this experiment in your transformation assays--E1A plus wild type p53 plus MDM-2, because it seems like that would distinguish whether a loss of function or a gain of function is important?

E. White

Well we did that, namely, the E1A plus MDM-2 experiment. Our cells have wild type p53 in them, they are primary cells.

D. Livingston

I am always fascinated by why the virus needs to have two gene products to deal with ideally the same problem, so I guess the question would be: In a 19K minus virus, can you select for virus that replicates efficiently, that do not have the deg phenotype but are mutant in p55.

E. White

Well, I think the answer to your question has to do with the role of the 55K protein in infection. I think the reason why there are two E1B proteins; one reason is that getting rid of p53 is really important for the virus to do. So, having redundant ways of doing it would be the most efficient way to ensure that p53 is disposed of. In productively effected cells the E1B 55K protein is normally found in a complex with E4, an E4 protein, and is not available to bind a p53. So I think in infected cells, the need for the 19K protein is because the 55K protein is not available.

D. Livingston

Do you think one could substitute HPV E6 for the 19K gene and, if so, would that virus be as infectious as wild type?

E. White

That would be an interesting experiment to do. We have not done that.

E. Mihich

You told us that TNF can induce apoptosis in HeLa cells and then that *bcl*-2 will antagonize the apoptosis induced by p53. Is it known whether *bcl*-2 will induce a change in the apoptosis induced by TNF, and whether perhaps the change represents a mode of killing by TNF or any other such agent. In other words, by inhibiting the apoptotic process can you induce an alteration in the mode of killing, or is the apoptosis the only mode of killing of those cells?

E. White

As far as I know, it is the only mode of killing. Are you asking if p53 is required for a cell death by TNF?

E. Mihich

With *bcl*-2 can you induce a different kind of death by a cytotoxic agent such as TNF, which can indeed kill cells in different ways?

E. White

Yes, *bcl*-2 and E1B will both prevent apoptosis by TNF. What I should say though is that I do not think it is known if p53 is in the TNF pathway. Because if you treat HeLa cells that do not have p53 they die and there is induction of p53 by TNF, so that says that TNF is inducing apoptosis a different way than E1A does. And that is consistent with the idea that E1B 19K and *bcl*-2 act downstream, at the end of the pathway to block apoptosis past p53.

D. Hockenbery

That was basically my question. You have assayed for E1A function in a p53-less background to see if it is independent of p53 in inducing apoptosis? Do you have other examples of apoptosis induced in a p53-less background?

E. White

No, I decided a long time ago that we are at the limit of our ability to do experiments, and I have made reagents available to anyone who wants them, and I believe that there are many labs now putting E1B into your favorite apoptosis assay. But I have not done those experiments.

M. Oren

Just a comment to support what you answered to the previous question. There seem to be TNF induced pathways which are p53 independent. This is an experiment that we did with David Wallach at the Weizmann. The human breast cancer line MCF7 is induced to die in response to TNF, and has endogenous wild type p53. Overexpression of the negative dominant p53, which worked so nicely in the IL3 paradigm, had no effect on apoptosis, or had no effect on cell death. TNF killed those cells as efficiently as it did the parental cells. Thus, TNF can work very well in these cells irrespective of p53 function.

E. White

So that would be, then, one example of a totally independent pathway.

R. Schimke

Is there a pattern to be observed such that if E1A and E1B are expressed, it is then not necessary to have a mutated p53? If so, this would suggest that to generate a state of transformation, it is important to alter apoptosis, and this can result from either the presence of E1B or an altered p53.

E. White

Clearly, the HPV induced tumors which express E6 will have wild type p53. I hope that there are other things besides p53. I have trouble believing that p53 is a single gene and it is the only thing that is doing this. I mean, it is very possible that other tumor suppressors may also act at the regulation of apoptosis and getting in the right combinations would obviate the need to have a mutant p53.

R. Schimke

I guess these are tumors which are either adenomas or papillomas. Will those not have a mutant type p53?

E. White

In the papilloma virus tumors, that is absolutely true. What we have done, we have looked at the status of p53 in the E1A immortalized lines and I have told you it is mutant. In the E1A plus E1B 19K transformants and the *bcl*-2 transformants, by antibody studies it appears to be wild type. So not only do you block the ability of p53 to kill cells by having *bcl*-2 or E1B, but you overcome the need to genetically select for a p53 mutation.

R. Schimke

Is this true of real tumors too, as well as just cells?

E. White

You may want to ask Tim McDonnell that question.

T. McDonnell

Thanks, I would just like to mention what we are seeing in the transgenic mouse models where we have the Mμ-myc mice and the *bcl*-2 immunoglobulin transgenic mice. They both develop lymphoma spontaneously. If you look for incidence of p53 mutations in those tumors that arise spontaneously in the *bcl*-2 mice, you see absolutely no p53 mutation, so it appears that they are not at a selective advantage; whereas if you look in the Mμ-myc mice, about half of those tumors will have mutations in p53. The double transgenic mice have, as far we have looked, no p53 mutation, so it appears that either one of those events alone is sufficient for overriding apoptotic cell death and there is no selective advantage for having both of them simultaneously.

D. Green

This is a thought question, unless you have done the experiment. Since we know that FAS/APO-1 defects or a decreased expression of FAS/APO-1 in mice leads to a massive lymphadenopathy by about 12 weeks of age, the mutation that we call LPR, one might expect if E1B completely or dramatically inhibits the death induced by anti-Fas antibodies, that expression of E1B in a mouse would lead to the same phenotype. First, has anybody made an animal like this, that expresses 19K?

E. White

If anyone would be, it would be me. We are in the process of that sort of study.

D. Green

On the other hand though, expression of *bcl*-2 in peripheral T cells does not lead to that phenotype and that suggests that either there are other pathways or perhaps *bcl*-2 only delays FAS induced death and does not completely inhibit, or like Peter says, that there are additional mechanisms involved in that resistance to FAS/APO-1 induced death. I just wondered.

M. Fried

What about Ras? Does Ras have any effect on p53 because that is an oncogene that cooperates with E1A and myc.

E. White

Yes, it prevents the cell from dying, but I should also say that things seem to be a little bit strange and we do not know why yet.

M. Fried

Are there any reports that RAS has any effect on p53 expression?

E. White

It does not affect p53 expression, that I have seen, but it appears to prevent the cells from dying by p53, but there is also something that we do not understand yet. There is something different about Ras and I do not know what it is.

M. Oren

Given that the region in E1A which is needed to induce apoptosis covers the part that is involved in RB binding, what happens if you transfect cells with E1A and overexpress RB at the same time? Is there any effect?

E. White

I have not done that experiment. I do not know if anyone else has.

M. Oren

That is kind of driving towards the extreme that RB, unlike p53, does exactly the opposite and, in fact, inhibits apoptosis.

E. White

I can tell you that people that work on RB have been calling me and asking me questions about programmed cell death. My thought, from what I have discussed with them, is that they think it might be inducing programmed cell death. But from the work of the knockout mice, that may not be true. I think the answer on RB will be coming.

M. Kastan

I just wanted to make a couple of comments, partially in answer to Dr. Schimke's questions that he raised. Clearly, p53 is not necessary for chemotherapy or DNA damage induced apoptosis. One of the best models that people use is HL60 cells and etoposide-induced apoptosis and those have no p53 genes. So, there are many types of apoptosis that occur where p53 is not necessary. In addition, in terms of the question of whether p53 is necessary for transformation because of loss of apoptosis, that is probably cell type specific. For example, in the thymocytes it may be very important to lose apoptosis and that is why you get lymphomas in the knockout mice. But in other tumor types, loss of p53 may lead to transformation by other mechanisms such as genetic instability because of inability to cell cycle arrest following damage. So you do not need in all cell types to invoke loss of apoptosis as a transformation event.

C-*MYC*: ONCOGENE AND TUMOUR SUPPRESSOR GENE

Gerard Evan, Trevor Littlewood, David Hancock, Martin Bennett, Elizabeth Harrington, and Abdallah Fanidi

Biochemistry of the Cell Nucleus Laboratory, Imperial Cancer Research Fund, P.O. Box 123, Lincoln's Inn Fields, London WC2A 3PX, UK

INTRODUCTION

Eukaryotic cell proliferation in metazoans is regulated by a variety of positive and negative signals that serve to rigidly control cell division. This underscores a major problem facing multicellular organisms: namely, how to allow the rapid proliferation of component cells and at the same time never allow those component cells to engage in competition with each other. In principle, any cell that acquires a growth advantage through mutation should out-compete its siblings and generate a hyperplastic clone from which ever more rapidly proliferating mutants are likely to arise. Such an event is, however, extremely rare as demonstrated by the fact that cancer arises in only one in three individuals during the entire course of their lives. Part of the answer to this paradox appears to reside in the multifunctional nature of the components that mediate cell proliferation. One of these, c-Myc, is the subject of this paper.

The proto-oncogene c-*myc* was first identified by its homology to the v-*myc* viral oncogene of the avian myelocytomatosis viruses. c-*myc* is one of several *myc* genes in the human genome of which c-*myc*, N-*myc* and L-*myc* share the greatest homology. However, only c-*myc* appears to be expressed in differentiated cells in the adult. The c-*myc* proto-oncogene has long been implicated in the control of normal cell growth[1,2]. c-*myc* is one of the immediate early growth response genes that is rapidly induced upon mitogenic stimulation of quiescent cells[3,4], suggesting a role in normal cell growth. Conversely, instances of deregulated expression of c-*myc* are widely reported in tumours. For example, reciprocal chromosomal translocations in human Burkitt's lymphoma and murine plasmacytoma deregulate c-*myc* by placing it under the control of the immunoglobulin gene locus - a region transcriptionally active in B lymphoid cells from which these tumours arise. c-*myc* is also activated *in situ* by retrovirus insertion in avian B cell leukosis. Amplification of *myc* genes, with concomitant elevated expression, is relatively common in various tumours, for example neuroblastoma and small cell lung carcinoma. Finally, almost all tested

Apoptosis, Edited by E. Mihich and R.T. Schimke
Plenum Press, New York, 1994

tumour cells express substantially elevated levels of c-*myc*, even though most show no evidence of any gross cytogenetic lesion.

The c-*myc* gene encodes a short-lived, nuclear phosphoprotein[5,6,7,8] that exhibits sequence-specific DNA binding in association with a heterologous partner, Max[9,10,11,12,13,14]. Its nuclear location[8], sequence-specific DNA binding and possession of both *trans*-activation[15,16,17,18,19] and basic-Helix-Loop-Helix-leucine zipper (bHLH-LZ) domains all suggest that the c-*myc* protein, c-Myc, is a transcription factor. Thus, c-Myc is presumed to exert its biological effects by modulation of specific target genes. However, the identity of these genes remains unknown.

GROWTH-PROMOTING FUNCTION OF c-MYC PROTEIN IN UNTRANSFORMED CELLS

In sub-confluent human or rodent fibroblasts, both c-*myc* mRNA and protein are absent in arrested, growth factor-deprived cells. Upon addition of mitogens, c-*myc* is rapidly induced - c-*myc* mRNA peaks after about 2 hours[4] and c-Myc protein after about 3 hours[20]. Thereafter, both c-*myc* mRNA and protein levels fall to between 20 and 60% of the peak and are then maintained with no apparent fluctuation throughout the cell cycle[20,21,22,23]. The mitogenic induction of c-Myc, whilst rapid, is significantly delayed when compared with that of other immediate early growth response genes like c-*fos* or *egr*-1[20]. Moreover, expression of c-*fos* and *egr*-1 is transient and confined to the G_0-G_1 transition whereas c-*myc* is not, suggesting that c-*myc* has a role not only in initiating cell growth but also in maintaining it. This view is supported by analysis of c-*myc* expression in confluent fibroblasts following addition of mitogens. Confluent fibroblasts do not enter S-phase following mitogenic stimulation, presumably due to a dominant anti-proliferative signal arising from contact inhibition. Nonetheless, mitogenic stimulation of confluent fibroblasts leads to induction of c-*myc* with the same kinetics as in sub-confluent cells. However, instead of being maintained, expression of c-*myc* rapidly falls back to background quiescent levels[20]. Thus, entry into the cell cycle correlates with the sustained expression of c-*myc* (at least up to the late G_1 commitment point) rather than its transient induction at the G_0/G_1 transition.

If continuous c-*myc* expression is required for maintenance of cell growth, it is germane to ask how rapidly c-*myc* expression is down-regulated when cells receive signals triggering their growth arrest. Arguably the most physiological signal for growth arrest is mitogen depletion. When asynchronous exponentially proliferating fibroblasts are deprived of mitogens, both c-*myc* mRNA[24] and protein[22] rapidly disappear over a period of 60-90 minutes, irrespective of the position of each cell within its cycle. This down-regulation of c-*myc* is consistent with immediate cessation of new c-*myc* expression and decay of c-*myc* mRNA and protein at rates consistent with their short half-lives. In response to mitogen withdrawal, the fibroblasts eventually undergo growth arrest in G_1. One interesting corollary of this observation arises from the fact that cells deprived of mitogens after the late G_1 commitment point, although rapidly down-regulating c-*myc* expression[22], nonetheless divide on schedule before arresting in G_1 of the subsequent cycle[25]. Thus, cells can complete S-G_2-M in the absence of detectable c-Myc protein, arguing that c-Myc is not *required* for the post-commitment phases of cycle and that regulation of c-*myc* occurs at the time cells receive growth-regulatory signals, rather than when the cells actually arrest. This is most consistent with the notion that c-*myc* is engaged in a signalling pathway that renders cells permissive for continued growth.

Thus, c-*myc* expression is tightly and continuously dependent upon mitogenic stimulation throughout cell growth.

CONSEQUENCES OF DEREGULATED c-MYC EXPRESSION

The tight and continuous mitogenic dependence of c-*myc* expression prompted us to ask whether c-*myc* down-regulation is required for growth arrest in fibroblasts. To test this, human c-*myc* under the control of a constitutive promoter was introduced into primary and immortalised Rat-1 and Swiss 3T3 fibroblasts using a retrovirus vector. Individual clones were isolated that constitutively expressed a range of c-Myc protein, varying from the levels seen in a normal log-phase fibroblast (e.g. clone 21 at ~2,000 molecules per cell[mpc]) to the level seen in many tumour cells (e.g. clone 26 at ~20,000 mpc). These Rat-1/Myc clones were then tested for their abilities to arrest growth in the absence of mitogens (serum withdrawal). 48 hours after transfer into 0.1% serum, the proportion of cells in S-phase was determined by pulse-labelling for 1 hour with the S-phase metabolic label 5-bromo-2'-deoxyuridine (BrdU). Cells were then detached by trypsinisation, probed with fluorescent anti-BrdU antibodies and then assayed by flow cytometry. Representative data derived from three clones are shown in Table 1 and show that even very low (i.e. normal) levels of c-Myc significantly inhibited growth arrest of cells in low serum if constitutively expressed. At higher levels of c-Myc expression this effect is even more marked - essentially no change is observed in the growth rates of cells, whatever the serum concentration. Thus, inability to down-regulate c-Myc renders cells unable to arrest growth in response to mitogen deprivation.

Table 1. Effect of constitutive c-Myc expression on growth arrest of serum-deprived Rat-1 fibroblasts

Cell Line	*c-Myc protein level (molecules per cell)*	*Percent cells incorporating BrdU per hour*	
		10% FCS	**0.1% FCS**
Rat-1/*neo* control	-	47.2	3.7
Rat-1/ wt Myc clone 21	5,500	37.2	16.7
Rat-1/ wt Myc clone 19	12,000	27.6	23.2
Rat-1/ wt Myc clone 26	20,000	28.2	31.5

INDUCTION OF APOPTOSIS BY C-MYC

Although fibroblasts constitutively expressing a c-*myc* gene are unable to arrest in low serum, we made the surprising observation that their numbers do not increase because considerable cell death occurs (Figure 1). Death occurs by apoptosis: it is rapid (30-60 minutes), involves a characteristic process of membrane blebbing, condensation, fragmentation and nuclear pycnosis, and cellular DNA is fragmented into a characteristic nucleosome ladder[26]. To be sure that this apoptosis was caused by c-Myc, several further studies were conducted.

First, we made use of the conditional c-Myc allele, c-Myc-ER, in which the c-Myc protein been fused to part of the human oestrogen receptor[27,28]. c-Myc-ER is active only in the presence of exogenous ß-oestradiol. Rat-1 cells expressing c-Myc-ER (Rat-1/c-Myc-ER cells) are completely viable and arrest in low serum. In contrast, serum deprived Rat-1/c-Myc-ER cells in the presence of ß-oestradiol do not arrest growth and undergo substantial apoptosis[26]. Control cells expressing the inactive c-Myc mutant D106-143 fused to ER are viable and arrest in low serum irrespective of the presence of ß-oestradiol. Thus, apoptosis requires active c-Myc.

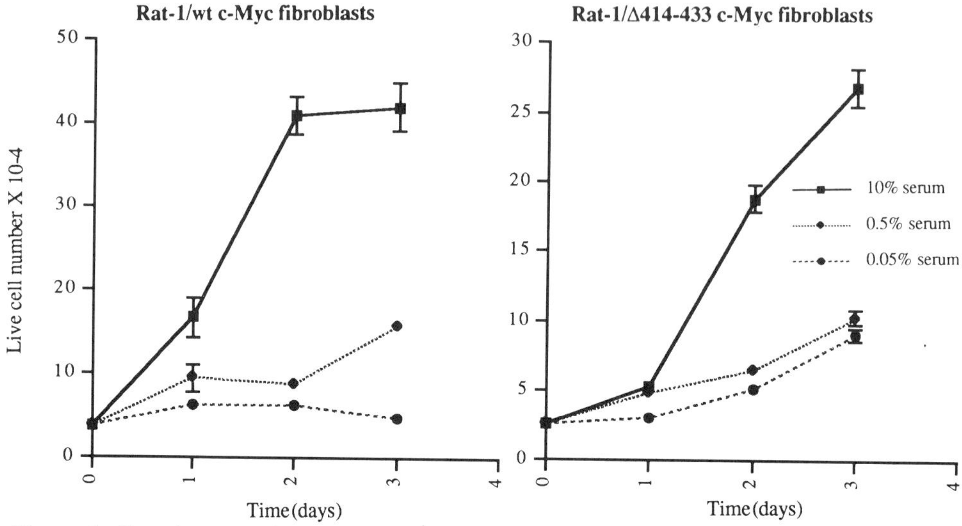

Figure 1. Growth curves in various concentrations of serum of Rat-1 fibroblasts consitutively expressing wild type c-Myc or control Rat-1 cells expressing the defective c-Myc mutant D414-433. Live cells were identified by trypan blue exclusion and cell numbers counted by haemocytometer. Quadruplicate counts were made for each point.

Second, we investigated the relationship between incidence of apoptosis and level of expression of c-Myc. Rat-1 clones expressing differing levels of c-Myc protein were examined for incidence of apoptosis under conditions of partial or complete serum deprivation (i.e. 10%, 2%, 0.5% and 0.1% serum). In clones expressing low levels of c-Myc (~2,000 mpc), apoptosis is only evident at the lowest serum concentrations after several days. In contrast, substantial apoptosis rapidly occurs in serum-deprived clones expressing high c-Myc levels and is evident even under modest serum deprivation (e.g. 2% serum)[26].

Third, we examined the abilities of a series of c-Myc mutants to induce apoptosis in low serum. Mutants lacking the N-terminal putative *trans*-activation domain (e.g. D106-143)[15,18] or the C-terminal DNA binding (e.g. 360 N>P and 364/6/7 R>A[15]) or dimerisation domains (e.g. D371-412, D414-433 or In414[29]) were completely inactive in inducing apoptosis[17,26]. In contrast, mutant D145-262, which lacks the entire central region of c-Myc, is fully active in inducing apoptosis. Thus, the same regions of the c-Myc protein that are required for co-transformation and inhibition of growth arrest in low serum are also required for apoptosis - in particular, those that confer upon c-Myc the ability to bind DNA in a sequence-specific manner and to *trans*-activate gene expression.

Finally, we examined whether Max, the heterodimeric partner for c-Myc, is required for apoptosis. To do this, we made use of mutants with reciprocally exchanged dimerisation specificity such that the c-Myc mutant (MycEG) can no longer interact with wild type Max but can dimerise with the corresponding MaxEG mutant[17]. Constitutive expression of either MycEG or MaxEG proteins alone did not induce apoptosis in serum-deprived Rat-1 cells. In contrast, if MaxEG and MycEG are co-expressed in Rat-1 cells, the cells undergo apoptosis in low serum[30]. Thus, Max is required for induction of apoptosis by c-Myc, just as it is for co-transformation and inhibition of growth arrest.

We conclude that the induction of apoptosis by c-Myc in serum-deprived fibroblasts requires sequence-specific DNA binding, *trans*-activation and dimerisation with Max and is an attribute that is genetically inseparable from its oncogenic and growth-promoting properties. Thus, c-Myc appears to promote two contradictory genetic programmes, growth and cell death, through the transcriptional modulation of target genes.

Table 2. Co-transforming and Apoptosis-inducing activities of Myc mutants

Myc mutant	*Nature of mutation*	*Active in co-transformation*	*Active in inducing apoptosis*
wild type	-	+	+
D7-91	Transcriptional modulation	-	-
D41-53	Loss of 1st Myc homology box	+	+
D106-143	Transcriptional modulation	-	-
D145-262	No observed effect	+	+
D265-317	No observed effect	+	+
D371-412	Deletion of part of helix-loop-helix	-	-
360 N>P	Destruction of DNA binding domain	-	-
D414-433	Deletion of leucine zipper	-	-
In414	Interruption of leucine zipper by insertion	-	-

SIGNIFICANCE OF INDUCTION OF APOPTOSIS BY c-MYC

Is the induction of apoptosis by c-Myc physiological or simply the fatal result of corruption of a normal cellular regulatory process? Various lines of evidence support the notion of a physiological role for c-Myc in apoptosis. Elevated expression of c-*myc* accompanies apoptosis in several physiological situations, for example involution of breast and prostate[31,32,33] and receptor-mediated cell death in embryonic thymocytes[34]. Moreover, a recent elegant study of receptor-mediated apoptosis in the A1.1 T cell hybridoma line showed that antisense inhibition of c-*myc* expression blocks apoptosis[35]. Thus, c-Myc appears to be a physiological regulator of apoptosis in certain circumstances.

Two simple models can be invoked to explain the induction of apoptosis by c-Myc in serum-deprived fibroblasts:

1. Apoptosis is a consequence of some catastrophic imbalance in proliferative pathways - in effect, a conflict of signals between the growth-promoting function of c-Myc (i.e. a "GO" signal) and the growth-inhibitory effect of low serum (i.e. a "STOP" signal). In this model, therefore, induction of apoptosis is not a normal function of c-Myc but a consequence of its ectopic expression. Arguably, this "imbalance" or "conflict" scenario is the most intuitively plausible of the two and it has been widely invoked to explain Myc-induced apoptosis (e.g.[36,37]). This model is depicted in Figure 2A. Here, the implementation of an apoptotic pathway is abnormal and is a specific pathological response to some internal conflict in growth signals (hypothesised as some form of internal cell growth clock). This model is most consistent with the idea that c-Myc implements its transcriptional apoptotic programme only

in response to the putative conflict of signals: the programme is not normally implemented. We call this the "*Conflict Model.*"

2. An alternative explanation for c-Myc-induced apoptosis[2] holds that induction of an apoptotic programme is a *bona fide* and normal physiological function of c-Myc that is always implemented by c-Myc, irrespective of growth conditions and not just during signalling "conflicts." However, once induced, this apoptotic pathway is not necessarily implemented because it is subject to independent modulation by anti-apoptotic "survival factors." According to this "*Dual Signal*" model (depicted in Figure 2B), cells expressing c-Myc survive in high serum because serum contains cytokines that specifically suppress apoptosis. Thus, induction of apoptosis by c-Myc in low serum is not due to a conflict of *growth* signals *per se* but rather to absence of anti-apoptotic cytokines whose actions are not directly linked to their abilities to promote cell cycle progression.

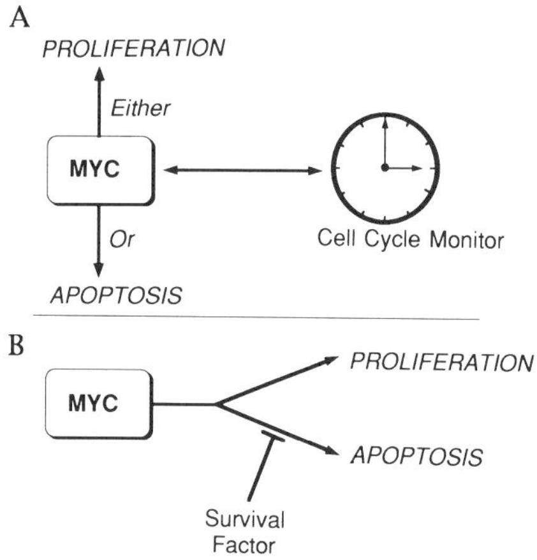

Figure 2. Alternative models for Myc-induced apoptosis

If the c-Myc-induced apoptotic programme is present but not active in cells growing in high serum, it is appropriate to ask what components of serum serve to suppress the programme. We therefore investigated whether specific cytokines could suppress c-Myc-dependent apoptosis in serum-deprived fibroblasts. To do this, we used Rat-1 cells that express the c-Myc-ER conditional allele of c-Myc. In the absence of ß-oestradiol , these Rat-1/c-Myc-ER cells arrest viably in low serum. Addition of ß-oestradiol induces entry into the cell cycle[26,28] and the rapid onset of apoptosis[26]. Accordingly, individual cytokines were added to Rat-1/c-Myc-ER cells in low serum, Myc activated by addition of ß-oestradiol, and apoptosis monitored by time-lapse videomicroscopy (Figure 3). Insulin-like growth factor-I (IGF-I) was found to be a potent inhibitor of apoptosis. IGF-II, insulin and PDGF AB were also fairly effective at blocking apoptosis. However, PDGF, basic FGF and the neuropeptide bombesin each had no anti-apoptotic effect, even though each is a potent mitogen for Rat-1 cells. Analogous results were obtained in experiments using Swiss 3T3 cells, rat embryo fibroblasts and rat primary vascular smooth muscle cells (not shown).

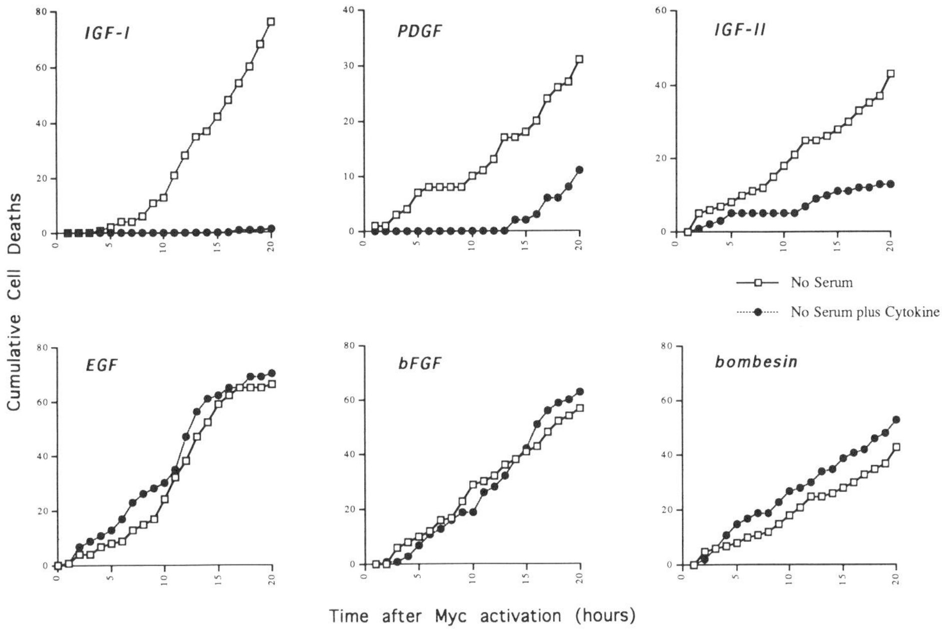

Figure 3. Protection of serum-deprived Rat-1 cells from Myc-induced apoptosis by cytokines. Identical cultures of Rat-1/c-Myc-ER cells were arrested in 0% serum for 48 hours, indicated cytokines added to each, and Myc then activated by addition of 2nM ß-oestradiol to the growth medium. Concentrations of cytokines used whereas follows: IGF-I at 100 ng/ml, PDGF AB at 10 ng/ml, IGF-II at 100 ng/ml, EGF at 10 ng/ml, basic FGF at 10 ng/ml and bombesin at 10 ng/ml. 100 cells of each type were picked at random and observed by time-lapse videomicroscopy for apoptosis.

IGF-I, like most cytokines, is a potent inducer of immediate early growth response transcription factors such as c-Fos, Egr-1 and c-Jun[38]. Thus, IGF-I initiates a signal transduction pathway that results in the modulation of specific genes. We were therefore interested to determine whether induction of immediate early genes was also required for IGF-I-mediated inhibition of apoptosis. Cycloheximide is a potent inducer of apoptosis in fibroblasts expressing c-Myc and, implicitly, such apoptosis occurs in the absence of new protein synthesis[26]. We initiated apoptosis in Rat-1/c-Myc-ER cells by activating c-Myc with ß-oestradiol and then triggered apoptosis by addition of cycloheximide to the cells at a final concentration of 10μg/ml. Apoptosis in the presence and absence of the anti-apoptotic cytokines IGF-I or PDGF AB was then monitored by time-lapse videomicroscopy (Figure 4). IGF-I exerted a profound anti-apoptotic effect on cycloheximide-treated cells whereas PDGF was ineffective. Thus, IGF-I blocks apoptosis by an epigenetic mechanism, whose signal transduction pathway is at least partially discrete from that triggering nuclear events.

What is the mechanism by which IGF-I blocks c-Myc-dependent apoptosis? According to the "Conflict Model," it is because IGF-I provides an obligatory auxiliary mitogenic signal that is required for cell cycle progression - without it cells sustain an imbalance and die. Alternatively, in the "Dual Signal" model, IGF-I acts as an anti-apoptotic survival signal independently of any mitogenic action. To discriminate between these two scenarios, we investigated whether IGF-I could suppress apoptosis in situations when it cannot act as a growth-promoting agent.

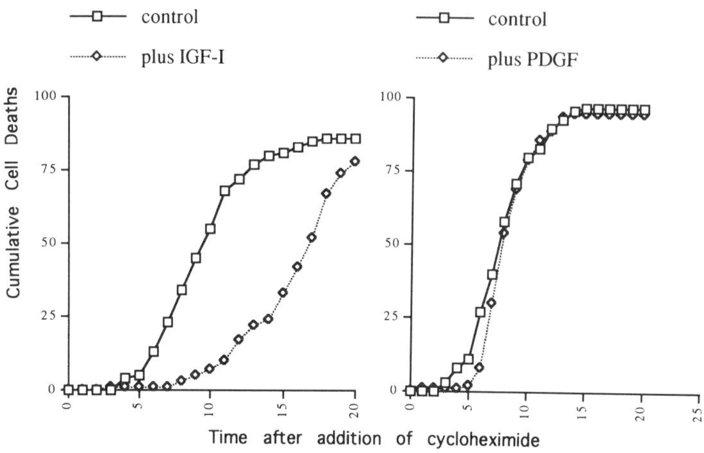

Figure 4. Modulation of cycloheximide-induced apoptosis by cytokines. Rat-1/c-Myc cells were transferred into serum-free medium for 1 hr and then cycloheximide added to a final conentration of 20 μg/ml. 1 hr later, cytokines were added as indicated (IGF-I to 100 ng/ml; PDGF to 10 ng/ml) and the cells observed by time-lpse videomicroscopy.

Activation of c-Myc induces apoptosis in Rat-1/c-Myc-ER cells treated with the drugs thymidine (a DNA synthesis inhibitor) or etoposide (an inhibitor of topoisomerase-II)[26,39]. According to the *Conflict Model*, such apoptosis arises because of a conflict between the growth-promoting effects of c-Myc and the growth-inhibitory effects of each drug, in which case the addition of an auxiliary mitogen like IGF-I might reasonably be expected to exacerbate cell death by further driving cell growth against the drug-induced block. In contrast, the *Dual Signal* model assumes that the function of IGF-I is to mitigate the apoptotic pathway and therefore predicts that IGF-I should suppress drug-induced apoptosis. Rat-1/c-Myc-ER cells were treated for 24 hours with either Thymidine (2 mM) or Etoposide (0.1μM) which both elicit profound cell cycle blocks in S-phase and S/G_2 respectively. c-Myc was then activated by addition of ß-oestradiol, either in the absence or presence of IGF-I and apoptosis monitored by time-lapse videomicroscopy. As can be seen from Figure 5, drug-dependent c-Myc-induced apoptosis is substantially inhibited and delayed by IGF-I. A less profound, but significant, inhibition was also provided by PDGF (not shown). Thus, IGF-I suppresses c-Myc-induced apoptosis under conditions in which it provides no growth-promoting function.

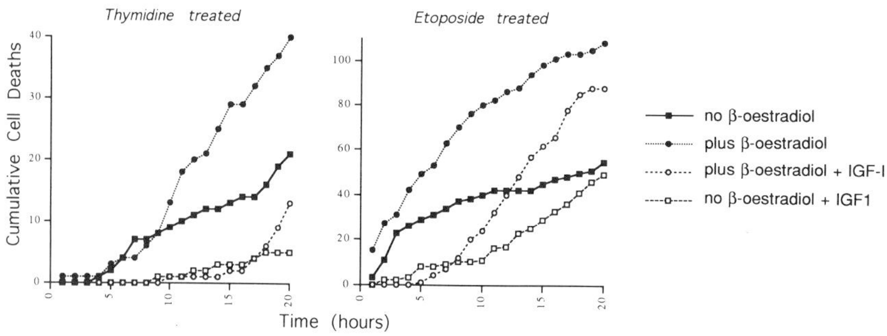

Figure 5. IGF-I-mediated protection from drug-induced apoptosis in Rat-1/Myc fibroblasts. Identical sub-confluent cultures of Rat-1/c-Myc-ER cells were arrested for 24 hours by either thymidine (2mM) or etoposide (0.1μM). IGF-I was then added to 100nM and c-Myc activated by addition of 2μM ß-oestradiol, as indicated. Cells were observed by time-lapse videomicroscopy.

Because of the possible complications in interpreting experiments with toxic drugs we sought a more direct and physiological way of examining the independent anti-apoptotic and mitogenic effects of IGF-I. Fibroblast proliferation is dependent upon mitogens only during the pre-commitment G_1 phase of the cell cycle. Once beyond this late G_1 point, fibroblasts complete their cycle on schedule independently of the presence or absence of mitogens[25]. Thus, transit of $S/G_2/M$ does not require growth factors. If c-Myc induces apoptosis in serum-deprived fibroblasts because of a conflict arising from lack of necessary auxiliary mitogens such as IGF-I, it follows that c-Myc-induced apoptosis should be restricted to the pre-commitment G_1 phase of the cell cycle when such mitogens are required. Thus, serum-deprivation of S-phase fibroblasts expressing c-Myc should result in death only after each cell has divided and re-entered the serum-dependent G1 phase of the cell cycle. In contrast, if the *Dual Signal* model is correct (Figure 6), activation of c-Myc in the absence of anti-apoptotic serum cytokines should induce apoptosis irrespective of cell cycle position. Furthermore, anti-apoptotic cytokines should block this apoptosis even during the mitogen-independent $S/G_2/M$ phases of the cell cycle. Accordingly, asynchronous exponentially proliferating Rat-1/c-Myc-ER cells were pulsed for 30 minutes with BrdU to label S-phase cells. Serum was then removed, c-Myc activated by addition of ß-oestradiol, and the cells incubated in serum-free medium for a further 2 hours. As G_2/M of Rat-1 cells is about 3 hours, it is impossible for any BrdU-labelled cell to traverse G_2/M and re-enter G_1 during this period. Cells were then fixed, and stained with propidium iodide to label cellular RNA and DNA and with FITC-conjugated anti-BrdU antibodies to label S-phase cells. 37% of all cells incorporated BrdU label and were therefore in S-phase at the time of labelling. Approximately 1.5% of BrdU positive cells exhibited pycnotic nuclei and cellular morphology indicative of apoptosis. The same percentage (~1.5%) of BrdU negative cells were also apoptotic. Apoptosis of BrdU positive cells was completely inhibited by addition of IGF-I to the growth medium during the period of serum deprivation. We conclude that c-Myc induces apoptosis in serum-deprived fibroblasts, and IGF-I inhibits it, independently of cell cycle position or of any requirement of growth factors for cell cycle progression. This result strongly argues against the idea that c-Myc-induced apoptosis arises from a non-specific conflict of signals and is in favour of the *Dual Signal* model.

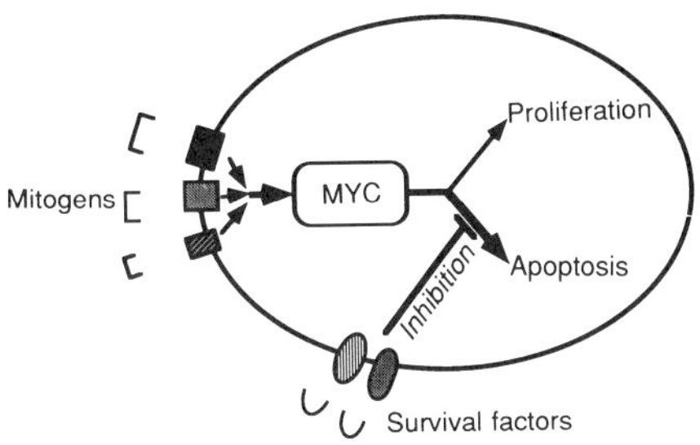

Figure 6. Dual signal model for Myc-induced apoptosis and its regulation by specific cytokines.

c-*myc* IN CANCER

The *Dual Signal* model, if correct, has important implications for our understanding of cell growth and carcinogenesis. If Myc couples the pathway of cell proliferation with that of cell death it means that every cell that proliferates automatically implements a potential "abort" pathway of programmed cell death. This implies that no mitogenic lesions in any proto-oncogene upstream of Myc can directly lead to neoplastic transformation because at the same time as it triggers cell proliferation it also triggers cell suicide. Only cells that continuously receive signals blocking apoptosis can survive mitogenic stimulation, implying that all cell *in vivo* exist on a "knife-edge" between survival and death. Although this may at first seem intuitively implausible, it does provide a very effective mechanism for limiting neoplasia by rendering intermediate stages of carcinogenesis intrinsically unstable. Deregulated or elevated expression of *myc* genes is so ubiquitous in tumour cells as to suggest that activation of *myc* genes is an essential step during carcinogenesis. However, because the proliferative and apoptotic functions of Myc are tightly coupled, it is impossible to select for one without also selecting for the other. Activation of c-Myc is therefore an unstable state from which almost all potential tumour cells will die unless they are rescued by an additional mutation that blocks the induction of apoptosis by c-Myc. This, then, implies that genetic mechanisms that block apoptosis are likely to be important during carcinogenesis.

The idea that anti-apoptotic lesions might co-operate with c-Myc during carcinogenesis is supported by various transgenic models of carcinogenesis - in particular, the strong oncogenic synergy between c-*myc* and the proto-oncogene *bcl*-2[40,41]. *bcl*-2 is activated by reciprocal translocation to the immunoglobulin heavy chain locus in a number of human B-cell follicular lymphomas and encodes two membrane-associated polypeptides. Although the molecular function of *bcl*-2 is unknown, its expression extends the life of lymphoid cell *in vitro*[42,43,44] and blocks apoptosis[45,46,47].

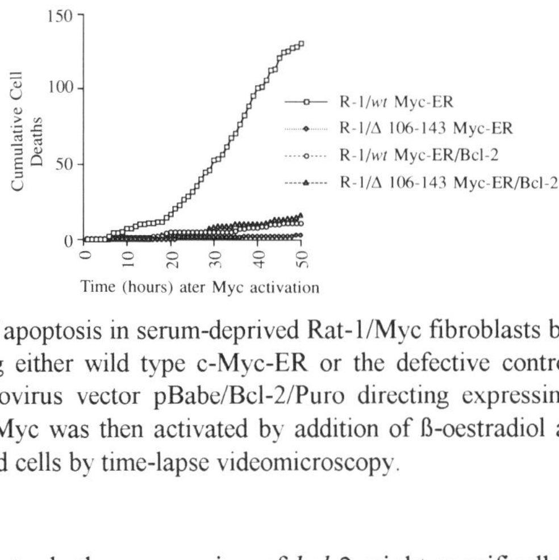

Figure 7. Inhibition of apoptosis in serum-deprived Rat-1/Myc fibroblasts by Bcl-2 expression. Rat-1 fibroblasts expressing either wild type c-Myc-ER or the defective control D106-143 c-Myc-ER were infected with retrovirus vector pBabe/Bcl-2/Puro directing expressing of Bcl-2a and serum deprived for 48 hrs. c-Myc was then activated by addition of ß-oestradiol and apoptosis monitored per 100 randomly picked cells by time-lapse videomicroscopy.

To formally test whether expression of *bcl*-2 might specifically block c-Myc-induced apoptosis in Rat-1 fibroblasts, Rat-1/Myc cells were infected with a retrovirus directing constitutive expression of human Bcl-2a protein[39]. Unlike fibroblasts expressing deregulated c-*myc* alone, fibroblasts expressing both c-Myc and Bcl-2 show markedly decreased incidence of apoptosis when serum deprived (Figure 7)[39,48,49] or treated with cytostatic drugs (Figure 8)[39,50]. However, such cells proliferate normally[39], demonstrating that *bcl*-2 inhibits only the apoptotic function of c-*myc* and not its growth-promoting one. The interaction between c-*myc* and *bcl*-2 is thus a paradigm for a novel form of oncogene co-operation in which one activated oncogene inhibits the lethal effects of another. Consistent with this, *bcl*-2 does not co-operate with c-*myc* in a classical focus-formation assay as does *ras*[39,51]. The oncogenic synergy between c-*myc* and *bcl*-2 therefore represents a novel form of oncogene co-operation distinct from the familiar c-*myc*/*ras* type. Many other oncogenes, including *pim-1, bmi-1, pal-1, and bla-1*, v-*abl* and N-*ras*, synergise with c-*myc* in transgenic mice[52,53,54,55,56] but do not co-operate with c-*myc* in focus forming assays. Potentially any of these proto-oncogenes might also function in an anti-apoptotic fashion analogous to *bcl*-2.

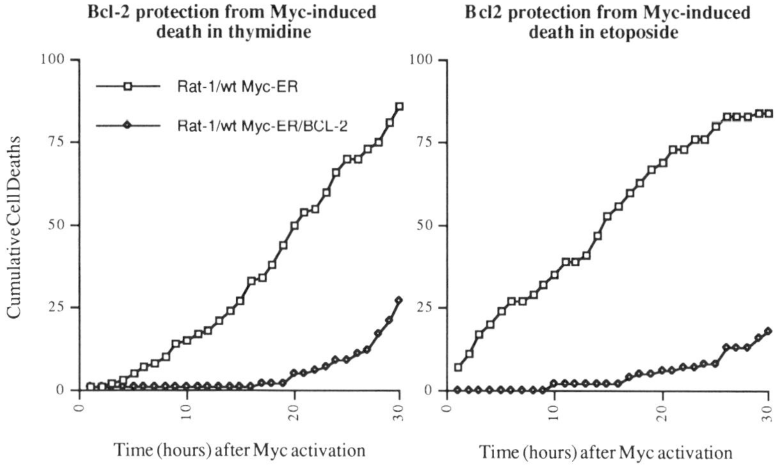

Figure 8. Protection from drug induced apoptosis in Rat-1/Myc cells by Bcl-2 expression. Rat-1/c-Myc-ER fibroblasts infected with either control pBabe/Puro retrovirus vector or pBabe/Bcl-2/Puro directing expressing of Bcl-2a were treated with drugs as indicated for 24 hrs and c-Myc activated by addition of ß-oestradiol. Apoptotic events were recorded per 100 randomly picked cells by time-lapse videomicroscopy.

COUPLING OF GROWTH AND DEATH PATHWAYS

Our data argue that the induction of apoptosis by the proto-oncogene c-*myc* is a normal physiological function of c-*myc* that serves to couple the pathways of cell proliferation with those mediating cell suicide. In this sense, c-*myc* act as both an essential component of cell proliferation and, simultaneously, as a potent suppressor of inappropriate cell proliferation. Thus, c-*myc* acts as both oncogene and tumour suppressor gene. *In vivo*, every cell that engages its proliferative machinery in response to mitogenic stimulation necessarily becomes absolutely dependent upon anti-apoptotic cytokines in its environment to keep it alive. This makes biological sense because each proliferating cell, whilst essential, presents a major neoplastic threat to the structural integrity of any multicellular organism. The coupling of proliferative and apoptotic pathways may thus act a major bulwark limiting the incidence and development of neoplasia. This model is an extension of that propounded recently by Raff[57] in which it is suggested that all cells require social signals for survival.

If cell suicide is the default pathway from which every cell must be actively rescued, mechanisms of evading apoptosis are certain to be important in evolution of neoplasms. A fuller understanding of how cell death pathways are modulated is therefore likely to provide a repertoire of novel molecular targets for the rational treatment of cancer.

REFERENCES

1. C.A. Spencer and M. Groudine, Control of c-myc regulation in normal and neoplastic cells. *Adv Cancer Res.* 56: 1-48 (1991)
2. G. Evan and T. Littlewood, The role of c-*myc* in cell growth. *Curr. Opin. Genet. & Dev.* 3: 44-49 (1993)
3. J.M. Almendral, D. Sommer, H. MacDonald-Bravo, J. Burckhardt, J. Perera and R. Bravo, Complexity of the early genetic response to growth factors in mouse fibroblasts. *Mol. Cell. Biol.* 8: 2140-2148 (1988)

4. K. Kelly, B.H. Cochran, C.D. Stiles and P. Leder, Cell specific regulation of the c-*myc* gene by lymphocyte mitogens and platelet-derived growth factor. *Cell*. 35: 603-610 (1983)
5. H. Persson, L. Hennighausen, R. Taub, W. DeGrado and P. Leder, Antibodies to human c-myc oncogene product: evidence of an evolutionarily conserved protein induced during cell proliferation. *Science*. 225: 687-693 (1984)
6. G. Ramsay, G.I. Evan and J.M. Bishop, The protein encoded by the human proto-oncogene c-myc. *Proc. Natl. Acad. Sci. USA*. 81: 7742-7746 (1984)
7. S.R. Hann and R.N. Eisenman, Proteins encoded by the human c-myc oncogene: differential expression in neoplastic cells. *Mol. Cell. Biol*. 4: 2486-2497 (1984)
8. G.I. Evan and D.C. Hancock, Studies on the interaction of the human c-myc protein with cell nuclei: p62c-myc as a member of a discrete subset of nuclear proteins. *Cell*. 43: 253-261 (1985)
9. E.M. Blackwood and R.N. Eisenman, Max: a helix-loop-helix zipper protein that forms a sequence-specific DNA-binding complex with Myc. *Science*. 251: 1211-7 (1991)
10. T.K. Blackwell, L. Kretzner, E.M. Blackwood, R.N. Eisenman and H. Weintraub, Sequence-specific DNA binding by the c-Myc protein. *Science*. 250: 1149-51 (1990)
11. G.C. Prendergast, D. Lawe and E.B. Ziff, Association of Myn, the murine homolog of max, with c-Myc stimulates methylation-sensitive DNA binding and ras cotransformation. *Cell*. 65: 395-407 (1991)
12. G.C. Prendergast and E.B. Ziff, Methylation-sensitive sequence-specific DNA binding by the c-Myc basic region. *Science*. 251: 186-9 (1991)
13. T. Littlewood, B. Amati, H. Land and G. Evan, Max and c-Myc/Max DNA binding activities in cell extracts. *Oncogene*. 7: 1783-1792 (1992)
14. E.M. Blackwood, B. Luscher and R.N. Eisenman, Myc and Max associate *in vivo*. *Genes Dev*. 6: 71-80 (1992)
15. B. Amati, S. Dalton, M. Brooks, T. Littlewood, G. Evan and H. Land, Transcriptional activation by c-Myc oncoprotein in yeast requires interaction with Max. *Nature*. 359: 423-426 (1992)
16. L. Kretzner, E. Blackwood and R. Eisenman, Myc and Max possess distinct transcriptional activities. *Nature*. 359: 426-429 (1992)
17. B. Amati, M. Brooks, N. Levy, T. Littlewood, G. Evan and H. Land, Oncogenic activity of the c-Myc protein requires dimerisation with Max. *Cell*. 72: 233-245 (1993)
18. G.J. Kato, J. Barrett, G.M. Villa and C.V. Dang, An amino-terminal c-myc domain required for neoplastic transformation activates transcription. *Mol Cell Biol*. 10: 5914-20 (1990)
19. G.J. Kato, W.M. Lee, L.L. Chen and C.V. Dang, Max: functional domains and interaction with c-Myc. *Genes Dev*. 6: 81-92 (1992)
20. J.P. Moore, D.C. Hancock, T.D. Littlewood and G.I. Evan, A sensitive and quantitative enzyme-linked immunosorbence assay for the c-*myc* and N-*myc* oncoproteins. *Oncogene Res*. 2: 65-80 (1987)
21. C.B. Thompson, P.B. Challoner, P.E. Neiman and M. Groudine, Levels of c-*myc* oncogene mRNA are invariant throughout the cell cycle. *Nature*. 314: 363-366 (1985)
22. C. Waters, T. Littlewood, D. Hancock, J. Moore and G. Evan, c-*myc* protein expression in untransformed fibroblasts. *Oncogene*. 6: 101-109 (1991)
23. S.R. Hann, C.B. Thompson and R.N. Eisenman, C-*myc* oncogene protein synthesis is independent of the cell cycle in human and avian cells. *Nature*. 314: 366-369 (1985)
24. M. Dean, R.A. Levine, W. Ran, M.S. Kindy, G.E. Sonenshein and J. Campisi, Regulation of c-*myc* transcription and mRNA abundance by serum growth factors and cell contact. *J. Biol. Chem*. 261: 9161-6 (1986)

25. A.B. Pardee, G1 events and regulation of cell proliferation. *Science*. 246: 603-608 (1989)
26. G. Evan, A. Wyllie, C. Gilbert, T. Littlewood, H. Land, M. Brooks, C. Waters, L. Penn and D. Hancock, Induction of apoptosis in fibroblasts by c-*myc* protein. *Cell*. 63: 119-125 (1992)
27. M. Eilers, D. Picard, K.R. Yamamoto and M.J. Bishop, Chimaeras of Myc oncoprotein and steroid receptors cause hormone-dependent transformation of cells. *Nature*. 340: 66-68 (1989)
28. M. Eilers, S. Schirm and J.M. Bishop, The MYC protein activates transcription of the alpha-prothymosin gene. *EMBO J*. 10: 133-41 (1991)
29. J. Stone, T. de Lange, G. Ramsay, E. Jakobvits, J.M. Bishop, H. Varmus and W. Lee, Definition of regions in human c-*myc* that are involved in transformation and nuclear localization. *Mol. Cell. Biol*. 7: 1697-1709 (1987)
30. B. Amati, T. Littlewood, G. Evan and H. Land, The c-Myc protein induces cell cycle progression and apoptosis through dimerisation with Max. *Submitted*. (1993)
31. V. Quarmby, W. Beekman, E. Wilson and F. French, Androgen regulation of c-myc messenger ribonucleic acid levels in rat ventral prostate. *Mol. Endocrinol*. 1: 865-874 (1987)
32. R. Buttyan, Z. Zakeri, R. Lockshin and D. Wolgemuth, Cascade induction of c-fos, c-myc, and heat shock 70K transcripts during regression of the rat ventral prostate gland. *Mol Endocrinol*. 2: 650-7 (1988)
33. N. Kyprianou, H.F. English, N.E. Davidson and J.T. Isaacs, Programmed cell death during regression of the MCF-7 human breast cancer following estrogen ablation. *Cancer Res*. 51: 162-6 (1991)
34. J.S. Riegel, E.R. Richie and J.P. Allison, Nuclear events after activation of CD4+8+ thymocytes. *J Immunol*. 144: 3611-8 (1990)
35. Y. Shi, J. Glynn, L. Guilbert, T. Cotter, R. Bissonette and D. Green, Role for c-*myc* in activation-induced apoptotic cell death in T cell hybridomas. *Science*. 257: 212-214 (1992)
36. J. Cohen, Overview: Mechanisms of apoptosis. *Immunol. Today*. 14: 126-130 (1993)
37. N. Touchette, Dying cells reveal new role for cancer genes. *J. NIH Research*. 4: 48-52 (1992)
38. R. Baserga, The double life of the IGF-1 receptor. *Receptor*. 2: 261-6 (1992)
39. A. Fanidi, E. Harrington and G. Evan, Cooperative interaction between c-*myc* and *bcl*-2 proto-oncogenes. *Nature*. 359: 554-556 (1992)
40. A. Strasser, A.W. Harris, M.L. Bath and S. Cory, Novel primitive lymphoid tumours induced in transgenic mice by cooperation between *myc* and *bcl*-2. *Nature*. 348: 331-3 (1990)
41. G. Klein, Comparative action of *myc* and *bcl*-2 in B-cell malignancy. *Cancer Cells*. 3: 141-3 (1991)
42. T.J. McDonnell, N. Deane, F.M. Platt, G. Nunez, U. Jaeger, J.P. McKearn and S.J. Korsmeyer, *bcl*-2-immunoglobulin transgenic mice demonstrate extended B cell survival and follicular lymphoproliferation. *Cell*. 57: 79-88 (1989)
43. G. Nunez, L. London, D. Hockenbery, M. Alexander, J.P. McKearn and S.J. Korsmeyer, Deregulated *bcl*-2 gene expression selectively prolongs survival of growth factor-deprived hemopoietic cell lines. *J Immunol*. 144: 3602-10 (1990)
44. D.L. Vaux, S. Cory and J.M. Adams, *bcl*-2 gene promotes haemopoietic cell survival and cooperates with c-myc to immortalize pre-B cells. *Nature*. 335: 440-2 (1988)
45. D. Hockenbery, G. Nunez, C. Milliman, R.D. Schreiber and S.J. Korsmeyer, *Bcl*-2 is an inner mitochondrial membrane protein that blocks programmed cell death. *Nature*. 348: 334-336 (1990)

46. C.L. Sentman, J.R. Shutter, D. Hockenbery, O. Kanagawa and S.J. Korsmeyer, *bcl*-2 inhibits multiple forms of apoptosis but not negative selection in thymocytes. *Cell.* 67: 879-88 (1991)
47. A. Strasser, A.W. Harris and S. Cory, *bcl*-2 transgene inhibits T cell death and perturbs thymic self-censorship. *Cell.* 67: 889-99 (1991)
48. R. Bissonnette, F. Echeverri, A. Mahboubi and D. Green. Apoptotic cell death induced by c-*myc* is inhibited by *bcl*-2. *Nature.* 359: 552-554 (1992)
49. A.J. Wagner, M.B. Small and N. Hay, Myc-mediated apoptosis is blocked by ectopic expression of *bcl*-2. *Mol Cell Biol.* 13: 2432-2440 (1993)
50. T. Miyashita and J.C. Reed, *bcl*-2 gene transfer increases relative resistance of S49.1 and WEHI7.2 lymphoid cells to cell death and DNA fragmentation induced by glucocorticoids and multiple chemotherapeutic drugs. *Cancer Res.* 52: 5407-11 (1992)
51. J.C. Reed, S. Haldar, C.M. Croce and M.P. Cuddy, Complementation by BCL2 and C-HA-RAS oncogenes in malignant transformation of rat embryo fibroblasts. *Mol Cell Biol.* 10: 4370-4 (1990)
52. M. van Lohuizen, S. Verbeek, B. Scheijen, E. Wientjens, H. van der Gulden and A. Berns, Identification of cooperating oncogenes in E mu-*myc* transgenic mice by provirus tagging. *Cell.* 65: 737-52 (1991)
53. M. van Lohuizen, M. Frasch, E. Wientjens and A. Berns, Sequence similarity between the mammalian bmi-1 proto-oncogene and the Drosophila regulatory genes Psc and Su(z)2. *Nature.* 353: 353-5 (1991)
54. Y. Haupt, W.S. Alexander, G. Barri, S.P. Klinken and J.M. Adams, Novel zinc finger gene implicated as *myc* collaborator by retrovirally accelerated lymphomagenesis in E mu-*myc* transgenic mice. *Cell.* 65: 753-63 (1991)
55. S. Verbeek, M. van Lohuizen, M. van der Valk, J. Domen, G. Kraal and A. Berns, Mice bearing the E mu-*myc* and E mu-*pim*-1 transgenes develop pre-B-cell leukemia prenatally. *Mol Cell Biol.* 11: 1176-9 (1991)
56. J.M. Adams and S. Cory, Transgenic models for haemopoietic malignancies. *Biochim Biophys Acta.* 1072: 9-31 (1991)
57. M.C. Raff, Social controls on cell survival and cell death. *Nature.* 356: 397-400 (1992)

Discussion

G. Guidotti

A comment and a question. In our laboratory we have results that fit your model. Using CEM cells, a $CD4^+$ cell line from human T-lymphoblastic leukemia that express c-*myc* continuously, the HIV-infection is followed shortly by the entry of a significant number of cells in the apoptotic pathway, as estimated by DNA fragmentation and confocal microscopy. Thereafter, the viral burden prevails and cell necrosis ensues. And the question is: Do Rat fibroblasts express IGF-1 receptors or, in your model, the protection by IGF-1 occurs by a spillover mechanism through insulin receptors?

G. Evan

All the mesenchyma cells that we use, Rat-1, Swiss 3T3 and rat embryo fibroblasts and primary rat vascular smooth muscle cells (VSMC), express copious amounts of IGF-1 receptor as do many other cell types. We get protection from apoptosis by IGF-1, IGF-II and insulin, presumably all mediated through the IGF-1 receptor. One interesting possibility that this raises is that endocrine levels of both insulin and IGF-I might be providing some form of basal anti-apoptotic support for many tissues. In insulin-dependent diabetics, endocrine levels of both insulin are low, as is that of IGF-I. This is because insulin and IGF-I synthesis are coordinately regulated. Perhaps this lack of basal anti-apoptotic factors has something to do with the degenerative pathologies that develop in long term diabetics? One further point is that IGF-I exerts its anti-apoptotic effect even in the presence of cycloheximide and therefore does not require any new protein synthesis or gene expression to be manifest. In contrast, the anti-apoptotic effects of PDGF appears to be indirect-perhaps by inducing expression of IGF-I in fibroblasts as has been sugested by Dr. Bengt Westermark of Uppsala.

M. Oren

1) Is wild type p53 required for myc-induced apoptosis? 2) What happens if you overexpess Max or Mad?

G. Evan

The second question first. Overexpression of either Max or Mad induces profound growth arrest in fibroblasts, presumably by competing out Myc function. Such cells do not die in low serum. As to the effect of activated *ras* on c-Myc-induced apoptosis in fibroblasts, we find *ras* has no detectable anti-apoptotic activity. Cells co-expressing *c-myc* and *ras* look morphologically transformed, grow rapidly, but die very quickly in low serum. However, out of these rapidly proliferating cultures, mutants seem to appear at low frequency that are blocked in apoptosis. These rapidly outgrow the other cells. As to the role of p53 in Myc-induced apoptosis, we find that co-expression of the adenovirus p55 E1B protein, which binds and inactivates p53, has no suppressive effect on apoptosis, suggesting that p53 is not involved in Myc-induced cell death. However, very recent experiments on fibroblasts obtained from the p53 knockout mice developed by Clarke *et al.* (Clarke, A.R., *et al.* (1993) Thymocyte apoptosis induced by p53-dependent and independent pathways. *Nature,* 362:849-52) show these cells to be resistant to Myc-induced apoptosis. Thus, we are at present confused by these contradictory results.

D. Livingston

Two questions: First, I did miss a point. When one deprives a replicating cell producing reasonable amounts of *myc*, of serum, for about three hours in S, and then re-feed, there is a delay in the next cycle. Where was the delay?

G. Evan

The delay is in G1. We do not know exactly where in G1, all we know is that the delayed G1 resembles a G0-S G1 (i.e., about 12-14 hrs) rather than an M-S G1 (6 hrs). Thus, we suggest that the delay is caused by the cells having to traverse G0-G1 in addition to G1 proper. These are hard experiments to do, and there is no effective biochemical definition of G0 versus G1, so most of this is inference. We have obtained essentially similar results using a range of primary and immortalized mesenchymal cells, including Rat-1, Swiss 3T3 and rat embryo fibroblasts and rat VSMC, all of which, incidently, possess wild-type p53. However, for technical reasons, most of our more conclusive studies have been conducted on early passage Swiss 3T3 cells.

D. Livingston

Was this effect specific to this particular line?

G. Evan

We have tried it in primary rat embryo fibroblasts, in Rat-1 cells, and in Swiss 3T3's, all of which, as I understand, have normal p53. So we have tried it in those cell lines and we have also done it in Rat vascular smooth muscle cells as well. That is all I know. The best experiments, the tightest ones where it has been done most carefully with the most time points have been with Swiss 3T3's.

D. Livingston

MC29 will kill a chicken if given a suitable dose in about three weeks, with massive leukemia. Is there a block to apoptosis in the virus-infected cells of these animals?

G. Evan

My understanding is that infection with v-*myc*-containing retroviruses is associated with massive apoptosis in the affected organs. Our experiments show that v-*myc* is a much more potent inducer of apoptosis than c-*myc*, as it is as an oncogene. I would suggest that this potent inductin of apoptosis may rapidly select for tumor cells with secondary lesions promoting survival, or that resultant tumors arise in protected micro-environments that are rich in anti-apoptotic cytokines. The rapid incidence of tumors induced by single-hit viral oncogenes is something of a mystery, however, because the same genes introduced into mammalian fibroblasts almost never induce complete transformation.

D. Livingston

Is this the case for MC-29?

G. Evan

It is for MC-29. But certainly if you do the same experiment, as far as I know, in the feline system, you do not get this massive tumor take within a few weeks.

D. Hockenbery

Is *myc*-induced apoptosis always associated with cell cycle progression?

G. Evan

When our fibroblasts are blocked with thymidine and c-Myc is then activated, they exhibit no cell cycle progression as judged by flow cytometry of time-lapse video microscopy.

D. Hockenbery

At what level does *myc* function, then, to enhance apoptosis if its cell cycle effects can be dissociated?

G. Evan

Let me describe what I think is going on. I believe that c-Myc regulates the transcription of a cadre of genes that regulate apoptosis. Either c-Myc activates part of the apoptotic machinery and/or it suppresses some anti-apoptotic function. Whatever, cells expressing c-Myc are now primed to undergo apoptosis and will do so with alacrity if deprived of factors or treated with genetoxic agents. Thus, an obligate component of cell proliferative machine is also an obligate activator of an apoptotic program. Every cell that divides is primed to kill itself. The biological rationale here is that it is better to have a dead cell than a potential tumor cell.

E. Mihich

In the shift of *myc* function on proliferation versus death, the agents that you quoted as being able to implement this shift, the interferons, etc., under particular conditions do not cause death or certainly not acute death, but rather differentiation. This differentiation may or may not develop into terminal death. How do you visualize, in your scheme, this pathway? Is it the same pathway which has different kinetics or is it a different pathway?

G. Evan

All our studies have been conducted in fibroblasts for one important reason--namely, that when deprived of growth factors, fibroblasts arrest without differentiating and, thus, turning into another type of cell. Subsequent re-addition of growth factors once again triggers their entry into cycle. This is not the case with most other cell types, such as haematopoietic cells, which tend to differentiate upon growth arrest. We therefore feel that fibroblasts give us a cleaner picture of the link between growth control and apoptosis, without the complexity of differentiation. In terminally differentiated cells that cannot proliferate, I imagine the situation is even more confused.

E. Mihich

Is it known whether *max* always inactivates *myc* or whether *myc* may inactivate *max*. What is known about the function of Max?

G. Evan

They appear to simply titrate each other, so it depends on how much of each a cell is expressing. At present, there is no evidence to suggest that Max has any positive function of its own: It may just act as a neutral vehicle allowing c-Myc to bind DNA. John Cleveland may have more to say on this. If overexpressed, Max competes with c-Myc-Max dimers for DNA binding sites. However, this still does not indicate whether the action of Max is negative (i.e., a transcriptional suppressor) or null (i.e., just acting to compete out c-Myc).

W. Giaretti

The model that you cited before, that is the human colorectal cancer progression characterized by *ras* mutations in early stages, and p53 in late stages, I think is quite potentially informative for all the hypothesis that have been discussed during your interesting presentation. If I recall correctly, you or a related group had performed Myc protein measurment in colorectal adenomas by flow cytometry. How do these data fit into your present model?

G. Evan

I am certainly not suggesting that the progressive or sequential models of carcinogenesis are wrong, merely incomplete. The Fearon/Vogelstein model (Fearon, E.R., Hamilton,S.R. and Vogelstein, B. (1987) Clonal analysis of human colorectal tumors. *Science* 238:193-197) makes little mention of the role of c-*myc* in colon carcinogenesis, perhaps for good reason. c-*myc* is overexpressed or deregulated in colonic tumors, but it is not clear at what stage the c-*myc* lesion occurs. My argument is simply that there are intermediate carcinogenic states that are intrinsically unstable or self-limiting. For example, deregulation of c-*myc* gives a proto-tumor cell a notional growth advantage because it drives cell cycle progression. However, it is also lethal because it drives apoptosis. The affected cell therefore has no time to progress and acquire new mutations that might rescue it from c-Myc-induced death. The only way a cell can pass through this log-jam is by simultaneous activation of c-*myc* and a death-suppressing mutation, i.e., a double mutation. In colon, it is not clear what degree of tissue regulation is mediated by apoptosis versus shedding of cells into the gut lumen. I suspect that the extent of apoptosis tends to be underestimated because each apoptotic event and its phagocytic clearance is so rapid. If apoptosis is already significant, it might be difficult to see any added effect of expressing c-Myc in such cells.

W. Giaretti

I would like to further comment on that. Aneuploidization takes place at a very high percentage already in the preneoplastic lesions, especially in adenoma regions with high dysplasia or early cancer; you may go probably up to 80% aneuploidy incidence. Probably, if apoptosis loses, aneuploidy wins. Two alternative routes appear to exist, at least at first approximation, and everything appears to derive from asymmetry in cell division. If apoptosis fails to stop an abnormal near-diploid G1 postmitotic cell, then aneuploidy may have its chance (*Int. J. Cancer* 45:597-603, 1990; *Int. J. Oncology* 2:165-171, 1993).

G. Evan

I think the gut is an excellent system for studying the physiological relevance of cell death and the consequences of its altered regulation. One obvious experiment to do is to transgenically introduce *bcl*-2 into gut, using a gut-specific promoter/enhancer, and ask what effects suppressing cell death in the gut may have.

D. Green

You have presented some somewhat paradoxical information about the effect of *myc* plus cyclohexamide. Just to add to that, we are looking at a system which is slightly different, where *myc* is inducible in CHO cells and expression causes apoptosis. We find that we get a high level of *myc* expression, which takes about one to two hours in our cells. If we block any subsequent transcription, we can get a pretty substantial inhibition of apoptosis, which is, of course, very consistent with what you are finding for the requirement for myc and max. I guess we all really believe that the real function of myc and max in this system is to drive expression of other genes that might be directly involved in these processes.

G. Evan

Yes, I think a point worth mentioning in this context is that searching for genes that are regulated as a cell undergoes apoptosis may be misplaced. If the trigger for apoptosis is epigenetic, as seems quite likely, then many, perhaps all, of the genetic mechanisms regulating apoptosis would not operate at the moment of suicide, but would be involved in setting up or suppressing the *potential* for apoptosis. Thus, if c-Myc is involved in establishing the potential for apoptosis, one should search for apoptosis-specific genes during c-Myc regulation, not at the time cells die.

J. Cleveland

A comment about Dr. Livingston's MC-29 question. MC-29 takes quite a bit longer to induce tumors than a dual oncogene retroviruses in the avian system such as MH2 which has both *myc* and *raf*. And if you take *myc*, put them into murine retroviruses and inject them into newborn mice, the tumors that arise are all late and they are all clonal.

D. Livingston

Yes, but if you used enough MC-29, you can kill an animal in three weeks with huge tumors.

J. Cleveland

These were with high titered murine viruses. In the mice system, if you use equal titers of v-myc-only retroviruses versus c-myc plus v-Raf retrovirus, there is quite a difference in how well they work.

G. Evan

In the transgenic experiments it is clear that rates of lymphomagenesis in lymphocytes continuously expressing c-*myc* are relatively low. Indeed, this has been used as a strategy to identify oncogenes that cooperate with c-*myc* because they enhance the rates of tumorigenesis (Haupt, Y., *et al.* (1991) Novel zinc finger gene implicated as myc collaborator by retrovirally accelerated lymphomagenesis in E mu-*myc* transgenic mice. *Cell* 65:753-63; van Lohuizen, M., *et al.* (1991). Identification of cooperating oncogenes in E mu-*myc* transgenic mice by provirus tagging. *Cell* 65:737-52). Some of these cooperating oncogenes act, I suggest, by suppressing the apoptotic effects of c-*myc*.

D. Livingston

Is it a prediction of your model that Burkitt's lymphoma should be resistant to certain chemotherapeutic agents?

G. Evan

The only thought I have had on Burkitt's lymphoma is a possible explanation for the obligate association between infection of B cells with Epstein Barr virus and subsequent activation of c-*myc* through reciprocal chromosomal translocation. It may be that prior infection by EBV introduces into B cells anti-apoptotic viral genes. For example, the *Bcl*-2 homologue BHRF1 or the EBV membrane oncogene LMP1 that induces expression of *bcl*-2 and the TNF-resistance gene A20 (Henderson, S., *et al.* (1993) Epstein-barr virus-coded BHRF1 protein, a viral homolog of Bcl-2, protects human B-cells from programmed cell-death. *Proc. Natl. Acad. Sci. USA* 90:8479-8483; Henderson, S., *et al.* (1991) Induction of *bcl*-2 expression by Epstein-Barr virus latent membrane protein 1 protects infected B cells from programmed cell death. *Cell* 65:1107-15; Laherty, C.D., *et al.* (1992) The Epstein-Barr virus LMP1 gene product induces A20 zinc finger protein expression by activating nuclear factor kappa B. *J. Biol. Chem.* 267:24157-60). This may protect the affected cell from subsequent induction of apoptosis when c-*myc* is activated by translocation. I know little about drug-resistance in Burkitt's lymphoma. One important point, however, is that it appears that tumor cells never lose the ability to undergo apoptosis, no matter how resistnat to it they become, say through acquisition of *bcl*-2 or loss of p53. Resistant cells merely require a higher level of damage to initiate the program. So the apoptotic mechanism is always present. This may indicate that it cannot be lost because it is comprised of components of other essential cellular functions.

D. Livingston

I was thinking mostly of American Burkitt's, which is almost exclusively EBV negative. It is a disease which is often quite sensitive to certain chemotherapeutic agents for a while.

G. Evan

I am afraid I do not know enough about American non-EBV-associated Burkitt's lymphoma.

M. D'Incalci

The data on the protection of drug-induced apoptosis were very impressive. Do you know if in the same conditions you really protect from cytotoxicity, or do you have other mechanisms of cytotoxicity?

G. Evan

We assay apoptosis by time-lapse video microscopy, which allows inspection of each cell death and verification that it is apoptotic rather than necrotic. We also have nucleosome ladders in our apoptotic cells. We feel that the best definition of apoptosis is the morphology of how they die--blebbing, shrinkage, fragmentation, and all of it very rapid. However, by time-lapse video microscopy it is difficult to examine large numbers of cells. One hundred is a typical number that might be followed in one of our experiments.

D. Papermaster

I am trying to understand the last point you have just made, namely, that the sister cells undergo apoptosis with a greater probability than other cells. I do not think I understand how that is consistent with the stochastic model you were describing earlier. What you are now describing would suggest a non stochastic process that is dependent on interaction of the cells near each other or the release of substances by the apoptotic cells that are triggering the death of the adjacent cells and not a stochastic model.

G. Evan

We find that some 60% of sister cells die within a few minutes of each other. That means that a lot (~40%) do not. Thus, the death clock can be re-set by external factors, as expected from the fact that cytokines like IGF-I and PDGF suppress apoptosis. Possible factors that might affect apoptosis of individual cells are cell-cell contact, micro-heterogeneity in the tissue culture plastic and localized depletion of cytokines. Nonetheless, the significant degree of simultaneous deaths in sister cells does suggest that, all things being equal, sister cells share some memory function derived from their parent that governs when they will die. That notwithstanding, apoptotic deaths in the culture as a whole appear random or stochastic.

R. Schimke

You have talked about a program which may be free running, a program that leads to proliferation but also to cell death. At the same time, you speak of damage, which suggests that a cell can sense the damage. A major question then is, are the genes that you are dealing with or those that others have described here, involved in the sensing, or are they involved in the transduction of that sensing information into a death program? And, if it is damage, what sort of damage? You mentioned DNA damage, and if so, do you have evidence for that?

G. Evan

I think there is no evidence to suggest that growth factor deprivation induces DNA damage *per se*. Factor-deprived fibroblasts expressing deregulated c-*myc* show no obvious signs of damage, it is just that a proportion of them die. Those that have not yet died look healthy and happy and divide normally. I therefore adhere to the model proposed by Raff (Raff, M.C. (1992) Social controls on cell survival and cell death. *Nature* 356:297-400) that argues that cell death is a default program, not one that needs to be initiated by damage.

P. Nicotera

My question is a bit more general. You have all these beautiful ways to induce apoptosis or to increase the susceptibility of cells to apoptosis by expressing different genes. In your particular case you have a number of different conditions such as for example the withdrawal of growth factor in the presence of c-myc expression and then etoposides which induce cell death. What do you think is the pathway that C-myc makes available for these cells to die in such diverse conditions. Etoposide is a DNA damaging agent for example. Growth factor withdrawal will probably work in a different way. Do you have any idea what could be the mechanism for this?

G. Evan

I believe that c-Myc regulates a transcriptional program that implements the potential to undergo apoptosis and so makes cells more sensitive to triggering its own demise. However, I have no idea what is involved in actually triggering each death event. In principle, c-Myc could act by inducing cell death components and/or by suppressing anti-apoptotic mechanisms. Both are logically equivalent. Nonetheless, there are clearly c-Myc independent apoptotic pathways--for example, the p53 system of Moshe Oren, the E1A system of Eileen White, and the programmed death of neurons (which express no detectable *myc* genes). The role of c-*myc*, as I see it, is to couple proliferation with the potential for apoptosis. Thus every cell that expresses c-*myc* (and hence is in cycle) has a built-in "abort pathway". Close examination of c-Myc induced apoptosis by video microscopy shows that apoptosis can never be completely suppressed by any amount of anti-apoptotic cytokines. Control of apoptosis seems never to be cut and dried. Under optimal conditions, some cells always die and even under severe pro-apoptotic conditions, some cells survive for long periods. The only absolute is the survival or death of an individual cell--when it is dead, it is dead. So what you have is a progressive population effects that has a catastrophic outcome at the level of each single cell.

C. Riccardi

I have a short question. You have shown protection with cytokines in your experiments. Have you ever seen induction of apoptosis using your cytokines?

G. Evan

No, we have never observed induction of apoptosis by any of the cytokines we use, except for a slight promotion of c-Myc-induced apoptosis by EGF. That may be fibroblast-specific because for other lineages (e.g., kidney) EGF appears to be a survival factor.

G. Guidotti

This comment is to remind us of some conditions in which to block the apoptotic pathway is insufficient to cause cell transformation. In transgenic mice for *bcl*-2 (placed under a strong enhancer) B-lymphocytes survival is enhanced and a lympho-accumulation occurs within one year. Lympho-accumulation of $CD4^-/CD8^-$ cells through inhibition of apoptosis also occurs in mice *lpr*, lpr^{CG} and *GLD* in which the receptor FAS/Apo1 is deleted (*lpr* mice), mutated (lpr^{CG} mice) of lacking a functional ligand (*GLD* mice). In all these mice, lympho-accumulation is not followed or accompanied by malignancies. Therefore, something more than just to stop apoptosis is required for cell transformation.

REGULATION OF APOPTOSIS IN INTERLEUKIN-3-DEPENDENT MYELOID PROGENITOR CELLS

John L. Cleveland[1], David S. Askew[1*], Concha Bello-Fernandez[1] and Graham Packham[1]

[1] Department of Biochemistry
St. Jude Children's Research Hospital
332 N. Lauderdale
Memphis, TN 38105

[*] Current address: Department of Pathology and Laboratory Medicine
University of Cincinnati College of Medicine
231 Bethesda Ave.
Cincinnati, OH 45267-0529

INTRODUCTION

Apoptosis plays an important role in the regulation of hematopoiesis by deleting growth factor-dependent progenitors following physiologic reductions in hemopoietins.[1,2] To examine signalling pathways which may influence apoptosis in hematopoietic progenitors we have utilized the Interleukin-3 (IL-3) dependent 32Dcl3 murine myeloid cell line. 32D cells are absolutely dependent upon IL-3 for proliferation and survival,[3,4] and have no propensity for developing factor-independent growth. In the absence of IL-3, 32D cells rapidly arrest in the G1 phase of the cell cycle and eventually undergo apoptosis.[5]

It is widely accepted that a key regulator of cell cycle progression is the c-*myc* proto-oncogene. MYC is a nuclear phosphoprotein having several domains common amongst transcription factors and, in conjunction with its heterodimeric partner *Max*, has been shown to possess sequence specific DNA binding activity (to an E-box sequence, CACGTG,[6]) and to activate transcription from heterologous promoters.[7,8] C-*myc* expression is universally induced by mitogens and rapidly down-regulated following growth factor deprivation or treatment with growth inhibitory agents.[9] By contrast, the identified c-*myc* partner, max, appears to be constitutively expressed.

We have previously shown that c-*myc* is an immediate early response gene following interleukin-3 (IL-3) stimulation of myeloid progenitor cell lines[5,10] and that transcription of c-*myc* is rapidly down-regulated following withdrawal of IL-3. This precedes, by several hours, the accumulation of cells in early G1. Down-regulation of

Apoptosis, Edited by E. Mihich and R.T. Schimke
Plenum Press, New York, 1994

MYC levels appears required for these (and other) cells to withdraw from the cell cycle, since enforced c-*myc* expression drives cell cycle progression into S phase in the absence of IL-3.[5]

The most remarkable consequence of this imbalance in mitogenesis is the rapid induction of apoptosis. Similar observations in fibroblasts and T cells[11,12] suggest that *myc* signalling pathways are involved in the induction of apoptosis in several cell types. However, since the programmed cell death of normal IL-3-deprived cells occurs in the absence of endogenous c-*myc* expression, this suggests the presence of multiple, yet distinct, signalling pathways which culminate in activation of apoptosis.

We have focussed our recent efforts on three issues. First, we have addressed whether the imbalanced mitogenesis provoked by *myc* expression is dominant to the suppression of programmed cell death by hemopoietins. Second, we have attempted to define the the mechanism by which *myc* expression derails cell cycle control and induces apoptosis. Since MYC has been shown to function as a transcription factor, our focus has been to identify genes which are transcriptionally regulated by MYC and address whether they may also influence the induction of apoptosis. Finally, since enforced c-*myc* expression promotes constitutive cell cycle progression, yet fails to substitute for the survival function provided by IL-3, we have addressed whether other oncogenes (other than *bcl-2*) which complement *myc* in transformation assays, such as activated *raf*, can also suppress apoptosis.

RESULTS

Enforced MYC Expression Promotes Inappropriate Cell Cycle Progression

In the absence of IL-3, 32D cells rapidly down-regulate *myc* expression, subsequently accumulate in G1 and eventually undergo apoptosis. By contrast, when *myc* is constitutively expressed, cells continue to display entry into S phase in the absence of ligand and have accelerated death.[5] Since growth control by hemopoietins involves signals which ensure cell survival, it is likely that IL-3-deprived cells fail to express genes required to promote viability. We therefore examined whether enforced *myc* expression also predisposes 32D cells to premature death under growth limiting conditions in the presence of excess specific hemopoietins which promote cell survival. Two conditions were evaluated. First, 32D cells, as is common for most hematopoietic cell cultures, undergo a density-dependent arrest in G1 when the cultures approach densities of 2-3 million cells/ml. Second, when deprived of IL-3 yet supplemented with erythropoietin (Epo), 32D cells display prolonged survival and arrest in G1 (Askew et al., submitted). Under both culture conditions, at times when cells noticeably accumulate in G1, c-*myc* expression is down-regulated (data not shown). By contrast, enforced *myc* expression continues to promote progression into S phase under both culture conditions (Askew et al., submitted). Thus, constitutive c-*myc* expression overrides normal cell cycle regulatory controls and promotes inappropriate cycling of these progenitor cells.

Activation of Apoptosis By Enforced MYC Expression Overrides The Suppression of Programmed Cell Death by IL-3 and Erythropoietin

In the absence of IL-3, inappropriate cell cycling of c-*myc* clones is associated with the rapid induction of apoptosis. This is evident by analyzing the rapid degradation of genomic DNA into oligonucleosomal fragments,[5] which first becomes detectable within two hours of ligand deprivation. Furthermore, characteristic changes in cellular morphology (cytoplasmic membrane blebbing, chromatin condensation and dissolution of the nucleus) are also rapid events following IL-3 deprivation (Figure 1). Interestingly, one feature which is consistently observed is that apoptosis is accompanied by the appearance of large vacuoles in the cytosol. These structures are also evident in ligand deprived 32D cells prior to other changes in cellular morphology or the degradation of genomic DNA; therefore, we believe that this is an early event which may indicate the initiation of the cell death program.

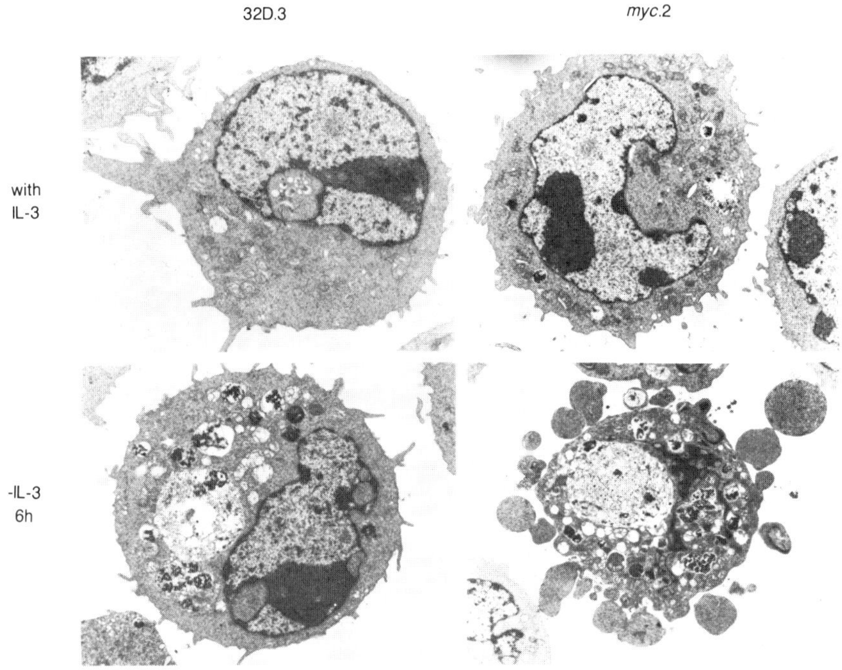

Figure 1. Electron photomicrographs of morphological changes in 32D cells and a cell which constitutively expresses c-*myc* (*myc*.2) following IL-3 withdrawal (for 6 hours). Magnification of cells was at 14,000X.

To address whether inappropriate cycling of *myc* clones was deleterious to cell survival at high cell densities, 32D cells and *myc* clones were initiated at equivalent densities and analyzed for their growth properties, viability and integrity of their genomic DNA with time in culture in the presence of excess IL-3. In contrast to parental cells, all c-*myc* clones lost viability as the cultures approached densities of between 1 and 2 $x10^6$ cells/ml. Viability showed a rapid decline between 48 and 96 hours and progressed rapidly to the death of the cells, despite the fact that the cultures were supplemented with an excess of IL-3. This was in marked contrast to control cells which maintained their viability for extended culture periods at densities of up to 3×10^6 cells/ml (Askew et al., submitted). The loss of viability in c-*myc* clones occurred over several days, consistent with the concept that the induction of programmed cell death is a stochastic process. To determine if enforced c-*myc* expression was associated with activation of apoptosis during density-dependent G1 arrest in IL-3-supplemented cultures, the integrity of genomic DNA in cells harvested on each day was analyzed by agarose gel electrophoresis. As shown in Figure 2, DNA degradation was undetected in control cells throughout the experiment. However, both *myc* clones revealed the classic oligonucleosomal genomic DNA ladder when cells reached densities of 1 X 10^6 cells/ml. Despite this fact, the few c-*myc* cells which remained viable at this time continued to show evidence of entry into S phase, yet eventually died.

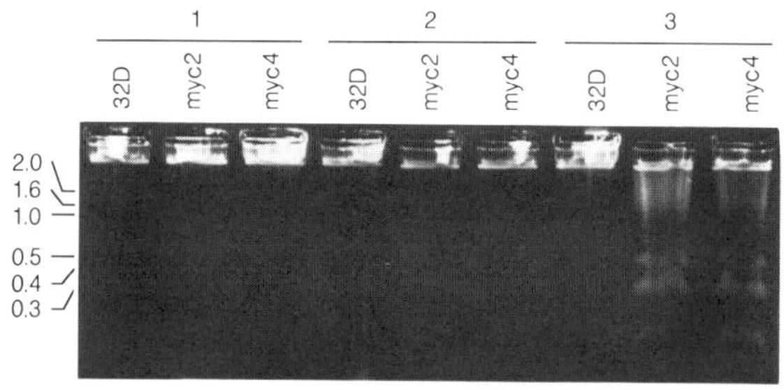

Figure 2. Electrophoretic Analysis of DNA degradation (apoptosis) with increasing cell density. The indicated cells were seeded at equivalent densities in medium supplemented with 100 U/ml IL-3 and genomic DNA analyzed by agarose gel electrophoresis.

Similar results were observed when cultures of control cells were compared to *myc* clones deprived of IL-3 yet supplemented with Epo (Askew et al., submitted). Therefore, down-regulation of c-*myc* expression appears to be required for cells to maintain viability in G1, even in the presence of factors such as IL-3 and Epo which normally suppress programmed cell death in myeloid progenitors.

The fact that enforced *myc* expression is associated with the induction of apoptosis in cells in IL-3, suggested that this program may be constitutively activated in cells which overexpress *myc* yet is only manifested when cells receive signals to G1 arrest. To address this issue we pretreated control cells and *myc* clones with cycloheximide and then deprived cells of IL-3 for 6 hours. If the apoptosis program was already initiated in these cells, a prediction would be that cells treated with the protein synthesis inhibitor would remain susceptible to programmed cell death. Analysis of genomic DNA in cultures treated under these conditions indicated that *myc* clones were susceptible to apoptosis even following pre-treatment with cycloheximide (Askew and Cleveland, unpublished results), suggesting that the endogenous suicide program is already primed in cells which have augmented MYC levels.

Ornithine Decarboxylase Is A Transcriptional Target of c-myc Required For Cell Cycle Progression

The precise mechanism by which c-*myc* promotes inappropriate cell cycle progression and apoptosis is unknown, yet we speculate that this may be due to the transcriptional activation of a set of genes which are themselves necessary for cell cycle progression. This hypothesis is consistent with the fact that MYC functions as a transactivator of reporter genes having a repeated array of synthetic CACGTG MYC binding sites.[8] In screening for alterations in IL-3 regulated gene expression in *myc* clones, we observed that the expression of one gene necessary for G1 progression,

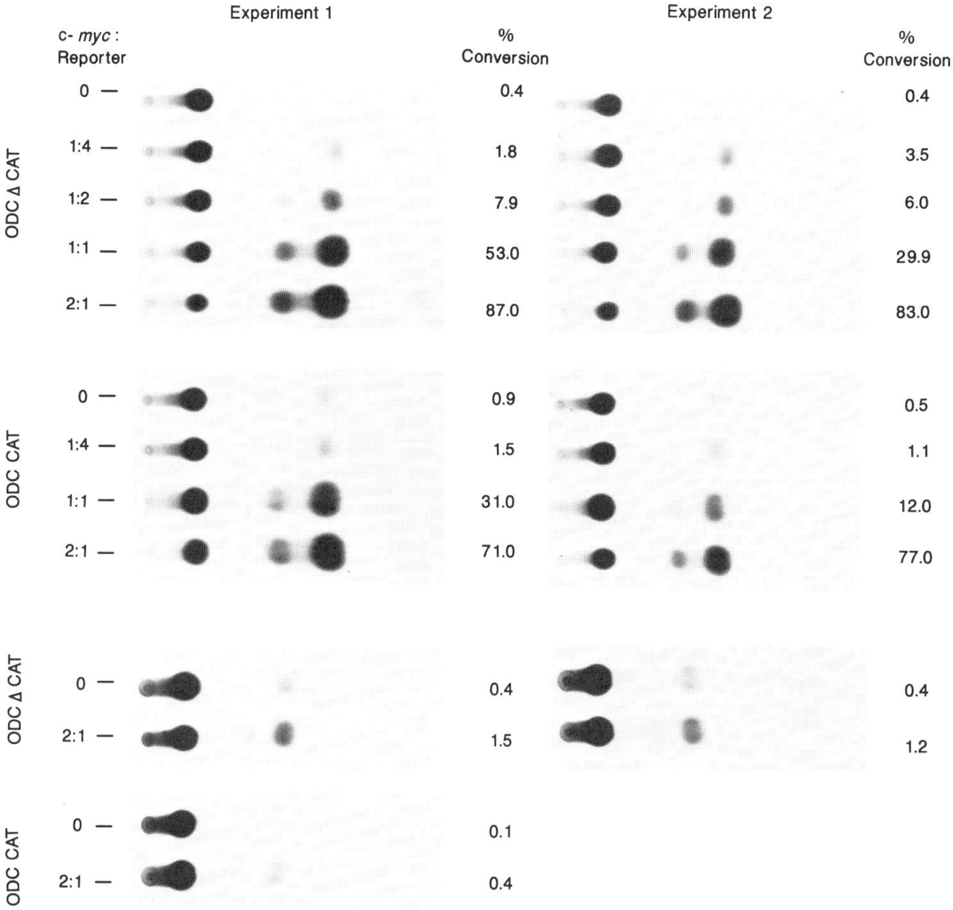

Figure 3. C-*myc* is a potent activator of ODC promoter/gene fusion constructs which requires the MYC binding (MB) sites in ODC Intron 1. (A). Representative CAT assays using standardized and balanced extracts from NIH 3T3 cells transiently transfected with two ODC promoter indicated reporter gene fusion constructs ODC CAT and ODC CAT and a c-*myc* expression plasmid. (B). Representative CAT assays in 32D myeloid cells. The quantitation shows the mean % conversion +/- deviation. ODCΔCAT contains 264 bp 5' of the ODC RNA cap site, whereas ODC CAT contain 2.5 kb of upstream sequence.

ornithine decarboxylase (ODC), was rendered independent of IL-3 in two different myeloid progenitor cells.[5,13] ODC is the rate-limiting enzyme of polyamine biosynthesis and specific inhibitors of ODC enzyme activity (such as DFMO), or polyamine depletion, arrests cells in G1.[14]

ODC expression is regulated as an early gene product in these cells (and in other cell types) and follows c-*myc* induction by IL-3 in a temporal fashion. Importantly, a conserved repeat of the MYC binding sequence, CACGTG, is present in ODC intron 1. In transient co-transfection experiments we have demonstrated that c-*myc* is a potent transactivator of ODC transcription in fibroblasts and myeloid cells (Figure 3) which requires these MYC binding sites.[7] Transactivation of the ODC promoter required functional MYC since point mutations which ablate MYC DNA binding abolish transactivation of this promoter.[7] Furthermore, when cloned upstream of the CAP site of a heterologous promoter, these MYC elements retain *myc*-responsiveness, suggesting that MYC activates ODC expression at the level of transcription.

ODC is an appealing transcriptional target for c-*myc* since the activity of this enzyme is required for G1 progression,[14] its expression is required for transformation[15] and, similar to *myc*, its overexpression augments the transformation of rat fibroblast cells in concert with activated *ras* oncogenes.[16] Our results indicate that ODC is a transcriptional target of c-*myc* and suggest that the ability of c-*myc* to promote cell cycle progression may be, in part, due to transcriptional activation of the ODC gene. However, the key issue which remains unresolved is whether this alteration in ODC expression is of consequence to MYC-mediated apoptosis. To address this question, we have generated clones of 32D cells which constitutively express the murine ODC gene. Preliminary data suggest that constitutive expression of ODC also renders 32D cells susceptible to premature apoptosis, although the kinetics of death are not as rapid as those seen in *myc* clones (Packham and Cleveland, unpublished results).

Activated RAF Synergizes With MYC To Promote Autonomous Growth And Suppresses Apoptosis

RAF is a cytosolic serine/threonine kinase whose activity is induced following ligand activation of several growth factor receptors, including those for IL-3 and Epo. Activated versions of *raf* have been shown to induce the promoters of the immediate early genes including c-*fos* and *egr*, yet notably fail to transactivate the c-*myc* promoter. It has been known for some time that both c-*myc* and v-*myc* synergize with *raf* in the transformation of several cell types including progenitors derived from primary bone marrow, fetal liver or B cell cultures. Moreover, transformation of fibroblast cells by *ras*, and their growth in serum and TPA, is *raf* dependent.[17] Therefore, in a naive sense, synergy could be considered as the result of complementary *raf*-signalling and *myc*-effector activities.

To test the effects of activated *raf* on apoptosis in 32D cells, cells were infected with a retrovirus which constitutively produces an activated form of the RAF kinase, v-*raf*. These clones display several interesting phenotypes versus *myc* clones (Cleveland et al., submitted). First, v-*raf*, but not c-*raf*, clones, have augmented growth in IL-3 with a shortened G1 phase and were able to proliferate in very low concentrations of IL-3. Second, in the absence of IL-3, v-*raf* clones rapidly accumulated in a G0/G1 state. However, the most striking effect was that all v-*raf* clones had a prolonged lifespan in the absence of IL-3 and this was associated with a delay in the onset of apoptosis (Figure 4).

In this regard, activated *raf* functions similar to *bcl-2*, which also prolongs myeloid cell survival, delays apoptosis and is also characterized by cells arresting in a G0/G1 state.[18,19] The specific mechanism by which activated RAF (or *bcl-2*) inhibits

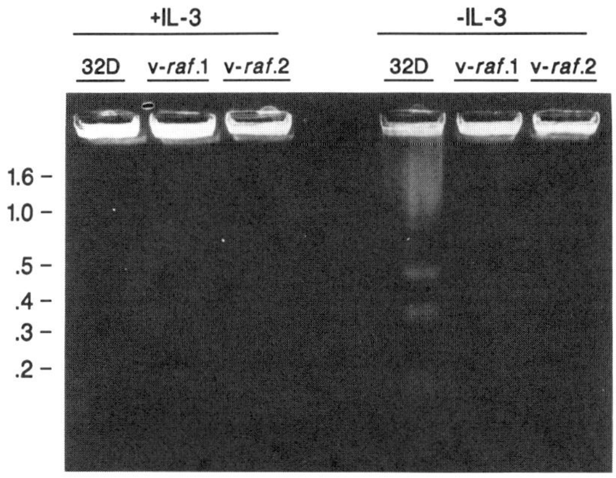

Figure 4. v-*raf* suppresses apoptosis in the absence of IL-3. Control 32D cells and two v-*raf* clones were washed of IL-3 and seeded at equivalent densities in media supplemented with (left panel) or lacking IL-3 (right panel). After 20 hours genomic DNA was prepared and analyzed.

the activation of apoptosis is however not clear. Suppression of apoptosis by activated RAF is not due to the secretion of factors which promote cell survival nor is it associated with the activation of known *raf* gene targets, as defined by co-transfection assays, such as *fos* or *egr*. Finally, when introduced in combination with v-*myc*, v-*raf* promotes IL-3-independent growth of 32D cells and renders them tumorigenic (Cleveland et al., submitted). Therefore, combined introduction of activated *raf* and constitutive *myc* expression appears to provide a balanced mitogenic signal which facilitates autonomous growth.

CONCLUSIONS

Current evidence indicates that hemopoietic growth factors suppress apoptosis in growth factor-dependent cells, and that this viability-promoting activity is distinct from effects on cell cycle progression.[1,2] Our results support this concept since enforced MYC expression clearly replaces, at least transiently, the requirement for hemopoietins for cell cycle progression, yet fails to provide survival functions necessary for suppression of apoptosis. By contrast, activated *raf*, which in concert with constitutive *myc* expression appears to provide a balanced mitogenic signal, does not alter cell cycle regulation in the absence of ligand yet suppresses apoptosis.

Our results, and those of others[11,12] clearly indicate that inappropriate *myc* expression induces apoptosis under conditions in which the cells receive signals which depress their proliferative activity. Such a system therefore provides an effective means to delete mutant cells which fail to respond to growth arrest signals. Notably hemopoietins, at least in our cell system, fail to suppress programmed cell death in cells which overexpress *myc* and these cells appear primed to die when they receive an antagonistic growth inhibitory signal. A key issue which remains unresolved is the mechanism by which *myc* executes the suicide program. Our bias is that *myc* promotes cell cycle progression and apoptosis by inducing the transcription of a common set of genes which are themselves ligand regulated and necessary for cell cycle progression.

To date we have identified one potential *myc* transcriptional target as ODC, yet at present have no direct evidence that this enzyme, or that its polyamine products, are integral components of MYC-induced apoptosis pathways, although both are clearly required for cell cycle progression.

Since c-*myc* activation is so widespread in neoplastic diseases, the apparent paradoxical effects of *myc* on promoting cell cycle progression and compromising cell survival suggests that cells harboring deregulated c-*myc* genes must express additional genes which can suppress apoptosis. In support of this concept, *bcl*-2 expression has been shown to suppress apoptosis that is associated with inappropriate *myc* expression in both fibroblasts and T-cells[20,21] and here we have shown that activated *raf* has similar effects. A logical prediction would therefore be that other oncogenes which also cooperate with *myc* in transformation assays, such as *ras*, would similarly suppress programmed cell death.[22] Finally, our results suggest the hypothesis that *in vivo*, where hemopoietins such as IL-3 are generally limiting, that the transformation of primary myeloid cells to an overt factor-independent leukemia may be an ordered process. This would first involve the activation of a gene such as *raf* or *bcl-2*, which supplies cell survival functions under limiting growth conditions, and would be followed by de-regulation of *myc* expression. Only under these conditions would constitutive c-*myc* expression be predicted to be advantageous and provide a growth advantage to transformed clones over their normal counterparts.

ACKNOWLEDGEMENTS

We are grateful to Joyce Galle and Kathy Fry for secretarial assistance. This work was supported by National Institute of Diabetes and Digestive and Kidney Diseases grant DK44158 (J.L.C.), the National Cancer Institute Center Support (CORE) grant PO CA21765 and by the American Lebanese Syrian Associated Charities (ALSAC).

REFERENCES

1. Koury, M.J., Programmed cell death (Apoptosis) in hematopoiesis, *Exp. Hematol.*, 20:391 (1992).
2. Williams, G.T., C.A. Smith, E. Spooncer, T.M. Dexter, and D.R. Taylor, Haemopoietic colony stimulating factors promote cell survival by suppressing apoptosis, *Nature (London)*, 343:76 (1990).
3. Greenberger, J.S., M.A. Sakakeeney, R.K. Humphries, C.J. Eaves, and R.J. Eckner, Demonstration of permanent factor-dependent multipotential (erythroid/neutrophil/-basophil) hemopoietic progenitor cell lines, *Proc. Natl. Acad. Sci. USA*, 80:2931 (1983).
4. Valtieri, M., D.J. Tweardy, D. Caracciolo, K. Johnson, F. Mavilio, S. Altmann, D. Santoli, and G. Rovera, Cytokine-dependent granulocytic differentiation. Regulation of proliferative and differentiative responses in a murine progenitor cell line, *J. Immunol.*, 11:3829 (1987).
5. Askew, D.S., R.A. Ashmun, B.C. Simmons, and J.L. Cleveland, Constitutive c-*myc* expression in an IL-3 dependent myeloid cell line suppresses cell cycle arrest and accelerates apoptosis, *Oncogene* 6:1915 (1991).
6. Blackwood, E.M., and R.N. Eisenman, *Max*: A helix-loop-helix zipper protein that forms a sequence-specific DNA-binding complex with *myc*, *Science* 251:1211 (1992).
7. Bello-Fernandez, C., G. Packham, and J.L. Cleveland, Ornithine decarboxylase is a transcriptional target of c-*myc*, *Proc. Natl. Acad. Sci. USA* (in press).

8. Kretzner, L., E.M. Blackwood, R.N. Eisenman, *Myc* and *Max* proteins possess distinct transcriptional activities, *Nature (London)*, 359:426 (1992).
9. Luscher, B., and R.N. Eisenman, New light on Myc and Myb. Part I: *myc*, *Genes & Devel*, 4:2025 (1990).
10. Cleveland, J.L., M. Dean, N. Rosenberg, J.Y.J. Wang, and U.R. Rapp, Tyrosine kinase oncogenes abrogate interleukin-3 dependence of murine myeloid cells through signalling pathways involving c-*myc*: Conditional regulation of c-*myc* transcription by temperature-sensitive v-*abl*, *Mol. Cell. Biol.*, 9:5685 (1989).
11. Evan, G.I., A.H. Wyllie, C.S. Gilbest, T.D. Littlewood, H. Land, M. Brooks, C.M. Waters, L.Z. Penn, and D.C. Hancock, Induction of apoptosis in fibroblasts by c-*myc* protein, *Cell*, 69:119 (1992).
12. Shi, Y., J.M. Glynn, L.J. Guilbest, T.G. Cotter, R.P. Bissonnette, and D.R. Green, Role for c-*myc* in activation-induced apoptotic cell death in T cell hybridomas, *Science*, 257:212 (1992).
13. Dean, M., J.L. Cleveland, U.R. Rapp, and J.N. Ihle, Role of Myc in the abrogation of Interleukin-3 dependence, *Oncogene Res.*, 1:279 (1987).
14. Bowlin, T.L., B. McKnown, and P. Sunkara, Ornithine decarboxylase induction and polyamine biosynthesis are required for the growth of Interleukin-2 and Interleukin-3 dependent cell lines, *Cell Immunol.*, 98:341 (1986).
15. Auvinen, M., A. Paasinen, L.C. Anderson, and E. Hotta, E., Ornithine decarboxylase activity is critical for cell transformation, *Nature (London)*, 360:355 (1992).
16. Hibshoosh, H., M. Johnson, and I.B. Weinstein, Effects of overexpression of ornithine decarboxylase (ODC) on growth control and oncogene-induced cell transformation, *Oncogene*, **6**:739 (1991).
17. Kolch, W., G. Heidecker, and U.R. Rapp, Raf-1 protein kinase is required for growth of induced NIH 3T3 cells, *Nature (London)*, 349:426 (1991).
18. Nunez, G., L. London, D. Hockenberry, M. Alexander, J.P. McKearn, and Korsmeyer, Deregulated *bcl*-2 gene expression selectively prolongs survival of growth factor deprived hemopoietic cell lines, *J. Immunol.*, 144:3602 (1990).
19. Vaux, D.L., S. Cory, and J.M. Adams, *Bcl*-2 gene promotes hemopoietic cell survival and cooperates with c-*myc* to immortalize pre-B cells, *Nature (London)*, 335:440 (1988).
20. Bissonnette, R.P., F. Echeverri, A. Mahboubi, and D.R. Green, Apoptotic cell death induced by c-*myc* is inhibited by *bcl*-2, *Nature (London)*, 359:552 (1992).
21. Fanidi, A., S.A. Harrington, and G.I. Evan, Cooperative interaction between c-*myc* and *bcl*-2 proto-oncogenes, *Nature (London)*, 359:554 (1992).
22. Wyllie, A.H., K.A. Rose, R.G. Morris, C.M. Steel, E. Foster, and D.A. Spandidos, Rodent fibroblast tumours expressing human *myc* and *ras* gene: Growth, metastasis and endogenous oncogene expression, *Br. J. Cancer*, 56:251 (1987).

Discussion

P. Nicotera

I found your data with ODC very interesting. We have published, a year and a half ago, that if you increase the intracellular content of spermine in thymocytes, you completely prevent apoptosis induced by glucocorticoid or calcium ionophore. We have a paper now which is submitted for publication where we show that in conditions of oxidative stress, you can induce cell proliferation or apoptosis, depending on the exposure levels. This has to do with the intra-cellular content of polyamines and the induction of ODC. The end point of this is that if you increase the intra-cellular content of spermine you protect cells from chromatin unfolding whereas if you decrease the content of spermine, then the cells will go to apoptosis. This is entirely dependent on induction of ODC dependent PKC activation.

J. Cleveland

Very nice, I would like to get a reprint.

M. Oren

Can you inhibit some of the apoptotic effects of *myc* by having difluoromethylornithine (DFMO) or some other ODC inhibitor around?

J. Cleveland

DFMO is tricky. Have you worked with it? It is a poor inhibitor of growth cell so when you do the experiment it takes almost three cell doublings to arrest the cells in G1. And when you do that experiment with a *myc* clone, they are not going to make it because they get to a point where they should really be arresting anyway and are dying. So we have tried the experiment and so far we have failed.

M. Kastan

I wanted to make sure I understood your deletion mutant experiments. When you knock out the *max* binding site you are still getting activation of ODC, is that right?

J. Cleveland

Yes, the *max* interaction domain, one of the *max* interaction domains, the leucine zipper.

M. Kastan

I thought when you do that you would no longer get good DNA binding with *myc*.

J. Cleveland

You do not in an *in vitro* assay. We believe there is a novel Myc dimerization partner out there that does not require the leucine zipper motif and that is one of the postdocs' projects, to molecularly clone other partners. And I will also mention in passing, it is a little controversial, but we also have Myc lacking the leucine zipper domain functioning in apoptosis, which is what we would predict, and these are different than other published results.

D. Livingston

I was wondering what would happen if you knocked out the ODC gene.

J. Cleveland

It would be lethal.

D. Livingston

Is there any pharmacologic evidence that would suggest that?

J. Cleveland

Phil Coffino could answer this question.

J. Isaacs

This is one of the problems with the irreversible inhibition of that enzyme. Originally it was hoped that it would be cytotoxic to many cells but there are polyamine transporters and if you inhibit the synthetic pathway, the cell upregulates its transporters, increasing uptake of the polyamines from the serum. Therefore, it is very hard actually to deplete the polyamines. It takes fairly heroic efforts with a series of inhibitors. There is an alternative way though. There are a variety of new polyamine analogs that are structurally related to polyamines but do not have the same affinities for DNA and are cytotoxic to cells. So it would be interesting in those systems to use those alternatives, the bis-ethyl-nor spermine (BES), for example, is one example of such polyamine antagonists. Actually Dr. Mihich could comment upon this better because this work is being done by Carl Porter at his Institution. Just simply inhibiting ODC is not a very good way to deplete polyamines at all because a cell will very rapidly respond and salvage polyamines some other source.

G. Evan

Have you looked at the E1A *myc* chimeras that Rawlston reported some years ago? For what it is worth, we have looked at the chimeras, all of them in fact, for induction of apoptosis in our system and the one that has the end terminus of *myc* and the C-terminus of E1A is a very potent inducer of cell death in serum deprived cells, and clearly that has no interaction with *max* whatsoever. Now, it may be something weird, but I just wondered what would happen in your system.

J. Cleveland

We have used Far-Westerns to see what else is out there that Myc will bind to. I do not really like these assays but there are several other proteins that you can detect with a C-terminal Myc probe and there are a lot of proteins you can identify with an amino-terminal probe. So my bias is that there is a lot of other interacting factors, and there is good evidence for this with other kinds of regulatory networks. The other point that should come out of all of this is that here we are really separating a *myc* function from transformation, so you clearly can induce cell death with a Myc gene lacking a domain that is required for transformation. So those two phenomena, I think, are starting to get separated.

E. Mihich

Commenting to Dr. Nicotera, I would like to mention that MGBG has a number of effects in addition to affecting polyamine pathways. One prominent one is to affect mytochondria and that could be a reason for apoptosis apart from the polyamine interference. As far as the comment of Dr. Isaacs is concerned, indeed Carl Porter has now an analog, which closely resembles spermidine, which has multiple activities on cells in the sense that it leads to a transcriptional induction of the acelytase which is the first step in degradation of polyamine pools and post-transcriptionally inhibits the synthetic decarboxylases and, in addition, it competes with polyamines for cellular uptake. And so through these three mechanisms, this analog causes a drop in polyamine pools which, as you pointed out, is very difficult to achieve otherwise, certainly not with ODC inhibitors.

J. Cleveland

Right.

E. White

I just had a comment on your separation of transformation from apoptosis. With E1A there are two domains that are required for the induction of DNA synthesis and mutations of either one will block transformation. In other words, both of them are required for transformation, but either one is sufficient for effects on the cell cycle. So, my question is: You have looked at transformation with the mutants, could they possibly induce DNA synthesis and apoptosis but be transformation defective because of lacking the second function?

J. Cleveland

Yes, we are repeating a lot of these experiments and then, on top of it, we are making double and triple mutants that you have to make to really test the model. For example, a leucine zipper deletion plus a transactivation domain minus dominant negative mutant. We are just starting those experiments with REFs.

R. Schimke

Have you considered placing the ornithine decarboxylase gene under the control of a regulated promoter? This would allow you to test the hypothesis that altered regulation of ornithine decarboxylase is causal in events leading to apoptosis. Most results of the type showing some time-related dependent of events beg the question of whether the events are causally related.

J. Cleveland

There is also a cell line out there that is a putrescine auxotroph and we have thought of doing rescues with ODC mini genes containing either Myc binding sites or mutated Myc binding sites to see if these are really relevant *in vivo*. That is one experiment we will be doing.

P. Nicotera

Just to reply to the comment of Dr. Mihich: Spermine is very efficient in lowering the affinity of mytochondria for calcium uptake for example. By definition, apoptosis does not involve depletion of ATP, or measured changes in energy. At least in our hands, in thymocytes, we do not have such an effect. But I think the most direct evidence for this is when you have a system where you can measure the polyamine directly. And we do have a depletion of polyamines in the system that we use. We measure directly spermine, putrescine and spermidine. So we can actually measure their intra-cellular count.

E. Mihich

But MGBG also has marked effects on DNA and those also complicate mechanistic interpretations.

ION SIGNALLING IN APOPTOSIS

Pierluigi Nicotera[1a], Boris Zhivotovsky[1], Giorgio Bellomo[2], and Sten Orrenius[1]

[1]Institute of Environmental Medicine, Division of Toxicology, Karolinska Institute, Box 210, S 171 77 Stockholm, Sweden
[2]Department of Experimental Medicine and Oncology, Institute of General Pathology, University of Torino 10125, Torino, Italy

INTRODUCTION

Apoptosis (Kerr et al., 1972; Wyllie et al., 1980), a genetically controlled cell deletion process modulates organ development (Ellis and Horvitz,1986) and immune cell ontogenesis (Smith et al., 1989) in a predestined fashion. In the adult organism, the apoptotic program is activated to kill neoplastic cells (Trauth et al., 1989) influencing the rate of tumor cell population expansion (Arends and Wyllie, 1991). On the other hand, apoptosis can be evoked in a number of pathological states. Thus, immuno-competent cells are inappropriately eliminated by apoptosis in HIV infection (Groux et al., 1992), whereas viruses (Rao et al., 1992) and bacterial toxins (Fiorentini et al, 1993) can directly elicit an apoptotic response in target cells. Finally, exposure to ionizing radiation (Zhivotovsky et al., 1981 and 1993b) or chemical toxicants (McConkey et al., 1988; Aw et al., 1990) can accelerate apoptosis in immature T cells.

Apoptosis seems to develop in four different phases (Fig.1). The presence of genes that activate or repress the cell death program is a prerequisite for the occurrence of apoptosis in developing organisms (Schwartz et al., 1990; Ellis and Horvitz, 1986) as well as in cultured cell systems (Hockenberry et al., 1990, Yonish-Rouash et al., 1991; Evan et al., 1992). The apoptotic program can be initiated by specific signalling molecules (e.g. glucocorticoid hormones) (Wyllie, 1980), by imbalanced signalling, such as the lack of a growth factor (Kyprianou and Isaacs, 1988), or by a dysfunction in a signalling pathway caused by toxicants (Aw et al., 1990; Dypbukt et al., 1993). Degradative processes are subsequently activated in the cell, which include alteration of cell adhesion molecules with loss of cell-cell contact, cytoskeletal modifications, cell shrinkage, rearrangement of the chromatin structure with DNA fragmentation, formation of apoptotic bodies and death (Arends and Wyllie, 1991). Finally, phagocytosis of the whole cell or cell fragments terminates the process (Savill et al., 1993).

---

[a] Address correspondence to this author at the Institute of Environmental Medicine, Division of Toxicology, Karolinska Institute, Box 210, 171 77, Stockholm, Sweden

Apoptosis, Edited by E. Mihich and R.T. Schimke
Plenum Press, New York, 1994

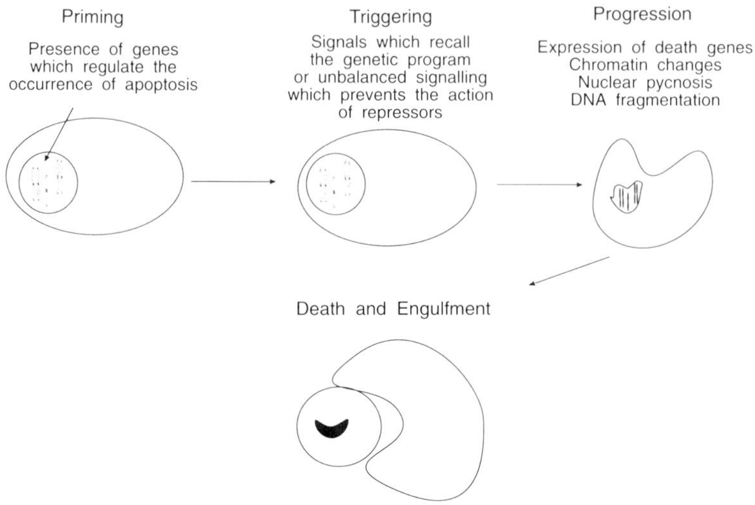

Figure 1. Schematic representation of possible phases in the development of apoptosis

Thus, regardless whether deterministically activated or triggered by accidental conditions, apoptosis requires a complex combination of signals that can either originate within the cell, or be delivered by exogenous molecules. The effective signals may differ between different organisms and tissues and in many cases remain poorly understood. For example, while it is clear that the occurrence and expression of different sets of genes decide the activation of apoptosis at a precise stage of development or cell maturation, the signals that trigger either their expression or repression are largely unknown. Further, it is unclear how biological mediators or xenobiotics initiate apoptosis accidentally in cells that possess the adequate genetic machinery.

Nevertheless, the past few years have seen a rapid development of the understanding of the signals that trigger apoptosis. Rather than the involvement of a single signalling system, the imbalance of multiple signalling pathways appears to be required to activate apoptosis in committed cells. In particular, the imbalance between intracellular Ca^{2+} signals and growth or mitogenic stimulation can result in apoptosis in many cell systems (Nicotera et al, 1992). While this proposal has been supported primarily by studies in lymphoid cell systems, more recent studies demonstrate that such paradigm also applies to various non-lymphoid cells. The following sections will describe some of the most recent findings concerning the role of Ca^{2+} in the activation of apoptosis in various cell systems.

INTRACELLULAR CA^{2+} SIGNALS ACTIVATE APOPTOSIS

A sustained intracellular Ca^{2+} overload can trigger several lethal processes. Proteolysis and disruption of the cytoskeletal organization, mitochondrial dysfunction and DNA damage can all be caused by Ca^{2+} accumulation within the cytoplasm or other intracellular compartments (Nicotera et al, 1992). While in many cases sudden and massive intracellular Ca^{2+} overload can rapidly lead to cell necrosis, there is now convincing evidence that disturbances in Ca^{2+} signalling can also trigger apoptosis. The finding that removal of extracellular Ca^{2+}, or buffering intracellular Ca^{2+} increases can prevent the nuclear changes typical of apoptosis (i.e., DNA degradation and apoptotic body formation) and cell killing, demonstrates the Ca^{2+} requirement in apoptosis (Nicotera et al., 1992). Further evidence for a critical role of Ca^{2+} accumulation in triggering apoptosis has been provided by Dowd et al. (1992), who have shown that transfection of WEHI7.2 thymoma cells with calbindin, a Ca^{2+}-binding protein, prevents apoptosis induced by glucocorticoids, cAMP or the calcium ionophore, A23187. The effect of calbindin expression appears to be post-translational and is likely related to the ability to buffer increase in intracellular Ca^{2+}.

At least three different Ca^{2+}-dependent mechanisms may be linked to the triggering of the nuclear changes seen in apoptosis. These include the activation of a Ca^{2+}-dependent endonuclease activity responsible for DNA fragmentation, the modification of chromatin conformation and its susceptibility to cleavage by the former or other nucleases, and alterations in gene expression.

Ca^{2+}-dependent endonuclease activation

Many *in vitro* models of apoptosis are associated with a loss of intracellular Ca^{2+} regulation and activation of a Ca^{2+}- dependent endonuclease activity. The Ca^{2+} ionophore A23187 stimulates DNA fragmentation and apoptosis in thymocytes (McConkey et al., 1989), and characteristic endonuclease activity in isolated nuclei is dependent on Ca^{2+} (Jones et al., 1989). In addition, Ca^{2+}-mediated endonuclease activation appears to be involved in the cytotoxicity of TCDD and tributyltin in thymocytes (McConkey et al., 1988; Aw et al., 1990).

Also in cell-mediated defence against malignant and virally infected cells Ca^{2+} can stimulate endonuclease activity and cause apoptosis. This is the case, in natural killer (NK) cell-mediated cell killing, where the lethal hit is delivered by a NK-stimulated Ca^{2+} increase in the target (McConkey et al., 1990). A key role for endonuclease-mediated DNA cleavage in target cell killing has been more recently confirmed by Helgason et al. (1993). In both cases, the critical event activated by Ca^{2+} seems to be DNA fragmentation. Although most of the evidence for Ca^{2+}-dependent endonuclease activation has come from studies in lymphoid cells, it appears that this process may also be important in a variety of other tissues. We have recently shown in collaboration with Dr. P.O. Berggren ´s laboratory, that exposure of pancreatic RINm5F cells to sera of patients affected by type I, insulin-dependent diabetes mellitus increases L-type Ca^{2+} channel activity, resulting in Ca^{2+} overload and apoptosis (Juntti-Berggren et al.,1993). If the Ca^{2+} increase is precluded by treatment of the cultures with the L-type Ca^{2+} channel blocker, verapamil, DNA fragmentation is prevented and apoptosis cannot ensue. Ca^{2+} increases resulting in the triggering of endonuclease activity can also be compartmentalized. For example, exposure of human adenocarcinoma cells to tumor necrosis factor α causes an initial intranuclear Ca^{2+} accumulation, endonuclease activation and apoptosis (Bellomo et al.,1992). This suggests that a compartmentalized nuclear Ca^{2+} overload may be sufficient to trigger DNA fragmentation.

Oligonucleosomal DNA fragmentation caused by endonuclease activation remains one of the most recurrent hallmarks of apoptosis. However, the endonuclease(s) responsible for apoptosis has not been conclusively identified. To date, several studies have reported partial purification and characterization of apoptosis-linked nucleases. Most have a Ca^{2+} and Mg^{2+} requirement (Arends et al., 1990), others seem to have mainly Ca^{2+} (Gaido and Cidlowski, 1991; Seiliev et al., 1992) or Mg^{2+} sensitivity (Kawabata et al., 1993). DNase I (Peitsch et al., 1993) and acidic nucleases such as DNase II (Barry and Eastman, 1993) have also been implicated. However, the cleavage specificity of DNase II for 3'P-5'OH ends makes this enzyme an unlikely candidate for the formation of DNA fragments such as those found in apoptotic cells, which present 5'P-3'OH ends (Beletsky et al.,1989). Finally the involvement of topoisomerase II in a configuration that cuts but cannot religate the DNA has been suggested by other reports (Walker et al., 1991, Roy et al., 1992). These discrepant findings may suggest that rather than a single distinct nuclease, various DNases may cause genome cleavage in different experimental models of apoptosis. Thus, in addition, to direct activation of nucleases, the Ca^{2+} requirement may also be related to its ability to cause changes in the chromatin structure (see below). This would increase the DNA susceptibility to digestion by a variety of DNases.

Despite the evidence that nuclease-mediated DNA cleavage is required for apoptosis in many systems, several studies have suggested that formation of apoptotic bodies and the apoptotic process in general, may progress without internucleosomal DNA fragmentation (Cohen et al., 1992; Oberhammer et al., 1992 and 1993). Endonuclease activation would eventually occur only at a late stage to dispose of bulky aggregates of newly formed heterochromatin. Work by Brown et al., (1993) has indicated that the chromatin condensation which preludes to the formation of apoptotic bodies occurs in the presence of Zn^{2+} a known inhibitor of endonuclease activity. Chromatin condensation would then be initiated by a detachment of chromatin loops from the scaffolding operated by topoisomerase II (Alnemri and Litwack, 1990; Walker et al., 1991; Brown et al., 1993) by cytoskeletal changes (Peitsch et al., 1993) or by proteases (Oberhammer F., personal communication). The observation that high molecular weight (HMW) DNA fragments (300-50 kb) are generated during the early phases of apoptosis prior to internucleosomal cleavage (Elias and Berry, 1991; Brown et al., 1993) would support this idea.

The lack of endonuclease involvement in the formation of HMW DNA fragments was suggested by the observation that Zn^{2+}, at 1 mM concentration did not prevent their formation in glucocorticoid-treated thymocytes (Brown et al., 1993). Our more recent results are not in agreement with this conclusion. We have found that treatment of human thymocytes with methylprednisolone causes the formation of 300 and 50 kb fragments. However, the formation of these large DNA fragments is inhibitable by Zn^{2+} at concentrations of 100μM, by the drug aurintricarboxylic acid and by incubating cells in Ca^{2+}-free conditions (Zhivotovsky et al. 1993a). In agreement with Brown et al. (1993), Zn^{2+} at 1mM, did not inhibit the formation of HMW DNA fragments. Notably, exposure of untreated human thymocytes to 1mM Zn^{2+} is sufficient to cause the formation of these large fragments. This can be explained by a direct Zn^{2+} toxicity at this concentration (McCabe et al., 1993), or by the multiple roles of Zn^{2+} in the regulation of enzymes that can promote chromatin relaxation. These include topoisomerase I (Samuel et al., 1993) and poly-ADP-ribose polymerase (Gradwohl et al., 1990). The observation that the aurintricarboxylic acid and Ca^{2+} withdrawal can prevent the formation of large DNA fragments strongly suggests the involvement of Ca^{2+}-dependent nuclease activity also in these early phases of thymocyte apoptosis..

On the other hand, mechanisms other than endonuclease activation can most likely cause the detachment of chromatin loops from the scaffolding in other models of apoptosis. Formation of 50 kb fragments and possibly detachment of chromatin loops from the scaffolding may also be triggered by topoisomerase II (Walker et al. 1991; Käs, et al., 1993). Our finding that HMW DNA fragments and apoptotic bodies can be formed by topoisomerase II reactive chemotherapeutic drugs in the presence of Zn^{2+} also supports the idea that topoisomerase II activation may trigger a different mechanism of chromatin degradation. Notably, the formation of HMW fragments caused by topoisomerase II inhibitors in human thymocytes is most efficiently blocked under Ca^{2+} -free conditions.

Role of Ca^{2+} in the formation of apoptotic bodies

A more direct role for Ca^{2+} in the formation of condensed chromatin and apoptotic bodies in thymocyte apoptosis is demonstrated by recent work using thapsigargin (Jiang et al., 1993). Thapsigargin is a highly selective inhibitor of the endoplasmic and sarcoplasmic reticular Ca^{2+} pumps (Thastrup et al., 1990). In a variety of cells addition of thapsigargin causes Ca^{2+} release from the endoplasmic reticulum (and indirectly also from other pools) and stimulates Ca^{2+} entry by a capacitative pathway (Inesi and Sagara, 1992). In thymocytes, in the presence of extracellular Ca^{2+}, thapsigargin causes a sustained intracellular Ca^{2+} elevation, that is associated with the formation of apoptotic bodies. The duration of the Ca^{2+} increase can be modulated in this system by withdrawal of extracellular Ca^{2+} at different times. By this approach we have been able to establish that the duration of Ca^{2+} elevation necessary to induce DNA fragmentation and formation of apoptotic bodies is at least 90 min. Ca^{2+} chelation prior to this time is sufficient to prevent the formation of condensed chromatin patches and apoptotic bodies (Jiang et al., 1993).

Ca^{2+} and the regulation of chromatin conformation

In eukaryotic cells, compaction of the genetic information into nucleosomal and supra-nucleosomal structures creates a partition of the genetic information. To be accessible for transcription a specific active gene requires both a transcription factor and a local chromatin relaxation. Active chromatin normally displays a higher degree of unfolding than heterochromatin (Ericsson et al., 1990). Several factors are responsible for chromatin relaxation, including changes in nuclear ion composition (Guo and Cole, 1989) and binding of histone H1 with the linker DNA regions (Clark and Thomas, 1986). In particular, it is well recognized that modification of H1-H1 interactions, or a loss of H1, can produce chromatin relaxation and confer transcriptional competence.

On the other hand, exposed DNA regions can also be easily cleaved (Pfaffle et al., 1990). It is generally recognized that the organization of the chromatin fiber presents DNase I hypersensitive sites in correspondence of transcriptionally active genes (Wu, 1979). However, such sensitive sites may extend also beyond the domain of active genes and cause a true partition of chromatin regions (Käs et al., 1993). Thus, as recently demonstrated by the same authors, redistribution of the H1 histone with consequent chromatin relaxation causes an increased sensitivity of large sectors of the genome to degradation by topoisomerase II or nucleases. The importance of chromatin organization in the susceptibility of the DNA to cleavage and in the progression of apoptosis has been suggested by early observations indicating that the apoptotic chromatin is depleted in histone H1 and that transcriptionally active regions are preferentially cleaved by endonucleases during apoptosis (Arends et al., 1990). Studies in our laboratory have indicated that apoptosis can be prevented by agents that increase chromatin compaction (i.e polyamines) (Brüne et al., 1991). Conversely, induction of chromatin unfolding can sensitize cells to apoptosis (Alnemri and Litwack, 1990; Brüne et al., 1991). Among the factors that may cause chromatin relaxation, ions are known to play an important role. Thus, increases in pH within one pH unit can cause chromatin unfolding (Guo and Cole, 1989), whereas other ions such as Cu^{2+} can play an important role in chromatin stabilization (Prütz et al., 1990).

It is becoming increasingly clear that ion transport across the nuclear envelope is regulated. We have reported the existence of an active Ca^{2+} -sequestration system on the nuclear envelope (Nicotera et al, 1989; Nicotera et al., 1990). This system promotes Ca^{2+} uptake into the nucleoplasm and causes an increase in the intranuclear free Ca^{2+} concentration. These observations are now supported by a number of studies reporting that Ca^{2+} and other ions are actively transported into the nucleus (Hechtenberg and Beyersmann, 1993). More recent studies in our laboratory have shown that nuclear Ca^{2+} uptake is linked to intranuclear alkalinization and that this change in ion composition in the nuclear interior causes chromatin unfolding (Fig. 2) and DNA cleavage by endonucleases (Nicotera et al., unpublished results).

Nuclear Ca^{2+} and pH signals may therefore be sufficient to promote the chromatin conformational changes necessary for the subsequent chromatin digestion. This assumption is supported by the finding described above that formation of large molecular weight DNA fragments caused by either topoisomerase II or endonuclease-mediated cleavage is prevented in the absence of Ca^{2+} (Zhivotovsky et al., 1993a). This would also provide an

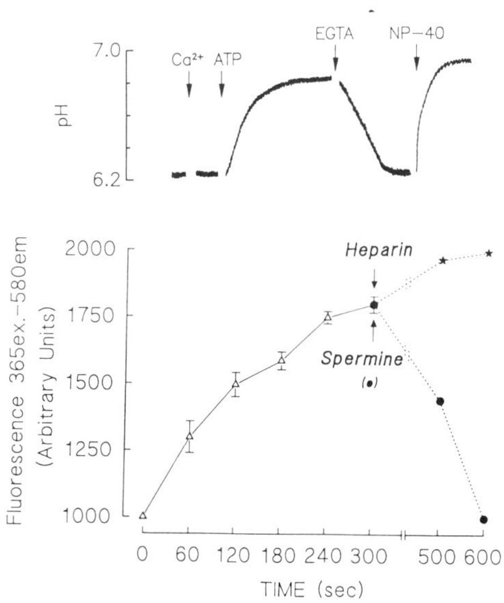

Figure 2 A: Increase in intranuclear pH in isolated liver nuclei. Nuclei were isolated as previously described (Nicotera et al., 1989) and loaded with 10 μM of the fluorescent pH indicator, BCECF-AM (Molecular Probes, Inc. Eugene, OR, USA). Intranuclear pH was measured essentially as described for intranuclear Ca^{2+} (Nicotera et al., 1989) except that the excitation ratio was 490/450 nm. and the emission was set at 510 nm. Addition of 1mM ATP promoted intranuclear alkalinization associated with Ca^{2+} uptake. Subsequent removal of Ca^{2+} with 2 mM EGTA caused intranuclear re-acidification. Permeabilization of the nuclear envelope with NP-40 allowed equilibration of BCECF with the incubation medium (pH 7.0). **B: Changes in chromatin folding induced by intranuclear Ca^{2+} and pH changes in isolated liver nuclei**. Nuclei were labelled with propidium iodide and incubated in the presence of 1μM Ca^{2+} and 1mM ATP, as described above. Ca^{2+} uptake and intranuclear alkalinization stimulated a time-dependent increase of propidium iodide fluorescence which indicated chromatin unfolding. Subsequently, the addition of 1 mM spermine (•) to cause chromatin compaction reduced the fluorescence intensity. In contrast, the polyanion heparin (⋆) caused a maximal increase in dye fluorescence, suggestive of maximal chromatin unfolding.

explanation for the inability to demonstrate the involvement of a single specific endonuclease in apoptotic DNA fragmentation. In this model, the role of ion signalling in apoptosis would be mainly related to the ability to promote chromatin structural changes.

It is apparent that prior to the formation of apoptotic bodies, the chromatin undergoes marked rearrangement. As illustrated in Fig.3, one hour after the exposure to methylprednisolone, thymocytes begin to display chromatin patches which tend to

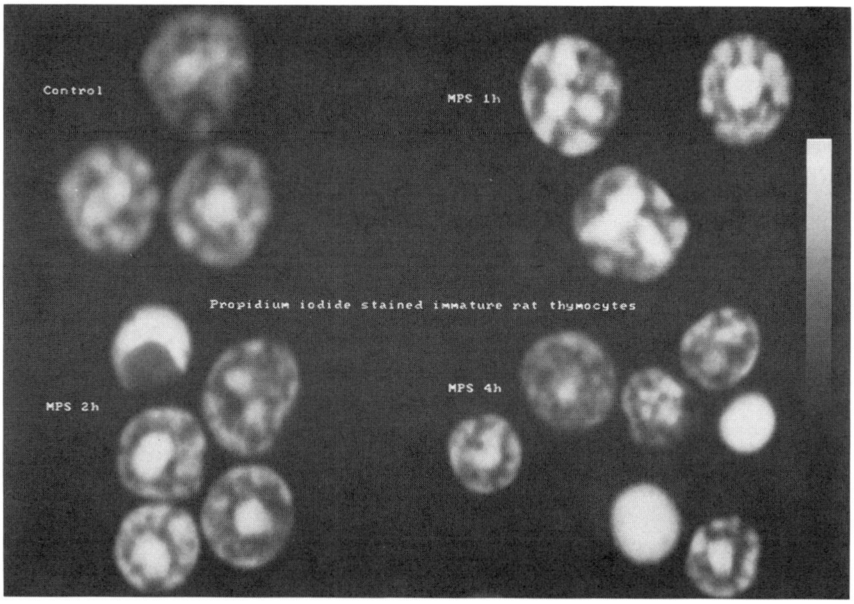

Figure 3. Confocal microscopy images of thymocytes undergoing apoptosis. Rat thymocytes were exposed to 0.1μM methylprednisolone (MPS). At the time indicated in the figure, cells were harvested, washed and seeded onto glass coverslips coated with poly-L-lysine. Cells were then stained with propidium iodide and examined with a BIO-RAD MRC-600 confocal microscope (excitation, 488 nm), was provided by a krypton-argon laser). Initial chromatin aggregation can be observed 1h after treatment. At 2h a few cells display apoptotic nuclei. Their number increases markedly between 2 and 4 h.

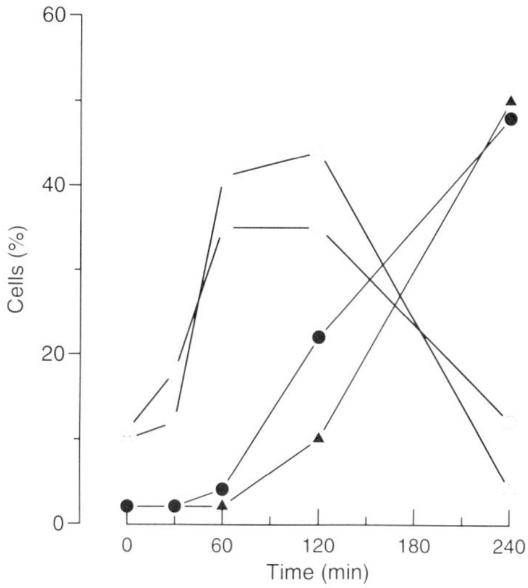

Figure 4. Relationship between early chromatin changes and formation of apoptotic bodies. Isolated rat thymocytes were exposed to 6 Gy X-irradiation (△,▲) or to 0.1 μM methylprednisolone (o,●). Cells were harvested at the times indicated and treated as described in legend to figure 3. Images were collected at the confocal microscope. The number of cells exhibiting early chromatin changes and apoptotic bodies was then scored. Open symbols indicate cells with patched chromatin. Closed symbols show cells with apoptotic nucleus.

progressively aggregate in a characteristic apoptotic body. The correlation between these early chromatin changes and apoptotic body formation is shown in Fig. 4 where it is clear that formation of these chromatin patches precedes the appearance of apoptotic bodies. If the number of cells displaying these early morphological changes is scored, it is apparent that a major chromatin rearrangement occurs in about 40% of the cells within 1h. As the formation of apoptotic bodies progresses, the cells with patched chromatin disappear suggesting that the early changes only precede the apoptotic body formation. This is also apparent in non-lymphoid cells. Exposure of human adenocarcinoma cells to TNF-α causes similar changes (Fig.5).

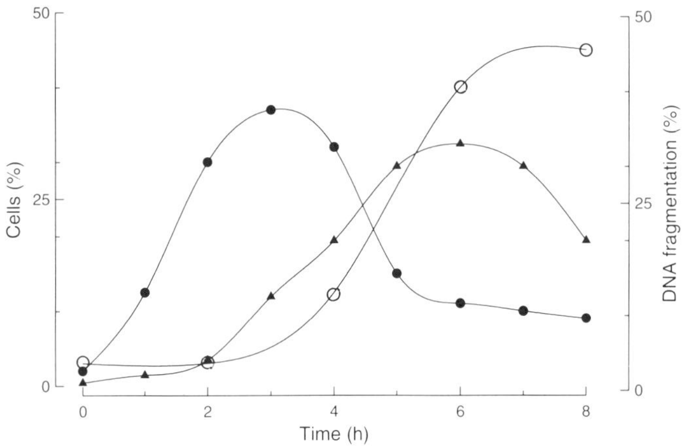

Figure 5. Early chromatin changes, apoptotic body formation and DNA fragmentation in human adenocarcinoma cells (BT-20) treated with tumor necrosis factor α. BT-20 cells were sensitized with 50 μg/ml cycloheximide and subsequently exposed to 50 units/ml of human recombinant TNF-α (Bellomo et al., 1992). At the time indicated cells were permeabilized and stained with propidium iodide or harvested and processed for DNA fragmentation. Cells stained with propidium iodide were imaged with a SPEX video-image analysis system. The percentage of cells displaying either patched chromatin (•) or apoptotic bodies (o) was scored. Detectable internucleosomal DNA fragmentation (▴) occurred after the appearance of patched chromatin and concomitantly with the formation of apoptotic nuclei. As discussed in the text, formation of large molecular weight DNA fragments precedes internucleosomal cleavage.

The importance of initial chromatin unfolding is suggested also by studies where increasing chromatin compaction is associated with protection from apoptosis. This is observed when thymocytes are loaded with a polyamine, spermine, and subsequently exposed to methylprednisolone or ionophore A23187 (Brüne et al., 1991). We have also observed that the polyamine level decreases immediately prior to endonuclease activation in pancreatic RINm5F cells exposed to oxidative stress. Pretreatment with the phorbol ester, TPA, which induces polyamine synthesis and restored the intracellular polyamine content, prevented DNA fragmentation and cell death (Dypbukt et al., 1993). Several observations suggest that in this system, polyamine depletion was responsible for the increased chromatin sensitivity to endonuclease cleavage; i) DNA fragmentation occurred only when cells were depleted of spermine; ii) pretreatment with TPA that restored spermine content prevented DNA fragmentation and restored the ability of the cells to grow; and iii) addition of spermine to a serum-free, incubation medium reduced DNA fragmentation. Because polyamines, in particular spermine cause chromatin compaction

(Snyder, 1989), it is conceivable that their increase induced by growth stimulation may counteract the chromatin relaxation required for DNA cleavage and formation of apoptotic bodies.

Ca^{2+} and gene expression

A role for Ca^{2+} and Ca^{2+} -dependent proteins (i.e. calmodulin) in the regulation of gene expression has long been known. Early work had demonstrated that the synthesis of prolactin and growth hormone in rat pituitary GH3 cells requires Ca^{2+} and subsequent studies showed that Ca^{2+} exerts its effect at the transcriptional level via calmodulin (White and Bancroft, 1987 and references therein). In addition Ca^{2+} can directly regulate the expression of certain genes, including c-fos (Morgan and Curran, 1986), and control the expression of others (i.e. interleukin-2) during cell activation (Crabtree, 1989). Moreover, according to the working hypothesis outlined above, ion changes may also favor the chromatin conformational changes required for gene activation. For example, initial chromatin unfolding initiated by ion changes in the nuclear interior may facilitate H1 redistribution or topoisomerase II activation. This would produce changes in the supercoiling of a particular chromatin domain (Villeponteau et al., 1986, Käs et al., 1993). The resulting torsional stress distorts the chromatin structure and imposes DNase sensitivity on a portion of that domain (Villeponteau, et al., 1986). A relaxation of chromatin high order structure may then occur and interaction of histone proteins (i.e., H1) with the nucleosomes may change further (Jin and Cole, 1986). These modifications have produced transcriptional competence but are not sufficient to produce transcription. Activation of specific gene(s) within the unfolded region may involve the activation and binding of a specific transcription factor protein, which would presumably not occur without the initial chromatin unfolding. A number of genes whose involvement in apoptosis has been suggested may be regulated by Ca^{2+}; examples include calmodulin (Dowd et al., 1991), *c-fos* (Smeyne et al., 1993) and possibly, growth arrest genes such as *gadd 153* (Bartlett et al., 1992). In addition, the role of calmodulin and Ca^{2+} in the regulation of the cell cycle is well established (Chafouleas et al., 1984). Expression of a constitutive form of Ca^{2+} /calmodulin- stimulated protein kinase II can cause cell cycle arrest in G_2 independently from phosphorylation of cdc2 as histone H1 kinase (Planas-Silva and Means, 1992). These changes causing cell cycle arrest may result in activation of the apoptotic process.

CONCLUDING REMARKS

During apoptosis, alterations of chromatin superstructure develop in both dividing and quiescent cells. Extensive rearrangement of chromatin loops and detachment from the nuclear scaffold proteins may be caused by a variety of signals including ion signals, phosphorylation/dephosphorylation reactions, proteolysis of linker proteins and nucleoskeletal alterations. In addition, marked chromatin redistribution occurs during the cell cycle. An imbalance in the signals that regulate chromatin structural assembly may increase the susceptibility of DNA to cleavage by endogenous nucleases. A possible working hypothesis would indicate that intranuclear Ca^{2+} elevations may have the dual effect of promoting chromatin changes and activating Ca^{2+}-dependent nucleases. In quiescent cells, such signals when dissociated from the activation of specific transcription factor proteins, may initiate the chromatin rearrangement that leads to genome fragmentation and apoptosis. Similarly, should such a signalling imbalance occur during mitosis the increased chromatin sensitivity to nuclease cleavage would result in extensive chromatin degradation and cell death. Work in progress in our laboratory is testing this hypothesis.

ACKNOWLEDGEMENTS

This work was supported by the Swedish Medical Research Council, (Proj. No. 03X-2471), the Swedish Natural Science Research Council (Proj. No. B-Bu 10173-300), the Royal Swedish Academy of Sciences and by Consiglio Nazionale delle Ricerche (Special project: Biology and Pathology of Calcium).

REFERENCES

Alnemri, E.S. and Litwack, G., 1990, Activation of internucleosomal DNA cleavage in human CEM lymphocytes by glucocorticoid and novobiocin, *J. Biol. Chem.*, **265**: 17323-17333.

Arends, M.J., Morris, R.G. and Wyllie, A.H., 1990, Apoptosis- The role of the endonuclease, *Amer. J. Pathol.*, **136**: 593-608.

Arends, M.J. and Wyllie, A.H., 1991, Apoptosis: mechanisms and roles in pathology, *Int. Rev. Exper. Pathol.*, **32**: 223-254.

Aw, T.Y., Nicotera, P., Manzo, L. and Orrenius, S., 1990, Tributyltin stimulates apoptosis in rat thymocytes, *Arch. Biochem. Biophys.*, **283**: 46-50.

Barry, M.A. and Eastman, A., 1993, Identification of deoxyribonuclease II as an endonuclease involved in apoptosis, *Arch. Biochem. Biophys.*, **300**: 440-445.

Bartlett, J.D., Luethy, J.D., Carlson, S.G., Sollott, S.J. and Holbrook, N.J., 1992, Calcium ionophore A23187 induces expression of the growth arrest and DNA damage inducible CCAAT/enhancer-binding protein (C/EBP)-related gene, *gadd 153.*, *J. Biol. Chem* **267**: 20465-20470.

Beletsky, I.P., Matyasova, J., Nikonova, L.V., Skalka, M. and Umansky, S.R., 1989, On the role of Ca^{2+}-Mg^{2+}-dependent endonuclease in the post-irradiation degradation of chromatin in lymphoid tissues, *Gen. Physiol. Biophys.*, **8**: 381-389.

Bellomo, G., Perotti, M., Taddei, F., Mirabelli, F., Finardi, G., Nicotera, P. and Orrenius, S., 1992, Tumor necrosis factor α induces apoptosis in mammary adenocarcinoma cells by an increase in intranuclear free Ca^{2+} concentration and DNA fragmentation, *Cancer Res.*, **52**: 1342-1346.

Brown, D.G., Sun, X. and Cohen, G.M., 1993, Dexamethasone-induced apoptosis involves cleavage of DNA to large fragments prior to internucleosomal fragmentation, *J. Biol. Chem.*, **268**: 3037-3039.

Brüne, B., Hartzell, P., Nicotera, P. and Orrenius, S., 1991, Spermine prevents endonuclease activation and apoptosis in thymocytes, *Exp.Cell Res.*, **195**: 323-329.

Chafouleas, J.G., Lagace, L., Bolton, W.E., Boyd, A.E. and Means, A.R., 1984, Changes in calmodulin and its mRNA accompany reentry of quiescent (G0) cells into the cell cycle., *Cell*, **36**: 73-81.

Clark, D.J. and Thomas, J.O., 1986, Salt-dependent co-operative interaction of histone H1 with linear DNA, *J. Mol. Biol.*, **187**: 569-580.

Cohen, G.M., Sun, X., Snowden, R.T., Dinsdale, D. and Skilleter, D.N., 1992, Key morphological features of apoptosis may occur in the absence of internucleosomal DNA fragmentation, *Biochem. J.*, **286**: 331-334.

Crabtree, G.R.,1989, Contingent genetic regulatory events in T-lymphocyte activation., *Science*, **243**: 355-361.

Dowd, D.R., MacDonald, P.N., Komm, B.S., Haussler, M.R. and Miesfeld,R.L., 1992, Stable expression of the calbindin-D28K complementary DNA interferes with the apoptotic pathway in lymphocytes, *Mol. Endocrin.*, **6**: 1843-1848.

Dowd, D.R., MacDonald, P.N., Komm, B.S., Maussler, M.R. and Miesfeld, R., 1991, Evidence for early induction of calmodulin gene expression in lymphocytes undergoing glucocorticoid-mediated apoptosis, *J. Biol. Chem.*, **266**: 18423-18426.

Dypbukt, J.M., Ankarcrona, M. Burkitt, M, Sjöholm Å, Ström, K., Orrenius S., and Nicotera, P., 1993, Different proooxidant levels stimulate cell growth, activate apoptosis, or produce necrosis in insulin-secreting RINm5F cells. *J. Biol. Chem. Submitted for publication.*

Elias, L. and Berry, C.O.A., 1991, Induction of differentiation by tumour necrosis factor In HL-60 cells is associated with the formation of large DNA fragments., *Leukemia*, **5**: 879-885.

Ellis, H.M. and Horvitz, H.R., 1986, Genetic control of programmed cell death in the Nematode C. Elegans, *Cell*, **44**: 817-829.

Ericsson, C., Grossbach, U., Björkroth, B. and Daneholt, B., 1990, Presence of histone H1 on an active Balbiani ring gene, *Cell*, **60**: 73-83.

Evan, G.I., Wyllie, A.H., Gilbert, C.S., Littlewood, T.D., Land, H., Brooks, M., Waters, C.M., Penn, L.Z. and Hancock, D.C., 1992, Induction of apoptosis in fibroblasts by c-myc protein, *Cell*, **69**: 19-128.

Fiorentini, C., Donelli, G., Nicotera, P. and Thelestam, M., 1993, *Clostridium difficile* toxin A elicits

Ca^{2+}-independent cytotoxic effects in cultured normal rat intestinal crypt cells, *Infect. Immun.*, **61**: 300-306.
Gaido, M.L. and Cidlowski, J.A., 1991, Identification, purification, and characterization calcium- dependent endonuclease (NUC18) from apoptotic rat thymocytes., *J. Biol. Chem.*, **266**: 18580-18585.
Gradwohl, G., Menissier de Murcia, J.M., Molinete, M. and Simonin, F., 1990, The second zinc-finger domain of poly(ADP-ribose) polymerase determines specificity for single-stranded breaks in DNA, *Proc. Natl. Acad. Sci. USA*, **87**: 2990-2994.
Groux, H., Torpier, G., Monté, D., Mouton, Y., Capron, A. and Amesien, J.C., 1992, Activation-induced death by apoptosis in $CD4^+$ T cells from human immunoficiency virus-infected asymptomatic individuals., *J. Exp. Med.*, **175**: 331-340.
Guo, X. and Cole, R.D., 1989, Chromatin aggregation changes substantially as pH varies within the physiological range., *J. Biol. Chem.*, **264**: 11653-11657.
Hechtenberg, S. and Beyersmann, D. , 1993, Differential control of frealcium and free zinc levels in isolated bovine liver nuclei., *Biochem. J.*, **289**: 757-760.
Helgason, C.D., Shi, L., Greenberg, A.H., Shi, Y., Bromley, P., Cotter, T.C., Green, D.R. and Bleackley, R.C., 1993, DNA fragmentation induced by cytotoxic-T lymphocytes can result in target cell death., *Exp. Cell. Res.*, **206**: 302-310.
Hockenbery, D., Nunez, G., Milliman, C., Schreiber, R.D. and Korsmayer, S.J., 1990, Bcl-2 is an inner mitochondrial membrane protein that blocks programmed cell death., *Nature*, **348**: 334-336.
Inesi, G. and Sagara, Y., 1992, Thapsigargin, a high affinity and global inhibitor of intracellular Ca^{2+} transport ATPases., *Arch. Biochem. Biophys.*, **298**: 313-317.
Jiang, S., Chow, S.C., Nicotera, P. and Orrenius, S., 1993, Intracellular Ca^{2+} signals activate apoptosis in thymocytes. Studies using the Ca^{2+}-ATPase inhibitor, thapsigargin, *J. Biol. Chem. Submitted for publication.*
Jin, Y. and Cole, R.D. , 1986, Exchange of H1 histone depends on aggregation of chromatin, not simply on ionic strength., *J. Biol. Chem.*, **261**: 15805-15812.
Jones, D.P., McConkey, D.J., Nicotera, P. and Orrenius, S. , 1989, Calcium-activated DNA fragmentation in rat liver nuclei, *J. Biol. Chem.*, **264**: 6398-6403.
Juntti-Berggren, L., Larsson, O., Rorsman, P., Ämmälä, C., Bokvist, K., Wåhlander, K., Nicotera, P., Dypbukt, J., Orrenius, S., Hallberg, A. and Berggren, P. , 1993, Increased activity of L-type Ca^{2+} channels exposed to serum from patients with type I diabetes, *Science*, **261**: 86-90.
Kawabata, H., Anzai, N., Masutani, H., Hirama, T., Yoshida, Y. and Okuma, M., 1993, Detection of Mg^{2+}-dependent endonuclease activity In myeloid leukemia cell nuclei capable of producing internucleosomal DNA cleavage, *Biochem. Biophys. Res. Commun.*, **191**: 247-254.
Käs, E., Poljac, L., Adachi, Y. and Laemmli, U.K., 1993, A model for chromatin opening: stimulation of topoisomerase II and restriction enzyme cleavage of chromatin by distamycin, *EMBO J.*, **12**: 115-126.
Kerr, J.F.R., Wyllie, A.H. and Currie, A.R. , 1972, Apoptosis: a basic biological phenomenon with wide ranging implications in tissue kinetics., *Br. J. Cancer*, **26**: 239-257.
Kyprianou, N. and Isaacs, J.T. , 1988, Activation of programmed cell death in the rat ventral prostate after castration., *Endocrinol.*, **122**: 552-562.
McCabe, M.J., Jiang, S. and Orrenius, S., 1993, Chelation of intracellular zinc triggers apoptosis in mature thymocytes., *Lab. Invest.*, **69**: 101-110.
McConkey, D.J., Chow, S.C., Orrenius, S. and Jondal, M., 1990, NK cell-induced cytotoxicity is dependent on a Ca^{2+} increase in the target, *FASEB J.*, **4**: 2661-2664.
McConkey, D.J., Hartzell, P., Duddy, S.K., Håkansson, H. and Orrenius, S., 1988, 2, 3, 7, 8-tetrachlorodibenzo-p-dioxin kills immature thymocytes by Ca^{2+}-mediated endonuclease activation, *Science*, **242**: 256-258.
McConkey, D.J., Hartzell, P., Nicotera, P. and Orrenius, S., 1989, Calcium-activated DNA fragmentation kills immature thymocytes, *FASEB J.*, **3**: 1843-1849.
Morgan, J.I. and Curran, T. , 1986, Role of ion flux in the control of *c-fos* expression., *Nature*, **322**: 552-555.
Nicotera, P., Bellomo, G. and Orrenius S., 1992, Calcium-mediated mechanisms in chemically-induced cell death, *Annu. Rev. Pharmacol. Toxicol.*, **32**: 449-470.
Nicotera, P., McConkey, D.J., Jones, D.P. and Orrenius, S. , 1989, ATP stimulates Ca^{2+} uptake and increases the free Ca^{2+} concentration in isolated rat liver nuclei, *Proc. Natl. Acad. Sci. U. S. A.*, **86**: 453-457.
Nicotera, P., Orrenius, S., Nilsson, T. and Berggren, P.O., 1990, An inositol 1,4,5-trisphosphate-sensitive Ca^{2+} pool in liver nuclei, *Proc. Natl. Acad. Sci. U. S. A.*, **87**: 6858-6862.
Oberhammer, F.A., Pavelka, M., Sharma, S., Tiefenbacher, R., Purchio, A.F., Bursch, W. and Schulte-Hermann, R., 1992, Induction of apoptosis in cultured hepatocytes and in regressing liver by transforming growth factor β 1, *Proc. Natl. Acad. Sci. USA*, **89**: 5408-5412.

Oberhammer, F., Wilso, J.W., Dive, C., Morris, I.D., Hickman J.A., Wakeling, A.E., Walker, P.R., and Sikorska, M., 1993, Apoptotic death in epithelial cells: cleavage of DNA to 300 and/or 50 kb fragments prior to or in the absence of internucleosomal fragmentation, *EMBO J.*, **12**: 3679-3684.

Peitsch, M.C., Polzar, B., Stephan, H., Crompton, T., MacDonald, H.R., Mannherz, H.G. and Tschopp, J., 1993, Characterization of the endogenous deoxyribonuclease involved in nuclear DNA degradation during apoptosis (programmed cell death)., *EMBO J.*, **12**: 371-377.

Pfaffle, P., Gerlach, V., Bunzel, L. and Jackson, V. , 1990, *In Vitro* evidence that transcription-induced stress causes nucleosome dissolution and regeneration., *J. Biol. Chem.*, **265**: 16830-16840.

Planas-Silva, M.D. and Means, A.R. , 1992, Expresson of a constitutive form of calcium/calmodulin dependent protein kinase II leads to arrest of the cell cycle in G_2., *EMBO J.*, **11**: 507-517.

Prütz, W.A., Butler, J. and Land, E.J. , 1990, Interaction of copper(I) with nucleic acids., *Int. J. Radiat. Biol.*, **58**: 215-234.

Rao, L., Debbas, M., Sabbatini, P., Hockenbery, D., Korsmeyer, S. and White, E., 1992, The adenovirus E1A protein induce apoptosis, which is inhibited by the E1B 19-kDa and Bcl-2 proteins, *Proc. Natl. Acad. Sci. USA*, **89**: 7742-7746.

Roy, C., Brown, D.L., Little, J.E., Valentine, B.K., Walker, P.R., Sikorska, M., Leblanc, J. and Chali, N., 1992, The topoisomerase II inhibitor teniposide (VM-26) induces apoptosis in unstimulated mature murine lymphocytes, *Exper. Cell. Res.*, **200**: 416-424.

Samuel, M., Zhu, C.X., Villanueva, G.D. and Tse-Dinh, Y.C., 1993, Effect of zinc removal on the conformation of Escherichia Coli DNA topoisomerase I., *Arch. Biochem. Biophys.*, **300**: 302-308.

Savill, J., Fadok, V., Henson, P. and Haslett, C. , 1993, Phagocyte recognition of cells undergoing apoptosis., *Immunol. Today*, **14**: 131-136.

Schwartz, L.M., Kosz, L. and Kay, B.K., 1990, Gene activation is required for developmentally programmed cell death., *Proc. Natl. Acad. Sci. USA*, **87**: 6594-6598.

Seiliev, A.A., Zvonareva, N.B., Zhivotovsky, B.D. and Hanson, K.P., 1992, Determination of some nuclear deoxyribonucleases in X-irradiated rat thymocytes, *Radiat. Environ. Biophys.*, **31**: 123-132.

Smeyne, R.J., Vendrell, M., Hayward, M., Baker, S.J., Miao, G.G., Schilling, K., Robertson, L.M., Curran, T. and Morgan, J.I., 1993, Continuous *c-fos* expression precedes programmed cell death *in vivo.*, *Nature*, **363**: 166-169.

Smith, C.A., Williams, G.T., Kingston, R., Jenkinson, E.J. and Owen, J.J.T. , 1989, Antibodies to CD3/T-cell receptor complex induce death by apoptosis in immature T cells in thymic cultures., *Nature*, **337**: 181-184.

Snyder, R.D., 1989, Polyamine depletion is asssociated with altered chromatin structure in HeLa cells, *Biochem. J.*, **260**: 697-704.

Thastrup, O., Cullen, P.J., Dröbak, B.., Hanley, M.R. and Dawson, A. 1990, Thapsigargin, a tumor promoter, discharges intracellular Ca^{2+} stores by specific inhibition of the endoplasmic reticulum Ca^{2+}-ATPase., *Proc. Natl. Acad. Sci. USA*, **87**: 2466-2470.

Trauth, B.C., Klas, C., Peters, A.M.J., Matzku, S., Moller, P., Falk, W., Debatin, K. and Krammer, P.H., 1989, Monoclonal antibody-mediated tumor regression by induction of apoptosis., *Science*, **245**:301-305.

Villeponteau, B., Pribyl, T.M., Grant, M.H. and Martinson, H.G., Novobiocin induces the in vivo cleavage of active gene sequences in intact cells, 1986, *J. Biol. Chem.*, **261**: 10359-10365.

Walker, P.R., Smith, C., Youdale, T., Leblave, J., Whitfield, J.F. and Sikoska, M., 1991, Topoisomerase II reactive chemotherapeutic drugs induce apoptosis in thymocytes, *Cancer Res.*, **51**: 1078-1085.

White, B.A. and Bancroft, C. (1987) Cheung, W.Y. (ed.) *Calcium and Cell Functions*, Academic Press, New York, pp. 109-132.

Wu, C., Bingham, P.M., Livak, K.J., Holmgren, R. and Elgin, S.C.R., 1979, The chromatin structure of specific genes: I. Evidence for higher order domains of defined DNA sequence, *Cell*, **16**: 797-806.

Wyllie, A.H., 1980, Glucocorticoid-induced thymocyte apoptosis is associated with endogenous endonuclease activation, *Nature*, **284**: 555-556.

Wyllie, A.H., Kerr, J.F.R. and Currie, A.R. , 1980, Cell death: The significance of apoptosis, *Int. Rev. Ctyol.*, **68**: 251-305.

Yonish-Rouach, E., Resnitzky, D., Lotem, J., Sachs, L., Kinchi, A. and Oren, M., 1991, Wild-type p53 induces apoptosis of myeloid leukaemic cells that is inhibited by interleukin-6, *Nature* , **352**: 345-347.

Zhivotovsky, B., Cedervall, B., Nicotera, P. and Orrenius, S., 1993a, Ca^{2+}-dependent formation of large molecular weight DNA fragments in thymocyte apoptosis. *Submitted for publication.*

Zhivotovsky, B., Nicotera, P., Bellomo, G., Hanson, K. and Orrenius, S., 1993b, Ca^{2+} and endonuclease activation in radiation-induced lymphoid cell death., *Exp. Cell. Res*, **207**: 163-170.

Zhivotovsky, B.D., Zvonareva, N.B. and Hanson, K.P., 1981, Characteristics of rat thymus chromatin degradation products after whole-body X irradiation, *Int. J. Radiat. Biol.*, **39**: 437-440.

Discussion

G. Lenaz

You have shown a Ca^{2+}/H^+ antiporter. Is this speculation or is it known? Also I would like to know, how do you think that ion gradients can be maintained in the nuclei where there are pores that should collapse the gradient over time?

P. Nicotera

The first question: We suppose it is a Ca^{2+}/H^+ antiporter. We do not yet have firm evidence for that. We know that when we add calcium in the presence of ATP, we see an intranuclear pH increase. That does not occur in the presence of calcium alone. As for the pore story, there has been a large debate since the publications of our data and there is now plenty of evidence that the nucleus can sequester calcium. The pores are not necessarily permeable and do not let all molecules or ions through freely. The last idea of the pore structure is that they are comprised of a central pore which is used for passage of large molecules and eight small channels around the central pore for the passage of ions. These channels were supposed to let ions through. However, there is recent evidence from a number of laboratories and old evidences from the sixties that the ion composition in the nucleus differs completely from the one in the cytoplasm and that nuclei have a membrane potential, suggesting that there are electrogenic pumps. In different experimental conditions, *in situ*, it is possible to observe variation in calcium sequestration within the nucleus at least during agonist stimulation. So a permeability barrier likely exists. Perhaps it is the opening or the closing of the ion pores, we do not yet know.

J. Ashwell

I want to clarify one point. You showed that methylprednisolone caused DNA fragmentation in isolated nuclei. Did you isolate the nuclei first and then treat them, or did you isolate nuclei from treated cells?

P. Nicotera

I am sorry I was probably not clear enough. With isolated nuclei we used only the calcium uptake model. Perhaps you are referring to the confocal microscopy images. Those experiments were performed in thymocytes. Cells are treated and then are permeabilized and stained with propidium iodide, at different times.

J. Ashwell

Another question concerns the calcium flux in nuclei. Have you dissected how much of that is extra nuclear calcium flux versus stores being released during agonist signaling?

P. Nicotera

We have not done very much in this respect but other groups have. For example, it has been observed in neural cells that caffeine increases nuclear calcium more than cytoplasmic calcium. Experiments in our laboratory have also shown that depolarization of superior cervical ganglion cells produces a selective calcium increase in the nucleus. There is work by a number of groups who have shown, that during agonist stimulation, alpha-1 agonist stimulation, you can have selective calcium increases in the nucleus. The whole issue began with Fain and Roger Tsien in 1988, when they published in Science that during muscle contraction the nuclear calcium would increase to a different extent than that in the cytoplasm.

G. Evan

Two questions: First, you showed this scheme where mild oxidative stress induced proliferation. How do you measure proliferation?

P. Nicotera

We measure it by colony forming efficiency and thymidine incorporation.

G. Evan

The second point is that Mike Jacobson in Martin Raff's laboratory has shown, I think pretty convincingly, that if you enucleate cells then they are quite capable of initiating everything that looks like apoptosis. That is, you treat them with staurosporin, kinase inhibitors, they show the whole process without any nucleus. How would that tie in with your results?

P. Nicotera

What do you think is apoptosis without the nucleus?

G. Evan

All I know from his experiments is that it looks like apoptosis except the cells have no nucleus. It is blocked by *bcl*-2 expression.

P. Nicotera

So, do you mean cell lysis or morphological changes of the cytoplasm?

G. Evan

There is no cell lysis. There is no release of intracellular components. It is exactly like apoptosis, there is just no nucleus.

P. Nicotera

What I can answer is that at least in the conditions where we keep the nucleus inside the cell, then our observations seem to apply. I do not know in other experimental systems. I am not trying to say that you need a nuclear activation of apoptosis. I think the events in the nucleus may very well be late. What our data show is that calcium or other signals can affect at least two different nuclear processes. One is chromatin conformation and eventually gene expression, and the other is endonuclease activation. However, I did not say that endonuclease activation is a causative event. Calcium can be involved in both apoptotic body formation and in the endonuclease activation. Should we discuss whether this is apoptosis or not, I think we are going to go into the largest debate in this field.

G. Evan

Sure, but it was an inference that I got from your talk that by blocking the DNA fragmentation you could block apoptosis.

P. Nicotera

Yes, if you block DNA fragmentation in thymocytes under those experimental conditions, you block apoptosis.

R. Schimke

Well, do you block cell death?

P. Nicotera

Yes, in these systems.

R. Schimke

Which kind of end point do you use to assess that cells are protected, dye uptake or other conditions?

P. Nicotera

Depending on the system you use, you have different end points. In the thymocytes, which are not dividing cells, we use trypan blue uptake. So when we treat thymocytes with thapsigargin in calcium free conditions, we prevent DNA fragmentation, apoptotic body formation, and prevent cell death as judged by trypan blue uptake. The same end point was used for the experiments with polyamines. Cell death is measured by trypan blue uptake. Now, calcium free conditions cannot be maintained for too long because Ca^{2+} withdrawal becomes toxic. So if you look at very late time points, you will start to see cells dying. However, in this case they do not die by apoptosis. The cell proliferation data which I showed at the very end of my talk are, of course, a different story. There we measure the ability of the cells to grow, and, as makers for apoptosis, the ability of the cells to form apoptotic bodies and to give ladders. We also used dye uptake and there was no dye uptake with 30 μM DMNQ for up to 24 hrs.

E. Mihich

Coming back to the polyamines for a moment, I have a question about the specificity of the effect. You did show differences in quantity between putrescine, spermine, and spermidine but, for a long time in the past, it was difficult to ascribe specificity of action to polyamines as some of their effects could be reproduced by magnesium for example. I am wondering: 1) how magnesium works in your system? and 2) do you have polyamines where their amino groups are blocked, for instance, methylated and do you see similar events or are those amino groups required?

P. Nicotera

No, we have not looked at magnesium effects. We have obtained some of these substituted polyamines and so we are going to try them but we have not done that yet. The specificity of effect, for spermine, spermidine, putrescine, is observed with two end point which are DNA fragmentation and trypan blue uptake for what concerns the thymocytes. We also correlated the ability of the three polyamines to produce chromatin compaction in isolated nuclei or permeabilized cells. And in these experiments you can see a clear difference of effect of the three of them, spermine being the most effective.

G. Bellomo

I would like to give an additional answer about the effects of magnesium and polyamines on chromatin conformation. You can still observe a clear spermine-induced chromatin compacting in nuclei incubated in media containing saturating concentrations of magnesium. So, I believe that the mechanisms underlying magnesium and spermine effects are different.

D. Green

You are probably aware of the paper by Makoto Iwata describing inhibition of glucocorticoid-induced apoptosis by ionophore by actually elevating the calcium level. How can that fit into this scheme?

P. Nicotera

We have actually similar results with other systems, for example, with cold shock-induced apoptosis. You asked me to give you an explanation for this. I probably will not be able to give you an explanation because we do not yet know. I would only speculate. On the other hand, the role of calcium in triggering apoptosis has been perhaps over emphasized. Calcium increases are one possible signal for apoptosis. We at least know it is a signal in certain experimental model systems. We know calcium is required but it is not a signal under other conditions. I will give you a very simple example with the same system: If you irradiate thymocytes or splenocytes with about 6 to 10 Gy, both cell types will undergo apoptosis. We also know that we have a transient calcium increase. This data are now in press in *Experimental Cell Research*. The calcium elevation is transient and occurs about 30 minutes after the exposure to irradiation. On the other hand, in the splenocytes we do not observe calcium increases and yet cells will die by apoptosis as well as the immature thymocytes. However, the resting calcium level of the splenocytes is about 3 times higher than that of thymocytes so presumably the resting level is sufficient to allow progression of apoptosis. Calcium chelation protects, however, from DNA fragmentation, apoptotic body formation and cell viability in both cell types. I think perhaps Dr. Bellomo has a better answer for your initial question because he has been working with a model system where protection of apoptosis by calcium can be observed.

G. Bellomo

What people generally believe is that the addition of calcium ionophores, such as A23187 or ionomycin, to cell suspensions or cultures invariably results in a marked and persistent increase in cytosolic Ca^{2+} due to the influx of the ion from the extracellular medium. This is not always true and mainly depends on the concentration of the ionophore employed. With low concentrations you exclusively obtain a depletion of intracellular Ca^{2+} pools. We have recently investigated apoptosis occurring in cultured human synovial cells exposed to "cold shock and rewarming". DNA fragmentation that takes place in these conditions is entirely Ca^{2+}-dependent since it is completely prevented by intracellular or extracellular Ca^{2+} chelators or by calmodulin inhibitors. Treatment of these cells with low concentrations of A23187 or ionomycin not only does not induce any appreciable DNA fragmentation, but completely inhibits apoptosis triggered by cold shock and rewarming. The reason for this effect is linked to the depletion of an intracellular Ca^{2+} pool (induced by ionophore treatment) that is essential for causing DNA fragmentation and apoptosis.

P. Nicotera

If I may add just a comment to this, the reason that induced us to use thapsigargin is to prevent the effects observed with A23187. When you use ionophores, depending on the gradients you have in the cells, you drive calcium from the highest concentration to the lowest so if we have any increase in calcium in the nucleus we drive it out. Thapsigargin has another advantage over the ionophores: It does not change intracellular pH. As you know, pH changes may be involved in apoptosis. Thapsigargin does not do that. Finally, thapsigargin-induced apoptosis can be blocked by chelating calcium at different time points. We know that you need about 90 minutes of sustained calcium increase to have apoptosis in that system and if you remove Ca^{2+} before, you prevent it. If Ca^{2+} is chelated beyond 90 min, you do not rescue the cells. Incidentally, thapsigargin is a good inhibitor of protein synthesis.

R. Schimke

The classical description of chromatin structure and apoptosis is one of chromatin condensation. Yet, in your studies, you describe the status of chromatin subject to degradation as one of chromatin decompaction. How do you reconcile these different descriptions of the nature of chromatin?

P. Nicotera

The word condensation has been used primarily from electron-microscopy studies. What we see with fluorescent dyes is not condensation, it is an aggregation; it is a different thing. The accessibility to a fluorescent chromatin dye, depends on the degree of chromatin, particularly when you are using intercalating dyes like propidium iodide. As you saw from some of the confocal pictures I showed, the fluorescence was markedly increased in all those conditions. Should the chromatin be compacted, then the dye would not access it. So it is not a compaction in the very specific sense of the word. I would call it rather an aggregation, based on this data. In terms of what happens in the early stages, yes, the ladders we showed there are clearly an end-point. We have very preliminary data that indicate the possible initial detachment of chromatin loops by endonuclease or topoisomerase II activation. Then you see the loop degradation that may as well be due to a number of DNAses. The overall mechanism, it is clearly calcium dependent, at least in thymocytes. I do believe that the early stages, like we see in many cases, are probably due to a signal for chromatin instability. As you know very well, mitosis is the condition during the cell cycle where you have the highest number of double strand breaks. Now if the right signals are not delivered, and I think we will hear more of this later on, then you may have apoptosis. For resting cells, and this is entirely an hypothesis, I see it as a disturbance in transcription. An initial opening or unfolding of a specific region of the chromatin, may probably occur close to the attachment of the loops to the scaffolding. This normally provides transcriptional competence and in the presence of a transcription factor the process will progress. However, if this does not occur, then you may observe H1 redistribution and exposure of DNAse hypersensitive sites. Laemmli has shown that they could be localized about 5kb apart. Such a situation may favor DNA fragmentation.

A. Eastman

Three brief questions. Thank you for bringing up the issue of pH and I will bring it up again in my talk. When you measured the pH in the nuclei, you were measuring in minutes whereas the DNA degradation is occurring at hours. Have you taken the pH measurement out longer?

P. Nicotera

If you expose isolated nuclei to calcium and ATP, fragmentation occurs within 20 minutes. The time frame we are talking about for pH measurements was 10 minutes. The traces I showed reflected short time measurements because the experiments were performed in single nuclei. If you take an entire population then you can see that these increases are sustained, actually for quite long, probably because we do not have the conditions to activate a rectifier in this system. So surely in cells, changes may occur within a longer time frame, depending on the model system you use.

A. Eastman

The second question relates to the presence of the calcium magnesium endonuclease. We have really only found it in thymocytes and you have used a whole variety of different cells. Did you find it in all the cells you have looked at?

P. Nicotera

It is highly expressed in liver nuclei.

A. Eastman

Also in the insulinoma and pancreatic cells?

P. Nicotera

That is correct.

A. Eastman

The third question really is more of a comment. You showed a correlation with PKC but PKC activation does many things. This poses the question as to whether this may be through ODC or some other pathway.

P. Nicotera

Yes, it can very well be through other pathways. I introduced PKC in relation to ODC because we have a specific experimental system where the two signals are associated. The pathways whereby PKC can protect in some systems or stimulate in others may be different. Again, it is a matter of balanced signaling. I think the point is not to focus on a single signal but to understand where the imbalance may be.

W. Giaretti

Resting G0 lymphocyte have a characteristic morphology similar to that of heterochromatin and have very little gene activity. Cleavage of this DNA would certainly be difficult in such a condensed chromatin state. If I understood you correctly, you said that a prerequisite of apoptosis is chromatin decondensation to allow a calcium-dependent endonuclease cleavage of DNA. I assume, then, that we never should observe apoptosis in resting G0 cells like unstimulated lymphocytes. On the contrary, we should observe apoptosis in stimulated lymphocytes. In fact, the model of lymphocyte transition from G0 to G1 is a good example of chromatin decondensation associated with a high influx of Ca^{2+} into the resting cells. Are there evidences that these predictions are true? Did people studying cell-cycle dependent apoptosis find this? Are there really findings that condensed G0 cells never go to apoptosis?

P. Nicotera

I would not say so because you have the condensation also in resting cells. Once you have chromatin decondensation, you may not even need a calcium-dependence in all cases. You can have a magnesium nuclease, as a matter of fact, which would have the same effect. Did I answer your question?

G. Guidotti

You have partially answered this question already saying that more than a signal transduction pathway can be involved in apoptosis. It has been reported that the effects of TNF-α can be mediated by sphingomyelin hydrolysis and ceramide generation. C_2-ceramide that enters the cell induces apoptosis in hours. Did you explore this transduction pathway in your model?

P. Nicotera

We have not tested this signaling pathway so C_2-ceramide may as well be a mediator of TNFα induced toxicity. We do not know what in the TNFα system mediates the selective increase in nuclear calcium and we do not know the role of plasma membrane signaling in this system. We observe that, like many others, TNF-induced apoptosis occurs only if you block protein synthesis. And the reason is that this may likely stop the induction of proteins which normally protect cells from TNFα induced toxicity. One of these is superoxide dismutase.

M. D'Incalci

Do you know of any study in which it has been looked at correlational between apoptosis and post translational modification of histone proteins, for example, phosphoration, acetylation, or poly ADP ribosylation of these proteins?

P. Nicotera

I think it is an issue which needs to be investigated. There is a number of modifications of the H1 histone which would affect chromatin structure. We have reported, in 1988, that if you do have increases in calcium and concomitantly poly-ADP-ribosylation, you favor DNA fragmentation in isolated nuclei. However, Samuil Umanski had some evidence a few years ago that poly-ADP-ribosylation in his experimental system would inhibit endonuclease-mediated DNA fragmentation. Again, we are talking about a very selective end point. I do not know how modification in H1 caused by ADP-ribosylation would result in apoptosis because, again, DNA fragmentation is not necessarily the only end-point in apoptosis.

MOLECULAR CONTROLS OF CELL CYCLE PROGRESSION FOLLOWING DNA DAMAGE: ROLES OF P53 AND ATAXIA-TELANGIECTASIA GENE PRODUCTS

Michael B. Kastan

The Johns Hopkins Oncology Center
Baltimore, Md. 21287

INTRODUCTION

It has been suggested that exposure to environmental DNA damaging agents contributes to the development of the vast majority of human tumors[1]. Therefore, an understanding of the molecular events involved in the cellular responses to such exposures should provide insights into mechanisms of human carcinogenesis. Much effort over the past 25 years has been focused on how altered nucleotide bases are removed and how the linear integrity of the DNA sequence is restored[2]. More recently, it has become clear that the timing of these repair processes relative to various critical cellular processes, such as DNA replication, may be important determinants for dictating the cellular consequences of the DNA damage. For example, a cell which continues to replicate its DNA prior to repairing lesions in the DNA may be more likely to result in daughter cells which have altered genetic information relative to the parental cell. It would be predicted that such altered daughter cells would be more likely to develop the genetic changes which contribute to a transformed phenotype (Figure 1).

There are many experimental examples of alterations in the progression of proliferating cells through the cell cycle following exposure to DNA damaging agents[3-8]. Such cell cycle "checkpoints" appear to be active cellular responses which permit optimal repair of DNA damage so that the cell will not replicate a damaged DNA template (G_1 and S-phase arrests) nor segregate damaged chromosomes (G_2 and M-phase arrests). The first DNA-damage responsive cell cycle checkpoint characterized at the molecular level was the RAD9 gene in yeast; abnormalities in the RAD9 gene result in a defect in the G_2 arrest following ionizing irradiation - such mutant yeast exhibit increased sensitivity and increased genetic abnormalities following exposure to ionizing radiation[6,9]. Hartwell and Weinert suggested four characteristics of cell cycle checkpoint gene products: 1) non-essential; 2) inhibitory; 3) feeds forward; and 4) transduces a signal. However, because of the biologic characteristics of yeast which

Apoptosis, Edited by E. Mihich and R.T. Schimke
Plenum Press, New York, 1994

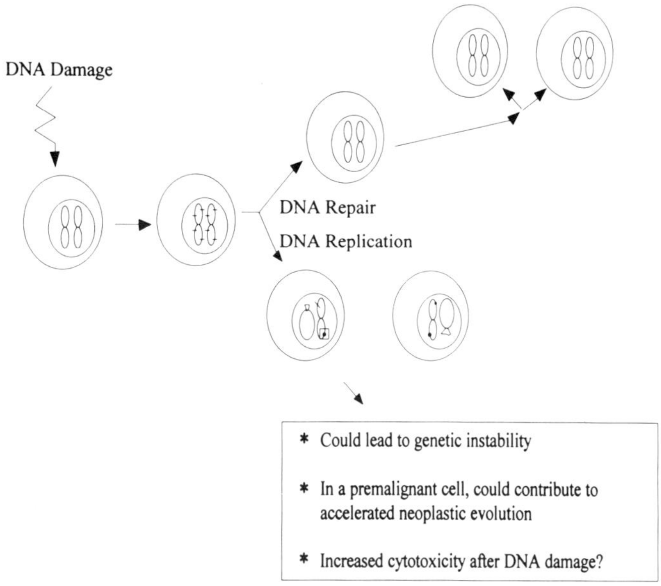

Figure 1. Schematic representation of alternative cellular responses to DNA damage. After DNA damage, cells face an important "decision fork" - whether to repair the DNA damage prior to progression through the cell cycle or whether to arrest in the cell cycle in order to repair the damage before replicating a damaged DNA template or segregating damaged chromosomes. Potential consequences of the lack of cell cycle arrests after damage include increased genetic abnormalities in daughter cells, potentially contributing to the genetic changes leading to a transformed phenotype, and increased sensitivity to the DNA damage. (Reproduced with permission from Environ. Health Persp., in press).

had lost RAD9 gene function, it has generally been thought that the physiologic consequences of defects in these checkpoints would include decreased cell survival and increased propagatable genetic abnormalities following DNA damage. As discussed below, however, we already know that cell survival following DNA damage is not always affected by the loss of cell cycle checkpoint function. We recently began to investigate some of the molecular mechanisms which control cell cycle progression following DNA damage in mammalian cells. These investigations have begun to elucidate a novel signal transduction pathway which responds to certain types of DNA damage, is critically dependent on the normal functioning of the p53 tumor suppressor gene, the most commonly mutated gene in human cancer, and appears to influence both the stability of the genome and cell survival signals, including radiation-induced apoptosis, in certain situations.

RESULTS AND DISCUSSION

p53 is a G1 cell cycle checkpoint determinant: In agreement with observations from many other laboratories, we found that exposure of mammalian cells to ionizing radiation (IR) can result in transient arrests of the cells at both G_1 and G_2 phases of the cell cycle. While characterizing molecular events which might control such arrests, we observed that levels of the "tumor suppressor" nuclear protein, p53, increased in temporal association with the decrease in replicative DNA synthesis[10] (Figure 2). Levels of p53 mRNA did not change following IR and increases in p53 protein occurred despite the use of actinomycin D (at transcription-inhibiting doses) as the damaging agent; these observations demonstrated that the increase in p53 protein following irradiation was due to post-transcriptional changes. This made teleologic sense - transcription utilizing a damaged DNA template would be almost as undesirable as replication using a damaged template. Since p53 is the most commonly mutated gene in human malignancies[11,12] and since environmental DNA damaging agents have been implicated in the development of up to 80% of human cancers[1], linking p53 function to the cellular response to DNA damage had significant implications in terms of understanding mechanisms of human carcinogenesis.

A functional role for the p53 gene product in the cellular response to certain types of DNA damage was strongly suggested by the observation that cells with wild-type p53 genes arrested in G_1 following IR, while cells with abnormal p53 genes lacked this checkpoint[10,13]. Since all cells evaluated, irrespective of p53 status, retained the G_2 arrest checkpoint, it appeared that p53 protein was involved in the G_1, but not in the G_2, arrest following ionizing radiation. A cause and effect relationship between p53 function and the arrest in G_1 following IR was then established by demonstrating: 1) acquisition of the G_1 arrest after ionizing radiation following transfection of wild-type p53 genes into cells lacking endogenous p53 genes[13]; 2) loss of the G_1 arrest after irradiation following transfection of mutant p53 genes into cells with wild-type endogenous p53 genes[13] (Figure 3); and 3) loss of the G_1 arrest in normal cells obtained from mice in which both p53 alleles were disrupted by homologous recombination[14].

Exposure of cells to caffeine potentiates the cytotoxicity of DNA damaging agents[3]. The mechanism of this potentiation appears to be at least partially due to inhibition of the cell cycle arrests[5,15]. We observed that both the G_1 and G_2 arrests following ionizing irradiation were abrogated by the presence of caffeine and that the

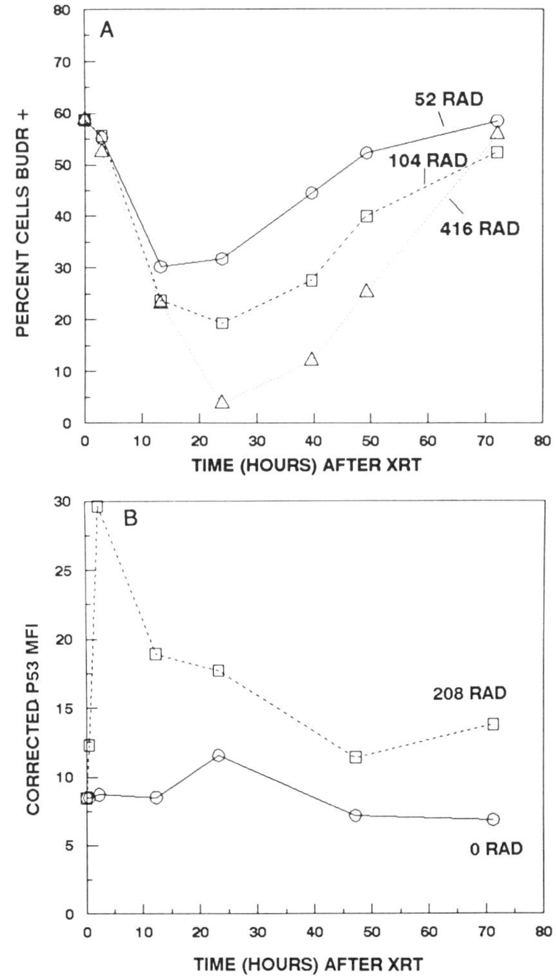

Figure 2. Time courses of changes in replicative DNA synthesis and p53 protein levels following exposure of cells to ionizing radiation. A. DNA synthesis was assessed by flow cytometric analysis of BrdUrd (BUDR) incorporation during a 4-hour pulse at various times after exposure to 52, 104 or 416 cGy ionizing radiation. B. The corrected mean fluorescence intensity (MFI) for p53 protein staining at various times following exposure to 208 cGy radiation. (Similar data is obtained when p53 protein levels are assessed by immunoblot or immunoprecipitation; data not shown). (Reproduced with permission from reference 10).

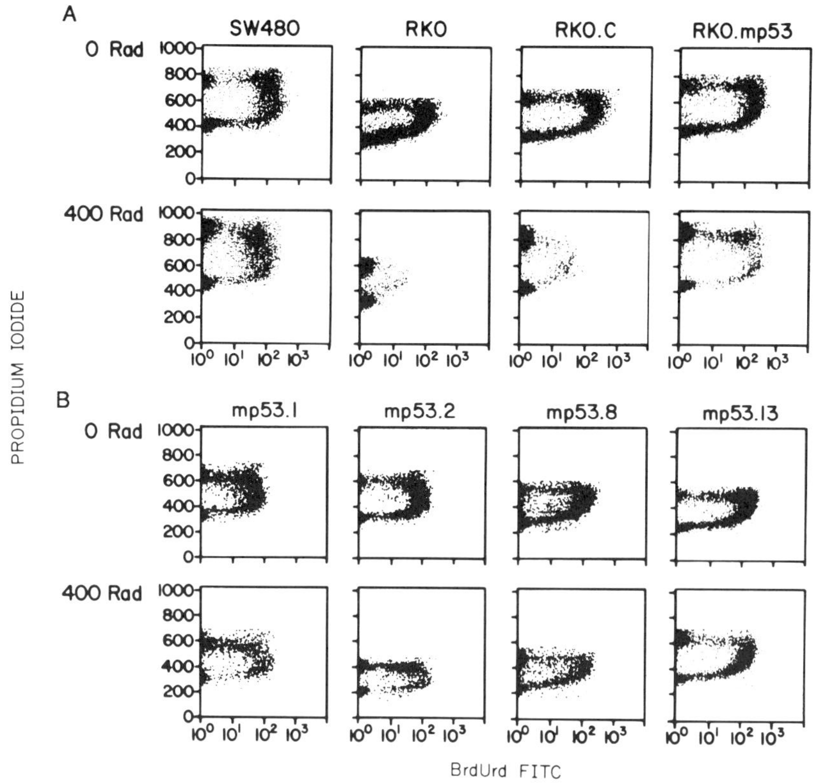

Figure 3. Cell cycle progression as a function of p53 status after irradiation in RKO colorectal carcinoma cells. A. Seventeen hours after exposure to 0 cGy (upper panels) or 400 cGy (lower panels) IR, SW480 colorectal carcinoma cells (endogenous mutant p53), parental RKO cells (wtp53), a polyclonal population of RKO cells transfected with a control vector (RKO.C), and a polyclonal population of RKO cells transfected with a vector containing a mp53 gene (RKO.mp53), were pulsed with BrdUrd for 4 hr and analyzed simultaneously for DNA synthesis (BrdUrd FITC) and DNA content (PI). B. Colonies of cells from RKO.mp53 were isolated by limiting dilution and screened for increased p53 protein expression by flow cytometry. The cell cycle status of selected clones was analyzed as described above 17 hr after exposure to 0 cGy (upper panels) or 400 cGy (lower panels) IR. Examples of four of these clones are shown. (Reprinted with permission from reference 13).

increase in p53 protein levels was also blocked[10]. Inhibition of protein synthesis by cycloheximide treatment blocked the G_1, but not the G_2, arrest and also abrogated the increase in p53 protein levels following IR. These results with caffeine and cycloheximide demonstrated that the G_1 arrest after IR is an active physiologic response to the DNA damage and does not result simply from a structural blockage of DNA synthesis. Thus, it appears that the cell "senses" abnormalities in the DNA, initiates a signal transduction pathway which results in a post-transcriptional increase in the levels of p53 protein, which subsequently causes cells to arrest in the G_1 phase of the cell cycle. The arrest in G_2 may result from the same signal in the DNA, but appears independent of p53 function.

Characterization of the molecular mechanisms of the induction of p53 protein and the mechanism by which this induction causes the G_1 arrest following IR are intense areas of investigation at this time. The inhibition of the p53 induction by caffeine provides one insight into these mechanisms - perhaps a cyclic nucleotide-dependent process is required for p53 induction following IR. Poly(ADP-ribose) synthesis following strand breaks has been implicated as a critical participant in the DNA repair process[16]. However, prior reports suggested that 3-aminobenzamide, an inhibitor of poly(ADP-ribos)ylation, does not affect the radiation-induced inhibition of DNA synthesis in human cells[17,18]. In agreement with this observation, we observed no alterations in cell cycle checkpoints or in p53 induction following IR when cells were treated with 3-aminobenzamide (Kastan and Plunkett, unpublished observations).

The A-T, GADD45 and MDM2 genes are involved in this signal transduction pathway: Ataxia-telangiectasia (A-T) is a human autosomal recessive disorder with many phenotypic characteristics, including: 1) hypersensitivity to ionizing radiation; 2) "radioresistant" DNA synthesis; and 3) a markedly increased incidence of cancer (for reviews, see references 19,20). Painter and Young[4] were the first to report an abnormality in the ability of A-T cells to cease replicative DNA synthesis following ionizing radiation. Several other laboratories reproduced this observation - interestingly, the cell cycle checkpoint defect in A-T cells[21] was reminiscent of the abnormality we were observing in cells with defective p53 function[10,13] (see Figure 3). Therefore, a link between the defect in A-T and the signal transduction pathway which utilizes p53 in causing the G_1 arrest following IR was investigated. We reasoned that if this A-T phenotypic characteristic were due to a defect proximal to p53 in this response pathway, A-T cells would lack the induction of p53 protein that is seen in normal cells following irradiation.

Therefore, the ability of a number of cell lines from A-T patients to increase p53 protein levels following IR was evaluated. Assessments of p53 protein levels 1 hour following irradiation by immunoprecipitation, immunoblot, and flow cytometry all demonstrated a deficiency in the induction of p53 protein levels in cells from A-T patients[14]. This defect was noted in both EBV-immortalized lymphocytes and in normal diploid fibroblasts from A-T patients. Though the complementation groups of some of the A-T cell lines utilized in these studies was not known, cell lines from patients which had been previously characterized as group A and group C were tested and were defective in the induction of p53 protein following IR. There was a suggestion, however, that the defect in cells from group C patients may not be as quantitatively severe as that seen in cells from group A patients. Since 83% of all A-T patients fall into complementation groups A or C[22], it appears that the vast majority of A-T patients are defective in this p53-dependent response to IR. More recent data from other laboratories have extended these observations to suggest that A-T cells are capable of increasing p53 protein following IR, but the induction is delayed and the level of induction appears quantitatively less than that seen in cells from normal individuals.

A gene distal to p53 in this signal transduction response pathway was also identified - transcription of the *GADD45* gene (a *g*rowth *a*rrest and *D*NA *d*amage-inducible gene) is enhanced by IR only in cells with wild-type p53 function[14]. A binding site for p53 was identified in the *GADD45* gene and wild-type, but not mutant, p53 protein binds to this site. This binding site can also activate transcription of a reporter CAT gene in the presence of wild-type, but not mutant, p53. Consistent with the observation mentioned above that A-T cells are defective in the induction of p53 protein following IR, Fornace and collaborators had previously shown that A-T cells are quantitatively defective in their ability to induce *GADD45* transcription following IR[23]. These experiments thus suggest the following steps in a signal transduction pathway which controls the arrest of cells at the G_1/S border following ionizing irradiation: following IR-type damage, the cell "recognizes" this damage and increases p53 protein levels via a post-transcriptional mechanism; optimal induction of p53 protein is dependent on the A-T gene product(s) at some step. Subsequently, this increased/altered p53 protein functions as a transcription factor to up-regulate the expression of *GADD45* and possibly other effector genes, which triggers the arrest (Figure 4). Further characterization of the various steps in this pathway, including elucidation of the signal in the DNA which initiates this pathway, is actively continuing.

Recent investigations have also demonstrated that the mdm2 oncogene is a participant in this signal transduction pathway. mdm2 was first identified as an oncogene in murine transformation system[24] and more recently the mdm2 protein has been shown to bind to p53 protein[25]. Certain human tumors appear to utilize amplification of the mdm2 gene as another mechanism of inactivating p53 function (rather than requiring p53 mutation)[26]. mdm2 appears to block p53 function (transcriptional activation of a reporter gene containing a p53 binding site) *in vitro*. In terms of the physiologic pathway in which we are suggesting p53 participates, this observation would predict that overexpression of mdm2 in cells would result in loss of the IR-induced induction of GADD45 and loss of the IR-mediated G1 arrest. These effects were noted: cells which overexpress mdm2 are defective both in their ability to induce GADD45 and to arrest in G1 following IR (Chen et al., manuscript submitted). Thus another way for cells to lose this checkpoint is to overexpress the mdm2 gene. Data from several laboratories have also suggested that expression of wild-type p53 results in increased transcription of the mdm2 gene[27,28]. Therefore, we asked whether exposure of cells to ionizing radiation resulted in increased levels of mdm2 expression and whether this induction was dependent on normal p53 function. Both of these predictions were also confirmed - mdm2 mRNA and protein levels increase in cells following IR, delayed from p53 induction by about 2 hours; no mdm2 induction by IR is noted in cells with abnormal p53 genes (Chen et al., manuscript submitted). These observations all taken together suggest a scenario in which ionzing radiation results in increased p53 protein levels, p53 subsequently induces increased mdm2 expression and mdm2 may then limit how long p53 mediates the growth arrest signal. This "feedback loop" hypothesis is currently being tested.

Physiologic consequences of loss of this cell cycle checkpoint function

Linkage of p53 and the A-T gene(s) in this pathway suggest that some previous observations about p53 and A-T and the development of human tumors are more than just coincidence. Lymphoid malignancies are the most common (or the first) tumors seen in both A-T patients[29,30] and mice in which both p53 alleles have been disrupted[31,32] - these are both situations in which the gene defect is *homozygous*. Interestingly, recent data from our lab suggets that the signal in the DNA which initiates this pathway is probably a DNA strand break (Nelson and Kastan, manuscript submitted). Since lymphoid cells develop a number of strand breaks during developments as they rearrange TCR or Ig genes, they may be particularly susceptible to developing genetic abnormalities (and thus to tumorigenesis) if this signalling pathway is not functioning optimally, as would be the case in cells with either homozygous p53 or A-T mutations.

In contrast to the homozygous situation, the risk of breast (and other non-lymphoid) tumors is markedly increased in situations in which individuals inherit defects in *only one allele* of either the p53 gene or the A-T gene(s)-Li-Fraumeni patients inherit an abnormality of one p53 allele and have an increased risk of breast, brain, lymphoid and other tumors[33,34] and A-T heterozygotes appear to have a significantly increased risk of developing breast cancers[35,36]. Interestingly, the increased incidence of breast cancer in A-T heterozygotes has been suggested to be related to exposure to low doses of IR[36]. The reasons for the

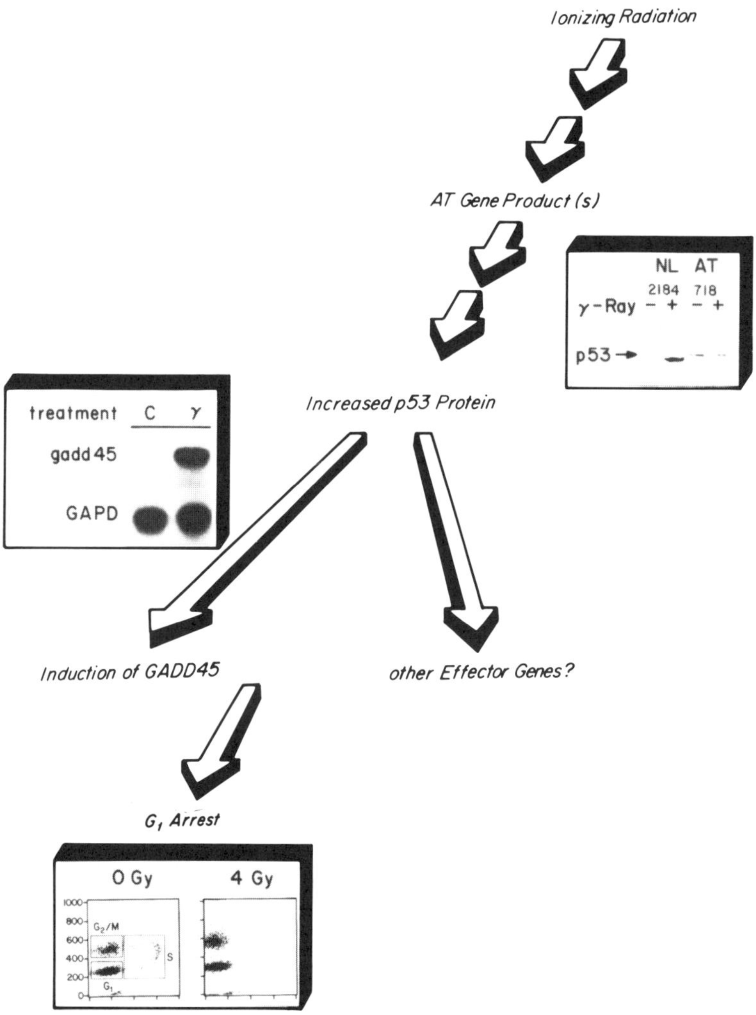

Figure 4. Schematic representation of IR-induced cell cycle checkpoint pathway in mammalian cells. Exposure of cells to IR results in an increase in p53 protein levels; optimal induction of p53 protein is dependent on normal function of the gene(s) defective in AT. The increase in p53 protein levels results in enhancement of GADD45 transcription and in an inhibition of progression of cells from the G1 into the S phase of the cell cycle. The role of the GADD45 gene product in mediating the G1 arrest is not yet clear.

development of these particular tumors with these particular gene defects remain to be elucidated.

As mentioned above, one predicted physiologic consequence of losing cell cycle checkpoints is the increased risk of developing genetic abnormalities following exposure to DNA damaging agents. Thus, defects at any step in this p53-dependent signal transduction pathway is likely to increase the chance that daughter cells will develop heritable genetic abnormalities. Such increases in genetic instability following IR have been observed in yeast with RAD9 mutations, which lack a G_2 checkpoint[9,37]. The observations that cells from A-T patients have increases in genetic abnormalities following IR[38,39] are consistent with these concepts and provide a direct example of genetic instability resulting from defects in this pathway. Finally, in mammalian cells, at least two types of genetic instability, gene amplification and aneuploidy, have already been correlated with loss of p53 function[40-42]. This model thus suggests that the high frequency with which tumor cells exhibit both p53 gene abnormalities[11] and gross chromosomal changes[43] is causally related through abnormalities in this response pathway to DNA damage.

Though abnormalities in this pathway are likely to be significant contributors to the development of malignancies, it appears that defects in this mammalian G_1 checkpoint do not lead to increased sensitivity to ionizing radiation. The observations that the G_1 arrest can be dissociated from the radiosensitivity of A-T cells during genetic manipulations[44,45] suggest that the radiosensitivity of AT cells is not dependent on loss of this checkpoint and may be due to some other manifestation of the deficiency in A-T. In light of the link established between p53 and the A-T gene product(s), the observation that SV40-transformed cells from normal individuals (which should have nonfunctional p53 gene products due to binding of SV40 T antigen to p53 protein) continue to be more radioresistant than SV40-transformed cells from A-T patients[46,47], would further support the notion of a dissociation between the G_1 checkpoint and cellular sensitivity to ionizing radiation since normal p53 function is required for the G_1 arrest. Finally, the observation that the failure to block replicon initiation in *S-phase* cells following irradiation also contributes to the radioresistant DNA synthesis in A-T cells provides another difference between A-T cells and cells with mutant p53 genes.

These observations were recently confirmed more directly by evaluating radiosensitivity in mammalian cells in which p53 function was disrupted by mutation, p53-inhibiting viral gene product overexpression, or p53 gene deletion. We found no decreases in clonogenic survival of fibroblasts or colorectal tumor cells following ionizing radiation when p53 and this checkpoint were disrupted[52]. Thus, it is feasible that the A-T genes function at a relatively "proximal" step in the cellular response to ionizing irradiation and leads to optimal p53 induction and the G_1 arrest by one pathway (which predisposes cells to genetic instability and tumor development when abnormal) and leads to increased cell death (i.e. "radiosensitivity") by a separate pathway when abnormal.

In terms of the subject matter of interest at this conference, as discussed in more detail by others at this conference, overexpression of p53 can lead to the induction of apoptotic cell death in certain situations[48]. More recently, it has been shown that ionizing radiation-induced apoptosis (but not other types of apoptosis) in thymocytes[49,50] and probably all hematopoietic cells[51] is dependent on p53. This is actually in agreement with our observation that loss of p53 function does not lead to increased radiosensitivity; in fact, these observations suggest that in certain cell types (e.g. hematopoietic cells), loss of p53 function can lead to relative radioresistance because of loss of an apoptosis signal. Cells in which IR does not lead to apoptosis do not have this part of the pathway and thus do not exhibit alterations in clonogenic survival following irradiation. (It should be pointed out, however, that loss of this checkpoint function may lead to alterations in cell survival following other types of DNA damage when other lesions are present in the DNA.)

CONCLUSIONS

We have defined several molecular events which control cellular responses to DNA damage. A critical participant in this signalling pathway is the p53 tumor suppressor gene, the most commonly mutated gene thus far identified in human cancers. Intriguingly, data suggests that p53 plays a critical role in both IR-mediated G1 arrest and IR-mediated apoptosis. A likely scenario based on these observations all taken together is that some of the steps in the IR-induced p53-dependent apoptosis pathway overlap with the steps in the p53-dependent cell cycle checkpoint pathway. In some cell types, the IR-induced signal which leads to p53 induction results in apoptosis and in other cell types it leads to a transient G1 arrest. Perhaps those cells which normally undergo apoptosis as part of the differentiation process also respond to IR by undergoing apoptosis, possibly because DNA strand breaks are the common lesion which signals this programmed cell death and which induces p53. Cells which do not normally undergo apoptosis in response to IR, use the DNA-strand break, p53-mediated signal to arrest in G1. Further characterizations of these pathways is certain to improve our understanding of both how tumors develop and how cells die-such insights should be useful in both cancer prevention and improved therapeutic options for treating tumors.

ACKNOWLEDGEMENTS

This work was supported in part by grants from the NIH (ES 05777) and the Council for Tobacco Research.

REFERENCES

1. R.Doll and R.Peto, The causes of cancer in the United States today, *J.N.C.I.* 66:1192, (1981).
2. P.C. Hanawalt, P.K.Cooper, A.K Ganesan, R.S. Lloyd, C.A.Smith, and M.E.Zolan, Repair responses to DNA damage: enzymatic pathways in E coli and human cells, *J. Cell. Biochem.* 18: 271 (1982).
3. L.J.Tolmach,R.W.Jones, and P.M.Busse, The action of caffeine on X-irradiated HeLa cells. I. Delayed inhibition of DNA synthesis, *Rad. Res.* 71: 653 (1977).
4. R.B.Painter and B.R.Young, Radiosensitivity in ataxia-telangiectasia: a new explanation. *Proc. Natl. Acad. Sci. USA. 77:* 7315 (1980).
5. C.C.Lau and A.B.Pardee, Mechanism by which caffeine potentiates lethality of nitrogen mustard. *Proc. Natl. Acad. Sci. USA 79:* 2942 (1982).
6. T.A.Weinert and L.H.Hartwell, The RAD9 gene controls the cell cycle response to DNA damage in saccharomyces cerevisiae. *Science* 241:317 (1988).
7. W.K.Kaufmann, J.C.Boyer, L.L.Estabrooks, and S.J.Wilson, Inhibition of replicon initiation in human cells following stabilization of topoisomerase-DNA cleavable complexes. *Mol. Cell. Biol.* 11: 3711 (1991).
8. P.M.O'Connor, D.K.Ferris, G.A.White, J.Pines, T.Hunter, D.L.Longo, and K.W.Kohn, Relationships between cdc2 kinase , DNA cross-linking, and cell cycle perturbations induced by nitrogen mustard. *Cell Growth and Diff. 3:* 43 (1992).
9. L.H.Hartwell and T.A.Weinert, Checkpoints: Controls that ensure the order of cell cycle events. *Science* 246:629 (1989).
10. M.B.Kastan, O.Onyekwere, D.Sidransky, B.Vogelstein, and R.W.Craig Participation of p53 protein in the cellular response to DNA damage. *Cancer Res*, 51:6304 (1991).
11. B.Vogelstein, A deadly inheritance. *Nature* 348: 681 (1991).
12. M.Hollstein, D.Sidransky, B.Vogelstein, and C.C.Harris. p53 mutations in human cancers. *Science* 253:49 (1991).

13. S.J.Kuerbitz, B.S.Plunkett, W.V.Walsh, and M.B.Kastan, Wild-type p53 is a cell cycle checkpoint determinant following irradiation. *Proc. Natl. Acad. Sci. USA 89:* 7491 (1992).
14. M.B.Kastan, Q.Zhan, W.S.El-Deiry, F.Carrier, T.Jacks, W.V.Walsh, B.S.Plunkett, B.Vogelstein, and A.J.Fornace,Jr., A mammalian cell cycle checkpoint pathway utilizing p53 and *GADD45* is defective in ataxia telangiectasia. *Cell*, 71:587 (1992).
15. R.A. Walters, L.R.Gurley, and R.A.Tobey, Effects of caffeine on radiation-induced phenomena associated with cell-cycle traverse of mammalian cells. *Biophys. J.* 14:99 (1974).
16. M.S.Satoh and T.Lindahl, Role of poly(ADP-ribose) formation in DNA repair. *Nature 356:*356 (1992).
17. M.R.James and A.R.Lehmann. Role of poly(adenosine diphosphate ribose) in deoxyribonucleic acid repair in human fibroblasts. *Biochemistry* 21:4007 (1982).
18. R.B.Painter, 3-Aminobenzamide does not affect radiation-induced inhibition of DNA synthesis in human cells. *Mut. Res. 143*: 113 (1985).
19. P.J.McKinnon, (1987). Ataxia-telangiectasia: an inherited disorder of ionizing -radiation sensitivity in man. *Hum. Genet. 75:* 197 (1987).
20. R.A.Gatti, E.Boder, H.V. Vinters, R.S.Sparkes, A.Norman, and K.Lange, Ataxia-telangiectasia: an interdisciplinary approach to pathogenesis, 70:99 (1991).
21. N.S.Rudolph and S.A.Latt, Flow cytometric analysis of x-ray sensitivity in ataxia-telangiectasia. *Mut. Res. 211:*31 (1989).
22. N.G.J.Jaspers, R.A.Gatti, C.Baan, P.C.M.L.Linssen, and D.Bootsma, Genetic complementation analysis of ataxia telangiectasia and Nijmegen breakage syndrome: a survey of 50 patients. *Cytogenet. Cell. Genet.* 49:259 (1988).
23. M.A.Papathanasiou, N.C.Kerr, J.H.Robbins, O.W.McBride, I.J.Alamo, S.F.Barrett, I.D.Hickson, and A.J.Fornace,Jr., Induction by ionizing radiation of the gadd45 gene in cultured human cells: lack of mediation by protein kinase C. *Mol Cell Biol 11:* 1009 (1991).
24. S. Fakharzadeh, R.S.Trusko, and D.L.George, Tumorigenic potential associated with enhanced expression of a gene that is amplified in a mouse tumor cell line. *EMBO J* 10: 1565, (1991).
25. J.Momand, G.P.Zambetti, D.C.Olson, D.George, and A.J.Levine, The mdm-2 oncogene product forms a complex with the p53 protein and inhibits p53-mediated transactivation. *Cell* 69:1237 (1992).
26. J.D.Oliner, K.W.Kinzler, P.S.Meltzer, D.L.George, and B.Vogelstein, Amplification of a gene encoding a p53-associated protein in human sarcomas. *Nature* 358:80 (1992).
27. Y. Barak, T. Juven, R. Haffner, and M. Oren, mdm2 expression is induced by wild type p53 activity. *EMBO J* 12: 461, (1993).
28. X. Wu, H. Bayle, D. Olson, and A.J. Levine, The p53-mdm2 autoregulatory feedback loop. *Genes and Development* (in press, 1993).
29. D.Morrell, E.Cromartie, and M.Swift, Mortality and cancer incidence in 263 patients with ataxia-telangiectasia. J. Natl. Cancer Inst. *77:* 89 (1986).
30. F.Hecht and B.K.Hecht, Cancer in Ataxia-telangiectasia patients. *Cancer Genet.Cytogenet.* 46:9 (1990).
31. L.A.Donehower, M.Harvey, B.L.Slagle, M.J.McArthur, C.A.Montgomery, J.S.Butel, and A.Bradley, Mice deficient for p53 are developmentally normal but susceptible to spontaneous tumours, Nature 356:215, (1992).
32. T. Jacks and R. Weinberg, unpublished observations.
33. D.Malkin, F.P.Li, L.C.Strong, J.F.Fraumeni,Jr., C.E.Nelson, D.H.Kim, J.Kassel, M.A.Gryka, F.Z.Bischoff, M.A.Tainsky, and S.H.Friend, Germ line p53 mutations in a familial syndrome of breast cancer, sarcomas, and other neoplasms. *Science 250:* 1233 (1990).
34. S.Srivasta, Z.Zou, K.Pirollo, W.Blattner, and E.H.Chang, Germ-line transmission of a mutated p53 gene in a cancer-prone family with Li-Fraumeni syndrome. *Nature 348:* 747 (1990).
35. M.Swift, P.J.Reitnauer, D.Morrell, and C.L.Chase, Breast and other cancers in families with ataxia-telangiectasia. *N. Engl. J. Med.* 316:1289 (1987).
36. M.Swift, D.Morrell, R.B.Massey, and C.L.Chase, Incidence of cancer in 161 families affected by ataxia-telangiectasia. *N. Engl. J. Med.* 325:1831 (1991).

37. T.A. Weinert and L.H.Hartwell, Characterization of RAD9 of saccharomyces cerevisiae and evidence that its function acts posttranslationally in cell cycle arrest after DNA damage. *Mol. Cell. Biol.* 10:6554 (1990).
38. F. Zampetti-Bosseler and D.Scott, Cell death, chromosome damage and mitotic delay in normal human, ataxia telangiectasia and retinoblastoma fibroblasts after X-irradiation. *Int. J. Radiat. Biol.* 39:547 (1981).
39. H.Nagasawa, S.A.Latt, M.E.Lalande, and J.B.Little, Effects of x-irradiation on cell-cycle progression, induction of chromosomal aberrations and cell killing in ataxia telangiectasia (AT) fibroblasts. *Mut. Res. 148:* 71 (1985).
40. L.R.Livingstone, A.White, J.Sprouse, E.Livanos, T.Jacks, and T.D.Tlsty, Altered cell cycle arrest and gene amplification potential accompany loss of wild-type p53. *Cell* 70:923 (1992).
41. Y.Yin, M.A.Tainsky, F.Z.Bischoff, L.C.Strong, and G.M.Wahl, Wild-type p53 restores cell cycle control and inhibits gene amplification in cells with mutant p53 alleles. *Cell* 70:937 (1992).
42. F.Bischoff,S.O.Yim, S.Pathak, G.Grant, M.J.Siciliano, B.C.Giovaenella, L.C.Strong, and M.A.Tainsky, Spontaneous immortalization of normal fibroblasts from patients with Li-Fraumeni cancer syndrome. *Cancer Res.* 50:7979, 1990.
43. E.Solomon, J.Borrow, and A.D.Goddard, Chromosome aberrations and cancer. *Science* 254:1153 (1991).
44. A.R.Lehman, C.F.Arlett, J.F.Burke, M.H.L.Green, M.R.James, and J.E.Lowe, A derivative of an ataxia-telangiectasia (A-T) cell line with normal radiosensitivity but A-T-like inhibition of DNA synthesis. *Int. J. Radiat. Biol. 49:* 639 (1986).
45. L.N.Kapp and R.B.Painter, Stable radioresistance in ataxia-telangiectasia cells containing DNA from normal human cells. *Int. J. Radiat. Biol.* 56:667 (1989).
46. C.Lambert, R.A.Schultz, M.Smith, C.Wagner-McPherson, D.McDaniel, T.Donlon, E.J.Stanbridge, and E.C.Friedberg, (1991). Functional complementation of ataxia-telangiectasia group D (AT-D) cells by microcell-mediated chromosome transfer and mapping of the AT-D locus to the region 11q22-23. *Proc. Natl. Acad. Sci. USA 88:* 5907 (1991).
47. N.Sullivan, and L.Lyne, Sensitivity of fibroblasts derived from ataxia-telangiectasia patients to calicheamicin γ_1. *Mut. Res.* 245: 171 (1990).
48. E. Yonisch-Rouach, D. Resnitzky, J. Lotem, L. Sachs, A. Kimchi, and M. Oren, Wild-type p53 induces apoptosis of myeloid leukaemic cells that is inhibited by interleukin-6. *Nature* 352: 345 (1991).
49. S.W. Lowe, E.M.Schmitt, S.W.Smith, B.A.Osborne, and T.Jacks, p53 is required for radiation-induced apoptosis in mouse thymocytes. *Nature* 362:847 (1993).
50. A.R. Clarke, C.A.Purdie, D.J.Harrison, R.G.Morris, C.C. Bird, M.L.Hooper, and A.H.Wylie. Thymocyte apoptosis induced by p53-dependent and independent pathways. *Nature* 362:849 (1993).
51. J.M.Lee and A.Bernstein, p53 mutations increase resistance to ionzing radiation. *Proc.Natl. Acad. Sci. USA* 90:5742 (1993).
52. W. J. Slichenmyer, W.G. Nelson, R.J. Slebos, and M.B. Kastan. *Cancer Res.* 53:4164 (1993).

Discussion

M. Oren

What is known about the response of other GADD genes to p53? Are there any that have been implicated in addition to GADD45?

M. Kastan

At first approximation the other GADD genes do not appear to be dependent on p53 for their induction, and I should also point out that GADD45 is induced by every type of DNA damaging agent imaginable and, in addition, by serum starvation. Those other inductions of GADD45 appear to be p53 independent; at least they may be aided by p53 being present but they are not dependent on p53 and appear to work through the GADD45 promoter. So there are many things that induce GADD45 but it is only the ionizing radiation induction of GADD45 that is dependent on p53 and is dependent on the intronic sequence.

M. Oren

Is it clear that the transcripts that are induced through p53 activation of the intron 3 element are the same as the ones that are induced when you trigger the promoter in a p53 independent manner?

M. Kastan

We have no evidence that there are different GADD45 transcripts induced by the different types of agents. They are the same size but there has really been no other work characterizing whether they have slightly different sequences or different protein products or anything like that. We are just in the process now of generating useful antibodies.

R. Horvitz

I think I got confused at one point, but what is it that put GADD45 downstream of the p53 and parallel to apoptosis as opposed to saying that GADD45 is downstream of p53 but before the branch that separates apoptosis from G1 arrest?

M. Kastan

GADD45 is downstream of p53 because its induction is dependent on normal p53 function. It is very possible that GADD45 is involved in an apoptosis branch of the pathway and not in the G1 arrest. We do not know that. At the time we made the initial slides showing where it was we did not even know about radiation induced apoptosis and the p53 dependence. So it is flapping in the wind right now in terms of where it is.

M. Fried

Have you actually deleted the sequence of GADD45 where the p53 binds? You just showed that they are important *in vitro* assays. If you delete the sequence and then put the deleted construct back in the cell, does p53 induce GADD45?

M. Kastan

That is a very good experiment that has not been done yet. We have done more than the *in vitro* assays. The induction of GADD45 is dependent on p53.

M. Fried

But it could still be dependent but a couple of steps away?

M. Kastan

Right, so that there are two separate pieces of information that we are putting together. At this point we have not knocked out the GADD45 intronic sequence in the intact cell and show that it loses the p53 induction.

M. Fried

In the p53 heterozygotes is there any difference in the response to radiation? Is it intermediate between the null and the wild-type?

M. Kastan

That is a very good question. We have been looking at that with many different end points: Is there a difference in survival? Is there a difference in genetic stability? Is there a difference in GADD45 induction? And the answer is: There appears to be some differences. The clearest difference actually is in the generation of aneuploidy. In Thea Tlsty's work in the heterozygote knockout mice she found the loss of one p53 allele greatly increased the incidence of aneuploidy, and in Tyler Jacks' work in the heterozygotes the thymocytes that had one p53 allele were intermediate in their apoptosis resistance in response to radiation. In terms of the cell cycle arrest, it has been much more difficult to quantitate differences. There are subtle differences when they are heterozygote but I am not convinced yet that they are physiologically significant.

D. Livingston

What about the other GADD genes? Are any of them under p53 control?

M. Kastan

No, that is what Moshe Oren asked. We do not have any evidence that the other GADD genes are dependent on p53.

D. Livingston

Has a GADD45 protein been detected?

M. Kastan

We are just now making antibodies, so we have not assayed the protein yet. The sequence of the gene predicts an 18-20 kilodalton protein. There is absolutely no function predicted by the sequence. It does not have DNA binding domains, nuclear localization domain, membrane spanning domains, it does not predict transcriptional activation or anything. There is nothing predicted about its function from the sequence.

D. Livingston

And the retinoblastoma gene product? Anything happen to it after irradiation in the presence and absence of p53?

M. Kastan

We know that there is not a change in the level of the retinoblastoma protein following irradiation. The arrest appears to be occurring prior to phosphorylation of RB. So that is marking where it is in G1. So it really is well into G1 and Steven Reid has discussed it and presented data at a couple of meetings, suggesting that the arrest is actually occurring prior to activation of the cyclin E/cdk2 complex. So again, that is consistent with it occurring prior to RB phosphorylation, but we do not know whether RB is physically in this pathway or not.

E. Mihich

I was intrigued by the differences in induction of p53 after DNA damage, and I was wondering whether these differences are related to number of DNA damage sites, site of DNA damage, or type of DNA? The reason why I am asking this is because there is evidence, for instance, that x-ray is equal in damaging chromatin DNA and episomal DNA whereas there are some topoisomerase II inhibitors that show major differences between damage to chromatin DNA versus episomal DNA. So the question is: Which are the differences, if any, in damage that would condition differences in p53 induction?

M. Kastan

Our data at this point just suggests that strand breaks are important. We do not know whether they are single or double strand breaks. In terms of the location of the breaks, we do not have any clue yet whether the location in the genome is important. How many breaks it takes? We do not know whether one break is enough, or it takes a thousand breaks, or a million breaks. There is a dose response that the more you irradiate the cells the higher the level of p53 is, within a certain range. And we have gone as low as 50 cGy and can see nice p53 induction. We are trying to do experiments now where we are inducing single breaks into exogenously added pieces of DNA to see if that is enough, but we do not have results yet.

D. Hockenbery

In Thea Tlsty's model, where it appears that wild-type p53 can have a dominant effect on the gene amplification phenotype, what also correlates with the gene amplification phenotype is an accelerated crisis of cells that go into S-phase in response to her selection conditions. Can you speculate on what the cell death occurring there is due to? It would suggest that wild-type p53 can suppress an accelerated cell death in that model. And also what wild-type 53 might be responding to in that system under PALA selection conditions.

M. Kastan

Actually, the effects of PALA on p53 induction is one of the slides that I skipped over because I thought it would be too complicated to go into. It takes a long time for PALA treatment to induce p53 and it appears to be occurring 24, 48, 72 hours after addition of the PALA to the cells, and it is at the time when strand breaks are occurring and this is true of other antimetabolites as well, like methotrexate, even though PALA seems to be more effective in inducing those strand breaks and inducing p53. So PALA alone does not seem to be inducing p53 just because of its addition. It is when you are getting the strand breaks. In terms of the early cell death that they get in those cells versus in cells with wild-type p53, they arrest and they become quiescent, but they do not die. We do not have an explanation for what the p53 dependence is, or whether there really is a good p53 dependence for that yet. So Thea and I talk about that a lot, we just do not have any answers to that at this point.

W. Giaretti

I would like again to discuss the issue of p53 and aneuploidization. You have cited some literature data in which loss of p53 increased aneuploidy. What type of cells did they use? In my opinion, it is probably true again that cell type here could really play a role. Take the example of epithelial colorectal cells. So far, it was not shown, to my knowledge, that in human colorectal adenomas you have abnormalities in p53 and still there you have aneuploidization to a large extent (Int. J. Cancer 45:597-603, 1990).

M. Kastan

Again, that gets to the caveat that I mentioned, that you do not need abnormalities in p53 to have genetic instability. In fact the RKO cells, the colorectal cells that have wild-type p53 are as unstable as any tumor cell line you want to work with. And that was shown, in fact, in Thea Tlsty's paper, that they amplified genes just fine. So there are ways to amplify genes that are independent of p53, there are ways to get genetic instability that are independent of p53. The aneuploidy related to p53 has been shown in fibroblasts from the knockout mice, either knocking out one allele or both alleles and there is a graded increase in the amount of aneuploidy and those in otherwise normal cells, at a time when they are starting off at very early passage and quite normal in culture. Curt Harris has shown some data in some talks recently, also true of his bronchial-epithelial cells, that he has manipulated with SV40 and with other gene products to show that at a time when p53 function is knocked out, that aneuploidy is increasing.

D. Livingston

Where does it map?

M. Kastan

It is on 1p about, I think 1p 31.2, it is in one of Al's first papers, I think maybe his MCB paper in '91 where they mapped it. I was very excited about the possibility that it might have been involved in *myc* amplification neuroblastoma because that 1p is where you get loss of heterozygosity and at the same time you get *myc* amplification in neuroblastoma, but the LOH is proximal to GADD45 on the chromosome.

M. D'Incalci

Do you still observe an induction of p53 if you inhibit DNA synthesis before or after radiation?

M. Kastan

We have not tried pharmacologically arresting cells prior to the irradiation, so I do not have any answer to that. I do know in peripheral blood T cells that are not replicating it is very difficult to see an induction. You see a little bit of induction, but we have to PHA stimulate them in order to get p53 levels up to the point where we can see the induction. We have not otherwise pharmacologically arrested cells prior to the irradiation to see if it would block p53 induction.

P. Nicotera

In the recent work that Andrew Wyllie's group published in Nature on p53 induction during irradiation, they proposed two distinct modes of cell death, one of which is activated by calcium and PMA, and the other one of which is depending on radiation induced DNA damage. The latter seems to be involving p53 induction. They say that there could be a downstream event common to both processes and they mention the GADD45 as a common event to the two modes of cell death. Do you have any comment on that?

M. Kastan

Right, that was the pathway I was showing at the end. In fact, I would suggest that the upstream events might be common between the two, from the radiation signal through to the p53 induction, and in some cell types you lead to the G1 arrest, in other cell types you lead to apoptosis. Which pathway GADD45 might be in, we are actively investigating.

P. Nicotera

Do you think you can have a number of different signals which could induce GADD45 expression like, for example, calcium and phorbol esters? Do you know if that is a signal which could induce GADD45 expression?

M. Kastan

TPA does not induce GADD45. In fact, of all the agents looked at it was the only one that had no effect on GADD45. That has not been done with calcium ionophore.

D. Livingston

If you slow the cell cycle in G1 by other means, even physiologic ones like serum deprivation, for example, or try to lower the concentration of any of the cyclin-D proteins, is GADD45 induced?

M. Kastan

GADD45 is induced by serum starvation alone in a p53 independent manner.

D. Livingston

So it might be expected that perturbations of G1, or altering the length of the G1 interval by a variety of means might result in activation of GADD45?

M. Kastan

Right, but that is in a p53 independent manner.

D. Livingston

Can you make a cell constitutively produce GADD45? If so, will it survive, and if it survives, what are its cell cycle characteristics?

M. Kastan

Al Fornace has some preliminary experiments, which he has given me permission to mention, that when he constitutively expresses GADD45 in cells, colony formation is markedly inhibited but, until we have good antibodies where we can really characterize changes in expression either with the anti-sense experiments where we knock out expression or with overexpression, I want to make sure protein levels are changing before we have any hard conclusions about that.

G. Evan

This is really a comment. The data of Mary Collins and others have shown quite clearly that high levels of specific cytokines can suppress cell death which is caused by irradiation or genotoxic damage. So I would just like to make the point that of course if one is growing cells under conditions where you are limiting or restricting or restraining programmed cell death, you are going to get strange results. So, for example, if you do experiments in fetal calf serum which have very high levels of IGF-1, as opposed to calf serum which does not, then if you irradiate those cells, you get much less cell death than if you do under conditions where they are not being hit with survival factors. So I think depending on your cell type and depending on what survival factors you have got there, you may get different results in terms of sensitivity to apoptosis, which is entirely due to the factors that surround the cells.

M. Kastan

That is a good point. For all the studies I have shown up to now, we tried to do it under maximal growth conditions, so we did not have other factors interfering with the pathways.

D. Green

I am not sure I understand the system itself, or how you think it works. Is p53 necessary and sufficient for the growth arrest, or is there an interaction with the DNA damage? Does DNA damage induce p53 and then the combination of DNA damage plus p53 produce a cell cycle arrest?

M. Kastan

No, if you transfect p53 into cells, it makes them stop growing.

D. Green

Is that normal p53, not mutant p53? The problem is that mutant cells will have accumulated DNA damage. Now you express p53 and the cells stop. Is that because there is DNA damage already existing in that cell, or is it because p53, once it is expressed, stops the cells cold?

M. Kastan

There are certainly systems in which wild-type p53 is put into cells which have wild-type p53 and growth is suppressed, though different people have had different results. Moshe, you are probably more aware of some of those experiments than I am.

M. Oren

There is quite a lot of evidence, including much unpublished data, that if you overexpress wild-type p53, you do not get growth arrest. So that would agree with what Doug Green brought up, and I really want to refer to a published study by Kathy Finlay and Arnie Levine in which they looked at primary rat embryo fibroblasts transfected with a p53 overexpression vector. If you look at short term proliferation, there is no effect of excess wild-type p53 and there cells proliferate happily, for that limited number of generations that they would have done it anyway. On the other hand, if you look at spontaneous immortalization, you get a marked reduction. So anything that would require apparently additional genetic damage, like spontaneous immortalization, would be inhibited. On the other hand, the ongoing cell cycle of cells which have been with wild-type p53 all that time apparently is much less sensitive, if at all. So, probably the general trend would be in agreement with what Doug Green suggested. As long as the cell is intact and is having functional p53, it can tolerate quite a bit more. Once you start putting any kind of pressure that would score for lack of genomic integrity, then p53 becomes inhibitory even if the cells to begin with had wild-type p53.

D. Livingston

But that phenotype could be epigenetic to p53 if it is like RB, where there are cells that are stopped and cells that are not stopped, and at least one difference between being stopped and non-stopped is whether or not the cell has the ability to phosphorylate the incoming product of the allele. So that could, indeed, be the case for p53 even. I agree that it is reasonable to be a little careful when one speaks about what p53 can and cannot do.

M. Oren

That is a good point. Unlike the case of RB, with p53 there is still no consensus on what type of phosphorylations would render the protein inactive. So there is really no formal way of disproving or proving the point you made so far.

PROLIFERATION INDEPENDENT ACTIVATION OF PROGRAMMED CELL DEATH AS A NOVEL THERAPY FOR PROSTATE CANCER

Yuzo Furuya, Richard S. Berges, Per Lundmo and John T.Isaacs

The Johns Hopkins Oncology Center and Department of Urology,
Johns Hopkins University School of Medicine, Baltimore, MD 21205

INTRODUCTION

Since the work of Charles Huggins in the 1940s, it has been known that prostatic cancer often retains an androgen responsiveness for stimulation of its growth. Prostatic cancer is thus often highly responsive to androgen ablation therapy. Based upon this, patients with non-organ confined disease eventually require systemic androgen ablation therapy. Nearly all men with metastatic prostatic cancer treated with surgically or chemically induced castration have an initial, often dramatic, beneficial response to such androgen withdrawal therapy[1]. While this initial response is of substantial palliative value, essentially all treated patients eventually relapse to an androgen-insensitive state and succumb to the progression of their cancer unless they die of intercurrent disease first; cures, if any, are rare[2]. Because of this nearly universal relapse phenomenon, the annual death rate from prostatic cancer has not decreased at all over the subsequent 50 years since androgen withdrawal has become standard therapy[3]. Over the last 50 years, the superficially benign nature of androgen withdrawal therapy has tended to disguise the fact that metastatic prostatic cancer is still a fatal disease for which no therapy is available which effectively increases survival[2,4].

Studies by a series of laboratories have demonstrated that a major reason for this universal relapse of metastatic prostatic cancer to androgen ablation is that prostatic cancer within an individual patient is heterogeneously composed of clones of both androgen-dependent and -independent cancer cells even before hormone therapy is begun[5-7]. Development of such tumor cells heterogeneity can occur by a variety of mechanisms [e.g. multifocal origin of the tumor, adaptation, or genetic instability[8]]. Regardless of the mechanism of development of such cellular heterogeneity, once androgen-independent cancer cells are present within individual prostatic cancer patients, the patient is no longer curable by androgen withdrawal therapy alone, no matter how complete, since this therapy kills only the androgen-dependent cells without eliminating pre-existing androgen-independent prostatic cancer cells. To effect all the heterogenous prostatic cancer cell populations within an individual cancer, effective chemotherapy, specifically targeted against the androgen-independent cancer cell must be simultaneously combined with androgen ablation to effect the androgen-dependent cells[9,10]. While the concept of early combinational chemo-hormonal therapy for prostatic cancer is valid, for such an approach to be therapeutically effective in humans, a chemotherapeutic agent which can effectively control the growth of the preexisting androgen-independent prostatic cancer cells must be available.

Apoptosis, Edited by E. Mihich and R.T. Schimke
Plenum Press, New York, 1994

There are presently no highly effective chemotherapeutic agents which can control the growth of androgen-independent prostatic cancer cells[4]. The inability to control androgen-independent prostate cancer cells in human and rodent tumors by standard chemotherapeutic methods, has lead to a search for new approaches.

Growth of a cancer is determined by the relationship between the rate of cell proliferation and the rate of cell death. Only when the rate of cell proliferation is greater than cell death does tumor growth continue. If the rate of cell proliferation is lower than the rate of cell death, then involution of the cancer occurs. Therefore a successful treatment of a cancer can be obtained by either lowering the rate of proliferation and/or by raising the rate of cell death so that the rate of cell proliferation is lower than the rate of cell death. Successful treatment of slow growing androgen independent prostatic cancer cells will probably require simultaneous anti-proliferative chemotherapy targeted at the small number of dividing cancer cells and some type of additional therapy targeted at increasing the low cell death rate of the majority of androgen-dependent cancer cells not proliferating within the prostatic cancer.

There are a variety of effective anti-proliferative chemotherapy presently available which can lower the rate of cell proliferation without increasing the rate of cell death (i.e. cytostatic agents) or agents which lower the rate of cell proliferation and also increase the rate of cell death (i.e. cytotoxic agents). Unfortunately, the cytotoxic agent presently available only lead to death of cancer cells if they subsequently undergo cell proliferation[11]. Therefore cancer cells not in cycle (i.e. cell in G_O of the cell cycle) at the time of exposure or not undergoing cell division soon enough after exposure to the cytotoxic chemotherapeutic agent can repair the damage induced by cytotoxic agent and are thus not killed by the therapy. Unfortunately, the majority (i.e., >90%) of prostatic cancer cells within an individual patient are not actively proliferating[12-15] and are thus resistant to standard cytotoxic chemotherapy[4]. Therefore, what is needed is some type of cytotoxic therapy which induces the death of androgen-independent prostatic cancer cells without requiring the cells to proliferate.

There are at least three cell proliferation independent methods to increasing the death rates of androgen-independent prostatic cancer cells. The first approach is to stimulate the host immune system to evoke/or enhance a cytotoxic anti-tumor response since immune killing of target cell death does not require the target cells to proliferate. The second approach is to block the host development of tumor blood supply. Both the growth and metastatic ability of cancers are critically dependent upon the ability of the cancer cells to induce the development of new blood vessels [i.e. termed angiogenesis][16]. If successful such an anti-angiogenic approach would limit the growth of androgen independent prostatic cancer via hypoxia induced tumor cell death. The third approach is to activate the "programmed cell death" pathway within these cells leading to their suicide. In programmed cell death, specific intracellular signals induce the cell to undergo an active energy-dependent process which does not initially require a change in the plasma membrane permeability[17]. Once initiated, programmed cell death leads to a cascade of biochemical and morphological events that result in the irreversible fragmentation of the genomic DNA and then the cell itself[17-20]. Both androgen-dependent normal prostatic glandular cells and androgen-dependent prostatic cancer cells can be induced to undergo programmed cell death following androgen ablation, and this death process does not require the cells to be in the proliferative cell cycle[21-24].

Summary of the temporal sequences involved in the programmed death of normal prostatic glandular cells following androgen ablation

The programmed death induced in the prostate by androgen ablation is cell type specific. Only the prostatic glandular epithelial cells and not the basal epithelial cells or stromal cells are androgen dependent and thus undergo programmed cell death following castration[22]. These glandular cells constitute approximately 80% of the total cells in the ventral prostate of an intact adult rat and approximately 70% of these glandular cells die by 7 days post castration[22]. Using the ventral prostate of the rat as a model system, the temporal sequence of events involved in the programmed cell death pathway induced by androgen ablation has begun to be defined, figure 1. In the androgen-maintained ventral prostate of an intact adult male rat, the rate of cell death is very low, approximately 2% per day, and this low rate is balanced by an equally low rate of cell proliferation, also 2% per day[25]. If animals are castrated, the serum testosterone level drops to less than 10% of the intact control value within two hours[21]. By 6 hours postcastration the serum testosterone level is only 1.2% of intact control[21]. By 12 to 24 hours following castration, the prostatic dihydrotestosterone (DHT) levels (i.e., the active intracellular androgen in prostatic cells) are only 5% of intact control values. This lowering of prostatic DHT leads to changes in nuclear androgen receptor function [i.e., by 12 hours after castration, androgen receptors are no longer retained in biochemically isolated ventral prostatic nuclei][21]. These nuclear receptor changes result in a major epigenetic reprogramming within the non-proliferating glandular cells (i.e. cell out of cycle in G_O) resulting in the activation phase, termed the D1 phase of the programmed death process, figure 2.

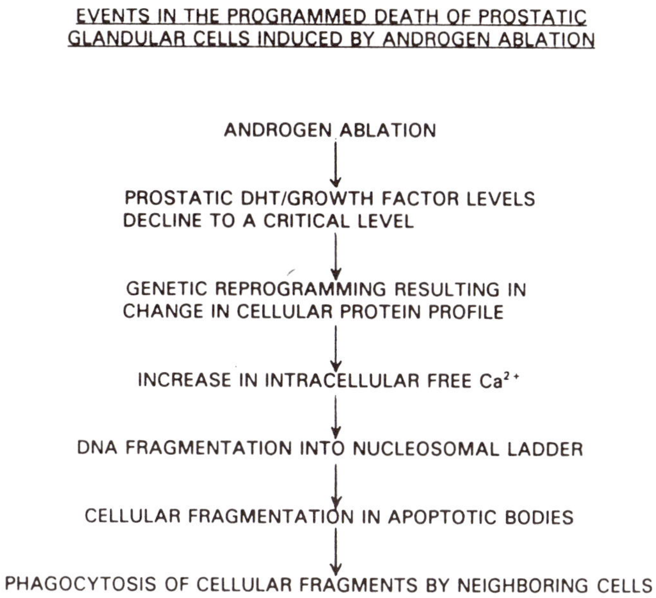

Figure 1. Overview of the temporal sequence of events in the programmed death of glandular cells induced by androgen ablation.

During this D1-activation phase, certain genes to be described later which were actively transcribed and translated before castration are rapidly turned off while other genes which initially were not actively transcribed and translated are rapidly induced when the program for cell death is activated by castration. The result of this epigenetic reprogramming is that during the D1-activation phase of the programmed death process, there is a change in the profile of proteins that are synthesized, which is coupled with an inhibition of glandular cell proliferation[26], a decrease in polyamine levels[27], and increase in intracellular free Ca^{2+} levels[23,28]. The increase in intracellular free Ca^{2+} occurring following castration is derived from the extra-cellular Ca^{2+} pool[23]. The mechanism for this induced elevation in intracellular free Ca^{2+} is not fully known. There are indications that enhanced expression of $TGF\beta_1$ mRNA and protein[29] as well as the receptor for $TGF\beta_1$[30] following castration are somehow involved in the elevation in the intracellular free Ca^{2+} (Ca_i) level[31].

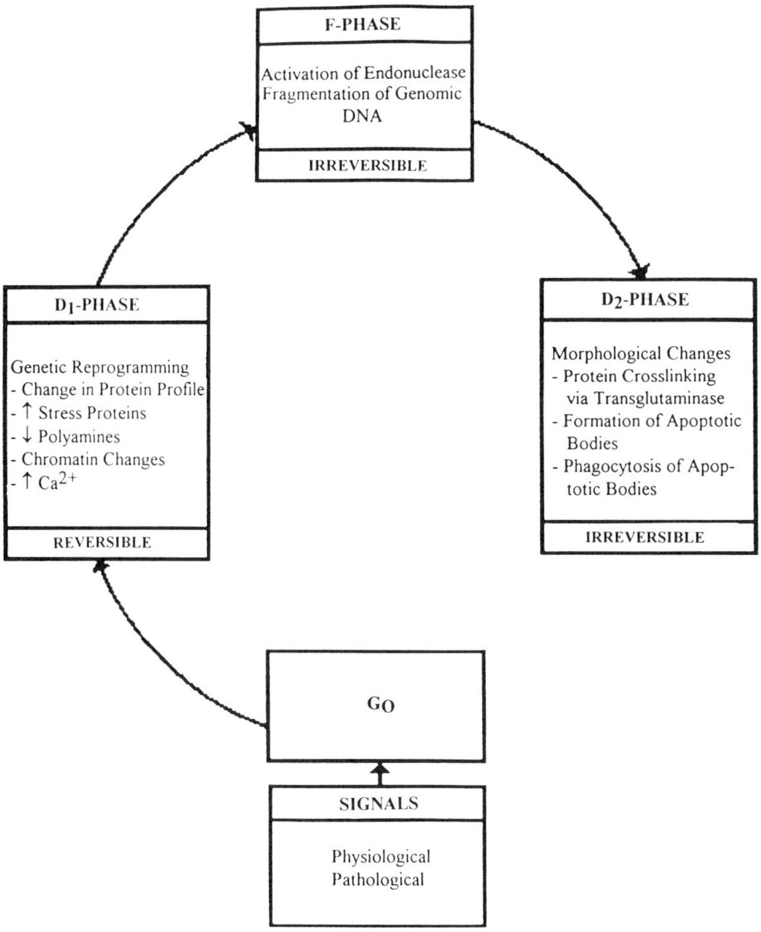

Figure 2. Summary of the phases through which a non-proliferating prostatic glandular cell traverses during its programmed cell death.

Once the Ca_i reaches a critical level, Ca^{2+}-Mg^{2+}-dependent endonucleases present within the nuclei of the prostatic glandular cells are enzymatically activated[23]. Normally, histone H_1 binds to genomic DNA in the linker region between nucleosomes and this binding is involved in packing of the DNA nucleosomes into solenoid structures. Likewise, DNA binding of polyamines, particularly spermine, due to their negative charge are involved in maintaining the spacial constraint of genomic DNA in a compacted form[32]. When the normal content of Histone H_1 and polyamines bound to genomic DNA, the DNA is compacted and is not an efficient substrate for the activated Ca^{2+}, Mg^{2+} endonucleases[33]. During this D1-activation phase, there is a decrease in polyamine levels[27] and the nuclear content of histone H_1[34]. Also during this phase there is a rise in the expression of the highly acidic [pI-3.5]α-prothymosin (to be described in next section). The combined results of these changes are that the genomic DNA conformation opens up in the linker region between nucleosomes in the glandular cells. This enhances the accessibility of the linker DNA to the activated $Ca^{2+}Mg^{2+}$-dependent endonuclease. Once this occurs, DNA fragmentation begins at sites located between nucleosomal units (i.e., F-phase of the programmed death process) and cell death is no longer reversible, figure 2. Recent unpublished studies using inverted pulse-gel electrophoresis has demonstrated the initial DNA fragmentation produces ≈ 100 kb size DNA pieces. Once formed, these 100 kb size pieces are further degraded into nucleosomal size pieces [i.e., > 1Kb].

During this F-phase, the nuclear morphology changes (i.e., chromatin condensation with nuclear margination) even though the plasma and lysosomal membranes are still intact and mitochondria are still functional[22]. During the subsequent portion of the death process, termed the D2 phase, the Ca^{2+}-dependent tissue transglutaminase actively crosslinks various membrane proteins (unpublished data) and cell surface blebbing, nuclear disintegration, and eventually cellular fragmentation into clusters of membrane bound apoptotic bodies occur, figure 2. Once formed, these apoptotic bodies are rapidly phagocytized by macrophages and/or neighboring epithelial cells[22,35]. Thus, within 7-10 days post castration ≈80% of the glandular epithelial cells die and are eliminated from the rat prostate[22].

Prostatic Gene Expression During Programmed Cell Death Pathway Induced by Androgen Ablation

The total RNA content per ventral prostatic cell decreased more rapidly than the loss of prostatic glandular cells (i.e., based upon prostatic DNA content) following castration[36]. Thus if accurate quantitation of the level of mRNA expression of genes on a per cell basis is to be obtained, correction for this temporal difference in the RNA/cell must be performed. Using total RNA extracts, equal amounts of total RNA were loaded per lane on to gels and Northern analysis used to quantitate the level of expression of a series of genes in the ventral prostate following castration. These results were then corrected for the substantial decrease in the total RNA content per cell occurring during prostatic regression induced by castration to allow the results to be expressed as the level of expression per cell for each gene. For comparison, the results were normalized to the expression per cell of each gene in the ventral prostate from intact (i.e., non castrated) male rats.

The expression of a series of genes are up-regulated during the period of programmed death by prostatic glandular cells induced by castration. These genes include c-*myc*[37], glutathione S-transferase subunit Yb_1[38], testosterone repressed prostatic message-2 (TRPM-2) [also called sulfated glycoprotein-2][39], transforming growth factor-$ß_1$[29], H-*ras*[40], calmodulin[36], α-prothymosin[36] and tissue transglutaminase[36]. TRPM-2[41], calmodulin[42], and tissue transglutaminase[43] previously have been demonstrated to be induced in a variety of other cell types undergoing programmed cell death. Several of the genes (i.e., c-*myc*, H-*ras*,) previously have been demonstrated to be involved in cell proliferation. Thus as a comparison, the relative level of expression of these same genes during the androgen-induced proliferation regrowth of the involuted prostate in animals previously castrated 1 week before beginning androgen replacement was determined. Previous studies have demonstrated that between 2-3 days post androgen replacement to 1 week castrated rats, the prostatic glandular cells are maximally undergoing DNA synthesis and cell proliferation[44].

These comparative results demonstrate that the expression of c-*myc*, H-*ras*, and tissue transglutaminase are enhanced in both prostatic cell death and proliferation[36]. In contrast, the expression of calmodulin[36], TRPM-2[36], TGFβ_1[36], glutathione S-transferase subunit Yb_1[38] and α-prothymosin[36] are enhanced only during prostatic cell death and not prostatic cell proliferation.

Additional analysis demonstrated that the expression of a series of genes are decreased following castration. For example, the C_3 subunit of the prostate in gene (i.e., the major secretory protein of the glandular cells) ornithine decarboxylase (ODC), histone-H_4, p53, glucose regulated protein 78, all decrease following castration[36]. In contrast to the decrease in the mRNA expression of these latter genes during programmed cell death in the prostate following castration, the expression of each of these genes is enhanced during the androgen induced prostatic cell proliferation[36].

Role of Cell Proliferation in the Prostatic Death Process Induced by Castration

Using the terminal transferase end-labeling technique of Gavrieli et al[45] to histological detect prostatic glandular cells undergoing programmed death and adjusting for the half-life of detection of these dying cells, the percent of glandular cells dying per day via programmed death in the prostate of intact and castrated rats was determined[46]. In intact (non castrated) rats, 1.2% of the glandular cells die per day via programmed death. Within the first day following castration, this percentage increases and between day 2-5 post castration, ~17-21% of these glandular cells die per day via programmed death. These results demonstrate that both the normal constitutive and androgen ablation induced elimination of glandular cells in the prostate is due to programmed cell death and not to cellular necrosis.

Using standard *in vivo* ^{3}H-thymidine pulse labelling, the percent of glandular cells entering the S-phase during the period of enhanced prostatic cell death occurring during the first week post castration was determined. In the prostates of intact, non castrated, adult male rats there is a low level of glandular cell proliferation. Using the percent of glandular cells in S-phase and the fact that S-phase is of 9 hr duration in these prostatic cells the daily rate of glandular cell proliferation is calculated to be ~1.3% per day in intact control rats[46]. This calculated daily rate of proliferation is essentially identical to the calculated daily rate of programmed death of prostatic glandular cells, which is consistent with the fact that the prostates of these adult, non-castrated, male rats are neither continuously growing nor regressing. Within 1 day following castration, there is an 80 % decrease ($p<0.05$) in the percent of glandular cells entering S-phase. By 4 days following castration, there is more than a 90% reduction in this value. Comparing the data demonstrates that greater than 98% of prostatic glandular cells die following castration without entering the proliferative cell cycle. These results confirm the previous studies of Stiens and Helpap[47] and Evans and Chandler[48] which likewise demonstrated a decrease in the percent of prostatic glandular cells in S-phase following castration.

Percent of ventral prostatic cells undergoing DNA repair following castration

Since double stranded fragmentation of genomic DNA is induced during programmed cell death, this raises the issue of whether DNA repair is activated during the process and whether such a futile process is required for G_O cell killing. In order to detect DNA repair in prostatic glandular cells, a high dose/long exposure bromodeoxyuridine (BrdU) labeling method was used. Instead of pulse labeling animals with a short exposure to a small dose of ^{3}H-thymidine, animals were given 50 mg/kg of BrdU and 6 hours later prostates were harvested for immunohistological detection of BrdU labeled prostatic glandular cells. This is a total dose of ~42 μmoles of BrdU per rat which is 3360 times higher than that the nucleotide precursor dose used in the ^{3}H-thymidine studies reported. When animals are injected with such a high dose of BrdU, incorporation of the nucleotide precursor is not limited to the first 30 min but continues for several hr following injection[46]. The use of such a high dose of BrdU and such a long period of exposure (ie 6 hr) before harvesting prostatic tissue coupled with the use of the highly sensitive immunocytochemical detection of BrdU maximizes the possibility of detecting both scheduled S-phase DNA synthesis and unscheduled DNA repair[46].

The percent of glandular cells incorporating BrdU into DNA was determined on prostatic tissue removed 6 hr after IP injection of 50 mg/kg of BrdU at various times following castration. These data demonstrate that there is a 3-4 fold increase in BrdU labeling by day 2 post castration[46]. By day 3 post castration, there is a >10 fold increase which peaks at a >20 fold increase in BrdU labeling on day 4 post castration before decreasing on day 5 post castration. The distinguishing feature between scheduled S-phase specific DNA synthesis and unscheduled DNA repair is that during S-phase DNA synthesis, there is a net accumulation of nuclear DNA content (i.e. cells have > diploid content of DNA). In contrast, during G_O DNA repair, no net accumulation occurs and the cell have a diploid content of DNA. To confirm that the majority of BrdU incorporation occurred as part of a futile G_O DNA repair process and not S-phase specific DNA synthesis, flow cytometer was used to sort the prostatic cells from 3 day castrated rate which have incorporated BrdU. These BrdU positively labeled prostatic cells were then propidium iodide staining and analyzed by flow cytometry for their DNA content. These studies demonstrated that 82.5 ± 6.9 % of these BrdU positive prostatic cells had a diploid (G_O) compliment of DNA[46].

The original single cell suspension used for these analysis included prostatic stromal cells, basal epithelial cells, and intra-epithelial macrophages in addition to glandular cells. Previous studies have demonstrated that these former cell types are not androgen dependent and that they continue to enter the S-phase following castration[49]. In addition, Evans and Chandler demonstrated that between 2-3 days post castration, there is a major increase in the proliferation of the intra-epithelial macrophages[48].

The continuing proliferation of these androgen independent prostatic cells following castration should be detectable in this analysis. Thus the detection of ~20% of the BrdU positively labeled cells having an increased compliment of DNA characteristic of S-phase cells is not unexpected. Regardless of the nature of these S-phase cells, the fact that >80% of the BrdU positively labeled cells on day 3 post castration have a diploid content of DNA demonstrates that the vast majority of BrdU incorporation into prostatic glandular cells following castration is due to G_O DNA repair and not to entrance into S-phase.

Neither progression through S-phase nor G_0 DNA repair is required for programmed death in the prostate induced by androgen ablation

The previous data demonstrates that during the programmed death of the prostatic glandular cells activated by castration, DNA fragmentation occurs which induces a futile process of DNA repair while these cells are in G_O. This raises the issue of whether such a futile G_O DNA repair process is associated with,but not causally required for prostatic cell death. To resolve this issue,rats were injected IP with 500mg of hydroxyurea(HU)/kg every 8 hr for 5 days. This dose of HU was chosen based upon demonstrating that this treatment inhibited both prostatic S-phase specific DNA synthesis and unscheduled G_O DNA repair by more than 90% for 8hr following an IP injection[46]. When intact male rats were treated with this tri-daily HU regimen for 1 week, there was no indication of an increase in programmed cell death in the prostate based upon the lack of an increase in morphologically detectable apoptotic bodies or terminal transferase end-labeled cells, or loss of DNA content.

Based upon these combined results, it is clear that tri-daily treatment with 500mg of HU/kg inhibits by at least 90% both scheduled S-phase DNA synthesis and unscheduled G_0 DNA repair without itself inducing programmed cell death in the prostate. Therefore, rats were castrated and either injected IP with 500mg of HU/kg every 8 hr or injected with the saline vehicle every 8 hr as a control. After 5 days of castration, the DNA content was reduced 51 ± 2% in the animals untreated with HU vs 47 ± 3% in castrated rats receiving the tri-daily HU treatment[46]. Histological analysis also demonstrated an identical atrophic morphology for the prostates from both groups and an identical frequence of glandular cell detected as apoptotic bodies in both groups of prostates. These data demonstrate that the programmed death of prostatic glandular cells induced by androgen ablation does not require either progression through S-phase or G_0 DNA repair.

p53 expression is not required for androgen ablation induced programmed death of G_0 prostatic glandular cell

To investigate the possible role of the p53 gene in the programmed cell death pathway induced by androgen ablation,the extent of programmed death of androgen dependent cells in the prostate and seminal vesicles following castration was compared between wild type and p53 deficient mice. The mutant mice were established using homologous recombination to produce null mutation in both of the p53 alleles[50]. These homozygous null mutations prevent any production of p53 protein in these mice[50]. Wild type (i.e., p53 expressing) mice and p53 deficient mice were castrated and after 10 days the animals were killed and their seminal vesicles and prostates removed, weighted, and DNA content determined. Histological sections were also prepared from each of these tissues. These analyses demonstrated that there is an identical decrease in the wet weight and DNA content in both the seminal vesicles and prostate from wild type and p53 deficient mice[46]. Histological analysis, likewise demonstrated an identical degree of cellular regression in these tissues in the two types of mice (i.e., similar percent of terminal transferase end-labeled prostatic glandular cells in the two groups of animals). These studies demonstrate that androgen ablation induced programmed death of androgen dependent cells does not require any involvement of p53 protein expression.

Redefining the Prostatic "Cell Cycle"

With the realization of the importance of programmed cell death, the older idea that prostatic cell number is determined by the proliferative cell cycle alone has been modified. Based upon this modification, a redefined "cell cycle", has been proposed[51]. The overall "cell cycle" controlling cell number is thus composed of a multicompartment system in which the prostatic glandular cell has at least three possible options, figure 3. The cell can be: 1) metabolically active but not undergoing either proliferation or death (i.e., G_0 cell); 2) undergoing cell proliferation (i.e., $G_0 \Rightarrow G_1 \Rightarrow S \Rightarrow G_2 \Rightarrow$ mitosis), or 3) undergoing cell death by either the programmed pathway (i.e., $G_0 \Rightarrow D_1 \Rightarrow F \Rightarrow D_2 \Rightarrow$ apoptotic cellular fragmentation) or by nonprogrammed (i.e., necrotic) pathway.

Therapeutic Implication of Programmed Cell Death for Prostatic Cancer

Using the human PC-82 prostatic xenograft system as a model, Kyprianou et al[24] demonstrated that androgen ablation activates the pathway of programmed cell death, not only in normal androgen dependent prostatic cells, but also androgen-dependent human prostatic cancer cells. Using bromodeoxyuridine incorporation into DNA to label human PC-82 prostatic cancer cells undergoing entrance into the S phase of the proliferative cell cycle, we have recently demonstrated that within 1 day following castration the number of PC-82 prostatic cancer cells entering the S phase declined from 8-10% to one-third this initial values (i.e., to a value 2-3%) and that after 2 days, the proliferative activity declined to below 1% (unpublished data). Combining these latter two studies demonstrated that the programmed death of androgen dependent human prostatic cancer cells induced by androgen ablation does not require these cells to go through a defective cell proliferation cycle but that these cells die without leaving G_0.

Additional studies have demonstrated that androgen ablation does not induce this programmed death process in androgen-independent prostatic cancer cells due to a defect in the initiation step[52]. Even with this defect, however, androgen-independent prostatic cancer cells retain the basic cellular machinery to undergo this programmed cell death pathway. This

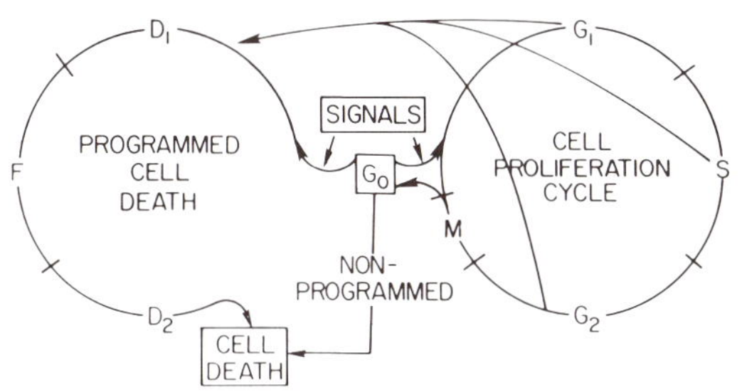

Figure 3. Revised "cell cycle" denoting the options of a G_0 prostatic glandular cell. D_1 denotes the period during which new gene and protein expression required for induction of the DNA fragmentation period (denoted F) occurs as part of the programmed cell death pathway. D_2 denotes the period during wich the cell itself fragment into apoptotic bodies as part of its programmed death.

was demonstrated by using a variety of chemotherapeutic agents which arrest proliferating androgen independent prostatic cancer cells in various phases of the proliferative cell cycle (e.g. G_1, S, or G_2) and which subsequently induces their programmed (i.e., apoptotic) death[52]. One explanation for the inability of androgen ablation to induce programmed death of androgen-independent prostatic cancer cells is that such ablation does not induce a sustained elevation in the intracellular free Ca^{2+} (Ca_i) levels in these cells.

To test this possibility, androgen-independent, highly metastatic Dunning R-3327 AT-3 rat prostatic cancer cells were chronically exposed *in vitro* to varying concentrations of the calcium ionophore ionomycin to sustain various levels of elevation in the their Ca_i[53]. These studies demonstrated that an elevation of Ca_i from a starting value of 35 nM to a value as small as only 3-fold above baseline (i.e., 100 nM) while not inducing immediate toxicity (i.e., death within $\leq$ hours) can induce the death of the cells if sustained for > 12 hr. Temporal analysis demonstrated that elevation in Ca_i results in these cells arresting in G_0 within 6-12 hr following ionomycin exposure. Over the next 24 hr, these cells begin fragment their genomic DNA into initially 100 Kb size pieces which are further degraded into nucleosome-sized pieces and during these next 24-48 hr these cells undergo cellular fragmentation in apoptotic bodies[53]. Associated with this programmed cell death is an epigenetic reprogrammed of the cell in which the expression of a series of genes (to be presented later) is specifically modified. These results demonstrate that even nonproliferating androgen-independent prostatic cancer cells can be induced to undergo programmed cell death if a modest elevation in the intracellular free Ca^{2+} is sustained for a sufficient time. Combining these latter ionomycin data with the chemotherapy data demonstrates that programmed death of androgen independent prostatic cancer cells can be induced in any phase of the cell cycle and does not necessarily require progression through the proliferation cell cycle (i.e., proliferation independent).

Ability of Thapsigargin (TG) to Activate Programmed Cell Death

Thapsigargin, figure 4, is a sesquiterpene γ-lactone isolated from the root of the umbelliferous plant, Thapsia garganica. Resin from this plant was used starting about 300BC as a medicine for rheumatic pains by the Greeks. The resin is a skin irritant and has been used in traditional Arabian medicine for centuries[54]. The active principles from the plant, a hexaoxygenated 6,7-guaianolide, was isolated in pure form and termed Thapsigargin (TG) in 1978 by the group of S. Brogger Christensen[55]. One of the mechanisms for its skin irritant effects is via its ability to induce mast cells to release histamine. A series of studies by the group of Christensen have demonstrated that this ability is due to an extracellular Ca^{2+} dependent effects which mimics the ability of calcium ionophores to induced mast cells to release histamine[54].

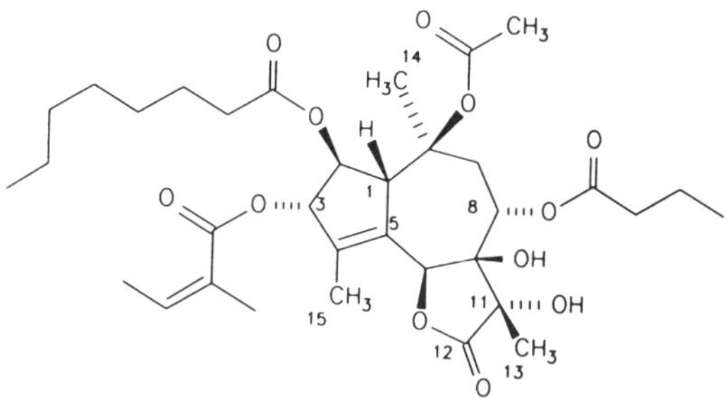

Figure 4. Chemical structure of the sesquiterpene laction, Thapsigargin.

More recent studies have demonstrated that the Ca^2 dependence for TG effects is due to the fact that this highly lipophilic agent enters cells and interacts with the Ca^{2+}-ATpase present in the endoplasmic reticulum (ER) and inhibits its enzymatic activity with a IC_{50} value of ≈30 nM[56]. Such inhibition is not only efficient but also highly specific since neither the plasma membrane or red blood cell Ca^{2+} ATpases are inhibited by TG even at μM concentrations[56]. Large pools of bound calcium are sequestered in the ER of cells even though the free intracellular Ca^{2+} (Ca_i) concentration is only 30-40 nM. This sequestered pool of bound Ca^{2+} can be specific and transiently liberated to elevate the Ca_i level from 30-40 nM to several hundred nM by a variety of intracellular signals. The best characterized of these signals is the production from the inositol phospholipids of inositol 1,4,5 triphosphate (IP_3). IP_3 binding to its specific receptor in the ER results in the release of the sequester Ca^{2+} and an elevation in the Ca_i. The elevation in Ca_i is usually transitory, however, since the elevated Ca^{2+} is rapidly pumped either out of cell via plasma membrane Ca^{2+} ATpase pumps or back into the ER via its Ca^{2+} ATpase pump. Recently, it has been demonstrated, however, that the sequestered Ca^{2+} in the ER is constantly "leaking" out into the cytoplasm of the cell and that the ER-Ca^{2+} ATpase is constantly pumping this free Ca^{2+} back into the sequestered stores of the ER[56]. Thus, when the cell permeable Thapsigargin inhibits the ER-Ca^{2+} ATpase pump, the leaking Ca^{2+} from the ER is no longer pumped back into a sequestered form resulting in the 3-4 fold elevation of the Ca_i without any requirement for IP_3 production. Such a primary elevation of Ca_i leads to a depletion of the ER Ca^{2+} pool results in many cell types of a signal being generated which induces a change in the permeability of the plasma membrane to extracellular divalent cations, particular Ca^{2+}. For many cell types this initial intracellular discharge of the ER sequestered calcium pools leads to a Ca^{2+} influx, in keeping with the prediction from the capacitance model of Ca^{2+} entry[57]. Thus, once these changes in the plasma membrane occurs, a Ca^{2+} influx into the cell occurs due to the high free Ca^{2+} concentration extracellularly (i.e., 1-3 mM)[56]. This produces a secondary elevation in the Ca_i which is sustainable (i.e., min-hours) if the TG inhibition is maintained[56]. Based upon this background, the ability of Thapsigargin to sustain an elevation in the Ca_i and to activate programmed cell death in androgen independent prostate cancer cells was tested.

Initially, *in vitro* testing was performed on a series of androgen independent prostatic cancer cell lines of rat (i.e., AT-3 cells) and human (i.e., TSU-pr, DU-145, and PC-3) origin. Microsomes from each of these 4 distinct cancer cell lines and assayed for their Ca^{2+} ATpase activity. These studies demonstrated that each of the lines possessed ER Ca^{2}-ATpase activity with a specific activity ranging between 12-40 nmoles of ATP hydrolyzed per min per mg of ER protein. Co-incubation of 500 nM TG with the microsomal preparation with the assay reagents resulted in $\geq$ 95% inhibition of the ER-Ca^{2+} ATpase activity of each of the cell lines. Based upon these results, each of these 4 cell lines was chronic exposed to 500 nM Thapsigargin. Using Fura-2 fluorescence ratio measurements, such Thapsigargin treatment resulted in a 2-3 fold elevation in the Ca_i levels from baseline values within 1-2 min of initial exposure. This elevation is only transient (i.e., returning to baseline by 5 min of TG exposure) if the cell culture media contained 5 mM EDTA to deplete the extracellular free Ca^{2+} level. In contrast, if no EDTA is present so that Ca_i level in the extracellular media is > 1 mM, then the response to TG treatment is a sustained (i.e., > 24 hours) 2-3 fold elevation in the Ca_i. These studies demonstrated that for all of the 4 cell lines tested, 500 nM TG treatment resulted in a sustained (i.e., > 24 hr) 2-3 fold elevation in Ca_i and that the major source of the calcium for this effect is extracellular.

Using 2 parameter flow cytometric analysis based upon DNA content and nucleus size, 500 nM TG treatment of each of the 4 distinct androgen independent prostatic cancer cell lines was found to arrest these cells in the G_0/G_1 phase of the cell cycle. This G_0/G_1 arrest was complete by 24 hours of continuous 500 nM TG exposure. Cells from each of the 4 prostatic cancer lines were incubated with ^{14}C-thymidine to uniformally pre-label their DNA. Pre-labeled cells were then treated with 500 nM TG and the percentage of the cells undergoing DNA fragmentation (i.e., to sizes $\leq$ 100 Kb) at various times of TG treatment was quantitated using inverted pulse gel electrophoresis. These results demonstrated that after a 24 hour lag period, the cells begin to fragment their DNA and that by ≈96 hours of treatment $\geq$ 95% of the cells have fragmented their DNA regardless of cell line tested.

Quantitative analysis of the DNA demonstrated the characteristic, nucleosomal ladder pattern of fragmentation, characteristic of programmed cell death. The temporal pattern of DNA fragmentation was tightly correlated with the loss of clonogenic ability by the cells for each of the 4 cell lines (i.e., 72 hours of TG treatment required for 50% of the cells to fragment their DNA and 50% loss of their clonogenic ability). In contrast to the high temporal correlation between DNA fragmentation and loss of clonogenic ability, there was more than a 24-48 hours lag period between the time required for 50% of the cell to fragment their DNA and loose clonogenic ability before 50% of the cells loss their cellular viability based upon plasma membrane integrity measure either by time-lapse by videomicroscopy or trypan blue extrusion. These time-lapse videomicroscopy studies did demonstrate, however, that morphological changes begin occurring within 3-6 hours of initial TG exposure. These changes initially involve round-up of cells. By 24 hours of TG treatment cells are smaller in size and rounded in morphology. Between 72-120 hr TG treatment, the cells undergo period of plasma membrane hyperactivity characterized by the production of plasma membrane blebbing. These surface blebs are highly dynamic, coming and going on the surface giving the appearance of the membrane boiling previously reported for ionomycin-induced programmed cell death of AT-3 prostatic cancer cells[53]. These combined results demonstrate that the initiation of DNA fragmentation is occurring in viable nonproliferating (i.e. G_0/G_1) cells from each of the 4 distinct androgen independent prostatic cancer cell lines tested, 24-48 hr before these cell lyse and that this DNA fragmentation is not the result of a loss of metabolic viability (i.e. loss of mitochondrial or plasma membrane function). In contrast, the data are consistent with the initiation of DNA fragmentation as the irreversible commitment step in the TG induced programmed death of nonproliferating androgen independent rodent or human prostatic cancer cells.

Analysis of a mRNA expression of the series of genes previously demonstrated to be enhanced during the programmed cell death of normal prostatic cells induced by androgen ablation demonstrated that TG treatment of androgen independent prostatic cancer cells likewise leads to an epigenetic reprogramming of the cells. AT-3 rat prostatic cancer cells were treated from 0-36 hours with either 500 nM TG, 10 μM ionomycin, or 100 μM 5-fluordeoxyuridine (5-FudU). Previously, we have demonstrated that prostatic cancer cells must progress through the proliferative cell cycle in order for 5-FudU to induce their programmed cell death[52]. In contrast, TG and ionomycin induce the proliferation independent programmed death of G_0 cells. These results demonstrate that within 1 hour of either TG or ionomycin treatment, expression of several of these genes is already elevated [eg. α-prothymosine, calmodulin, ornithine decarboxylase (ODC)] and that by 6 hours additional genes expression is enhanced (eg. glucose-regulated protein-78 (GRP), c-*myc*). Many of these enhancements are acute with expression decreasing at 24 hours of treatment. There are major differences in gene expression during the proliferation independent programmed death induced by TG or ionomycin and the proliferation dependent death induced by 5-FudU (eg. in the latter, c-*myc*, calmodulin, prothymosine are not induced while H-*ras* and TRPM-2 are). These results demonstrate that the programmed death induced by all of these agents involve an active epigenetic reprogramming of the cell and the pathway induced by TG is essentially identical to that induced by ionomycin, but distinct from that induced by 5-FU.

CONCLUSION

Metastatic prostatic cancer is a fatal disease due to the fact that presently there is no effective systemic therapy for the control of the androgen independent prostatic cancer cells within the patient. A major reason why androgen independent prostatic cancer cells are resistant to present therapies is due to their extremely low rate of proliferation. Androgen independent prostatic cancer cells can be induced to undergo cell proliferation independent programmed death if the intracellular free Ca^{2+} can be elevated for a sustained period. One method to sustain an elevated Ca_i is to inhibit the endoplasmic reticulum Ca-ATpase pump selectively in these cells. As a possible method to do this, thapsigargin is being derivatized to a peptide prodrug form which is only activatable by the protease activity of prostatic specific antigen [PSA][58]. Since only prostatic glandular cells and prostatic cancer cells secrete PSA locally[59], and since the PSA which reaches the blood serum is not enzymatically active[60], the cytotoxicity of this prodrug should be targetable to these prostatic cells selectively without activating non-specific programmed cell death systemically throughout the body.

REFERENCES

1. The Leuprolide Study Group, Leuprolide versus diethylstilbestrol for metastatic prostatic cancer, New Engl. J. Med. 311-1281-1286 (1984).
2. H. Lepor, A. Ross, and P.C. Walsh, The influence of hormonal therapy on survival of men with advanced prostatic cancer, J. Urol. 128:335 (1982).
3. C. Boring, T. Squires, and T. Tong, Cancer Statistics, 1992, Ca-a Cancer Journal for Clinicians 42:19 (1993).
4. D. Raghavan, Non-hormone chemotherapy for prostate cancer: principles of treatment and application to the testing of new drugs, Seminars in Oncology 15:371 (1988).
5. A.A Sinha, C.E. Blackhard, and U.S. Seal, A critical analysis of tumor morphology and hormone treatment in the untreated and estrogen treated responsive and refractory human prostatic carcinoma, Cancer 40:2836 (1977).
6. G.R. Prout, B. Leiman, J.J. Daly, R.A. MacLoughlin, P.P. Griffin, and H.H Young, Endocrine changes after diethylstilbestrol therapy, Urology 7:148 (1976).
7. J.T. Isaacs and D.S. Coffey, Adaptation vs selection as the mechanism responsible for the relapse of prostatic cancer to androgen ablation as studies in the Dunning R-3327 H adenocarcinoma, Cancer Research 41:5070 (1981).
8. J.T. Isaacs, Cellular factors in the development of resistance to hormonal therapy, In: Bruchovsky, N., Goldie, J, (eds), Drug and Hormone Resistance in Neoplasia I 139-156, CRC Press, Boca Raton (1981).
9. J.T. Isaacs, The timing of androgen ablation therapy and/or chemotherapy in the treatment of prostatic cancer, Prostate 5:1 (1984).
10. J.T. Isaacs, Relationship between tumor size and curability of prostate cancer by combined chemohormonal therapy, Cancer Research 49:6290 (1989).
11. S.E. Shackney, G.W. McCormack, and G.J. Cuchural, Growth rate patterns of solid tumors and their relationship to responsiveness to therapy, Annals of Internal Medicine 89:107 (1978).
12. B. Helpap, R. Steins, and P. Bruhl, Autoradiographic in vitro investigations on prostatic tissue with C-14 and H-3 thymidine double labeling method, Beitr. Pathol. anat. Allg. Pathol., 151:65 (1974).
13. J. S. Meyers, G. Sufrin, and S.A. Maring, Proliferation activity of benign human prostate, prostatic adenocarcinoma and seminal vesicle evaluated by thymidine labeling, J. Urol., 128:1353 (1982).
14. R. Nemoto, K. Hattori, K. Uchida, T. Shimazui, Y. Nishijima, K. Koiso, and M. Harada, S-phase fraction of human prostate adenocarcinoma studies with in vivo bromodeoxyuridine labeling, Cancer (Phila.), 66:509 (1990).
15. M.V. Sadi and E.R. Barrack, Determination of growth fraction in advanced prostate cancer by Ki-67 immunostaining and its relationship to the time to tumor progression after hormonal therapy, Cancer (Phila.), 67:3065 (1991).
16. J. Folkman, What is the evidence that tumors are angiogenesis-dependent? J. Natl. Cancer Inst. 82:4 (1990).
17. A.H. Wyllie, J.F.R. Kerr, and A.R. Currie, Cell death: the significance of apoptosis, International Reviews of Cytology 68:251 (1980).
18. J.F.R. Kerr, A.H. Wyllie, and A.R. Currie, Apoptosis: a basic biological phenomenon with wide ranging implications in tissue kinetics, Br. J. Cancer, 26:239 (1972).
19. S.R. Umansky, B.A. Korol, and P.A. Nelipovich, In vivo DNA degradation in thymocytes of γ-irradiated or hydrocortisone-treated rats, Biochim. Biophys. Acta, 655:9 (1981).
20. A.H. Wyllie, R.G. Morris, A.L. Smith, and D. Dunlop, Chromatin cleavage in apoptosis: association with condensed chromatin morphology and dependence on macromolecular synthesis, J. Pathol., 142:66 (1984).
21. N. Kyprianou and J.T. Isaacs, Activation of programmed cell death in the rat ventral prostate after castration, Endocrinology, 122:552 (1988).
22. H.F. English, N. Kyprianou, and J.T. Isaacs, Relationship between DNA fragmentation and apoptosis in the programmed cell death in the rat prostate following castration, Prostate, 15:233 (1989).

23. N. Kyprianou, H.F. English, and J.T. Isaacs, Activation of a Ca^{2+}-Mg^{2+}-dependent endonuclease as an early event in castration-induced prostatic cell death, Prostate, 13:103 (1988).
24. N. Kyprianou, H.F. English, and J.T. Isaacs, Programmed cell death during regression of PC-82 human prostate cancer following androgen ablation, Cancer Res., 50:3748 (1990).
25. J.T. Isaacs, Antagonistic Effect of Androgen on Prostatic Cell Death, The Prostate 5:545 (1984).
26. N. Kyprinaou and J.T. Isaacs, Biological Significance of Measurable Androgen Levels in the Rat Ventral Prostate Following Castration, The Prostate 10:313 (1987).
27. A.E. Pegg, D.H. Lockwood, and H.G. Williams-Ashman, Concentrations of Putrescine and Polyamines and their Enzymic Synthesis during Androgen-Induced Prostatic Growth, Biochem. J. 117:17 (1970).
28. P. Martikainen and J. Isaacs, Role of Calcium in the Programmed Death of Rat Prostatic Glandular Cells, The Prostate 17:175 (1990).
29. N. Kyprianou and J.T. Isaacs, Expression of transforming growth factor-ß in the rat ventral prostate during castration induced programmed cell death, Mol. Endocrinol. 3:1515 (1989).
30. N. Kyprianou and J.T. Isaacs, Identification of a cellular receptor for transforming growth factor-ß in rat ventral prostate and its negative regulation by androgens, Endocrinol. 123:2124 (1988b).
31. P. Martikainen, N. Kyprianou, and J.T. Isaacs, Effect of Transforming Growth Factor-$ß_1$ on Proliferation and Death of Rat Prostatic Cells, Endocrinol. 127:2963 (1990).
32. R.D. Synder, Polyamine depletion is associated with altered chromatin structure in HeLa cells, Biochem. J. 260:697 (1989).
33. B. Brüne, P. Hartzell, P. Nicotera, and S. Orrenius, Spermine prevents endonuclease activation and apoptosis in thymocytes, Exp. Cell Res. 195:323 (1991).
34. L.W.K. Chung and D.S. Coffey, Biochemical characterization of prostatic nuclei I. androgen-induced hanges in nuclear proteins, Biochimica. Biophysica. Acta 247:570 (1971).
35. J.F.R. Kerr and J. Searle, Deletion of cells by apoptosis during castration-induced involution of the rat prostate, Virchows. Arch. [Cell Pathol] 13:87 (1973).
36. Y. Furuya and J.T. Isaacs, Differential gene regulation during programmed death (Apoptosis) versus proliferation of prostatic glandular cells induced by androgen manipulation, in press (1993).
37. V.E. Quarmby, W.C. Beckman, Jr., E.M. Wilson, and F.S. French, Androgen regulation of c-myc messenger ribonucleic acid levels in rat ventral prostate, Mol. Endocrinol. 1:865 (1987).
38. C. Chang, A.G. Saltzman, N.S. Sorensen, R.A. Hiipakka, and S. Liao, Identification of glutathione S-transferase Yb_1 mRNA as the androgen repressed mRNA by cDNA cloning and sequence analysis, J. Biol. Chem. 262:11901 (1987).
39. M.L. Montpetit, K.R. Lawless, and M. Tenniswood, Androgen Repressed Messages in the Rat Ventral Prostate, Prostate 8:25 (1986).
40. R. Buttyan, Z. Zakeri, R. Lockshin, and D. Wolgemuth, Cascade induction of c-fos, c-myc and heat shock 70K transcripts during regression of the rat ventral prostate gland, Mol. Endocrinol., 2:650 (1988).
41. R. Buttyan, C.A. Olsson, J. Pintar, C. Chang, M. Bandyk, P-Y Ng, and I.S. Sawczuk, Induction of the TRPM-2 gene in cells undergoing programmed death, Mol. Cell Biol. 9:3473 (1989).
42. D.R. Dowd, P.N. MacDonald, B.S. Komm, M.R. Haussler, and R. Miesfeld, Evidence for early induction of calmodulin gene expression in lymphocytes undergoing glucocorticoid-mediated apoptosis, J. Biological Chemistry 266:18423 (1991).
43. L. Fesus, V. Thomazy, and A. Falus, Induction and activation of tissue transglutaminase during programmed cell death, FEBS Lett. 224:104 (1989).
44. D.S. Coffey, J. Shimazaki, and H.G. Williams-Ashman, Polymerization of deoxyribonucleotides in relation to androgen-induced prostatic growth, Arch. Biochem. Biophys. 124:184 (1968).
45. Y .Gavrieli, Y. Sherman, and S.A. Ben-Sasson, Identification of Programmed Cell Death In Situ via Specific Labeling of Nuclear DNA Fragmentation, J. Cell Biol. 119:493 (1992).
46. R.S. Berges, Y. Furuya, L. Remington, H.F. English, T. Jacks, and J.T. Isaacs, Cell proliferation, DNA repair, and p53 function are not required for programmed death of prostatic glandular cells induced by androgen ablation, Proc. Natl. Acad. Sci. (USA), (in press, 1993).
47. R. Stiens and Helpap, Regressive changes in the prostate after castration. A study using histology, morphometrics and autoradiography with special reference to apoptosis, Pathol. Res. Pract. 172:73 (1981).
48. G.S. Evans and J.A. Chandler, Cell proliferation studies in the rat prostate: II. The effects of castration and androgen replacement upon basal and secretory cell proliferation, Prostate 11:339 (1987).

49. H.F. English, J.R. Drago, and R.J. Santen, Cellular response to androgen depletion and repletion in the rat ventral prostate: Autoradiography and morphometric analysis, The Prostate 7:41 (1985).
50. S. Lowe, E.M. Schmitt, S.W. Smith, B.A. Osborne, and T. Jacks, p53 is required for radiation induced apoptosis in mouse thymocytes, Nature 362:847 (1993).
51. R. Berges and J.T. Isaacs, Programming events in the regulation of cell proliferation and death, Clinical Chemistry 39:2 (1993).
52. N. Kyprianou and J.T. Isaacs, Thymine-less death in androgen independent prostatic cancer cells, Biochem. Biophys. Res. Commun. 165:73 (1989).
53. P. Martikainen, N. Kyprianou, R.W. Tucker and J.T. Isaacs, Programmed death of non-proliferating androgen independent prostatic cancer cells, Cancer Res. 51:4693 (1991).
54. S.B. Christensen, E. Norup E and U. Rasmussen, Chemistry and structure-activity relationship of the histamine secretagogue Thapsigragin and related compounds, In: Natural Products and Drug Development (Editors: P. Krogsgaard-Larsen, S. Brogger Christensen, H. Kofod) pp. 405 (1993).
55. U. Rasmussen, S.B. Christensen, and F. Sandberg, Thapsigargin and thapsigargicin, two new histamine liberators from Thapsia garganica. L. Acta. Pharm. Suec. 15, 133 (1978).
56. O. Thastrup, P.J. Cullen, B.K. Dr_bak, M.R. Hanley, and A.P. Dawson, Thapsigargin, a tumor promoter, discharges intracellular Ca^{2+} stores by specific inhibition of the endoplasmic reticulum Ca^{2+}-ATPase, Proc. Natl. Acad. Sci. USA, 87:2466 (1990).
57. J. Lytton, M. Westlin, and M.R. Haley, Thapsigargin inhibits the sarcoplasmic or endoplasmic reticulum Ca-ATPase family of calcium pumps, J. Biol. Chem. 266:17067 (1991).
58. A. Christensson, C.B. Laurell, and H. Lilja, Enzymatic activity of the prostate-specific antigen and its reactions with extracellular serine proteinase inhibitors, Europe J. Biochem. 194:755 (1990).
59. L.D. Papsidero, M. Kuriyama, M.C. Want, J. Horoszewicz, S.S. Leong, L. Valenzuela, et al, Prostate antigen: a marker for human prostate epithelial cells, J. Natl. Cancer Inst. 66:37 (1981).
60. H. Lilja, C. Christensson, U. Dahlen, M-T. Martikainen, O. Nilsson, K. Pettersson, et al, Prostate-specific antigen in human srum occurs predominately in complex with α_1-antichymotrypsin, Clin. Chem. 37:618 (1991).

Discussion

T. McDonnell

Considering tumor progression and the response to Thapsigargin, have you looked in the context of p53 mutation or *bcl*-2 overexpression, whether or not this compound is still effective?

J. Isaacs

The cell line that I used *in vivo*, the TSU, was shown by my brother, Dr. William Isaacs, to have allelic loss of 1 allele for the p53 and the other allele is a null mutation so it does not express p53 and it undergoes programmed cell death induced by a variety of agents. In a series of studies that we have done with Maria Hartwick, at Hopkins, we have looked at the same androgen-independent prostatic cancer cells where a simbis-virus infection will activate programmed cell death. This is very interesting since simbis-virus does not integrate into the DNA, so it is not perturbing the DNA. When *bcl*-2 is transfected into those cells, the same type of virus is being produced but no cell death. The dose response curve for ionomycin, Thapsigargin and s-fluorodeoxyuridine in those *bcl*-2 transfectants suggests that *bcl*-2 expression is not preventing cell death, it is changing the kinetics of cell death. I am confused about some of the situations with *bcl*-2, whether it is a qualitative or a quantitative effect. And again this may be very important. The way we do our studies is that we give these drugs continuously, and that certainly is not mimicking what occurs in patients. So now we are going back and ask what happens if instead of exposing them chronically to these agents, we use pulse exposures. The *bcl*-2 may make them much more able to handle pulse exposures, whereas when you give it chronically, it is just too much for them to overcome.

A. Eastman

You give a long list of genes that are induced. Do you have some idea of how many of those genes might be induced in the dying cells and how many might be induced in the surrounding surviving cells? And also, which of those might be necessary for death and which ones are just passenger traits?

J. Isaacs

What we have been trying to do, is where we have antibodies for the various proteins, and where the antibodies are specific enough, to try to actually identify if enhanced protein expression is occurring also. TGF-β_1, for example, is upregulated in the epithelial cells themselves that are undergoing death. The receptor for TGF-β_1 which seems to be in the cells is also upregulated following castration. Calmodulin protein is demonstrable in the epithelial cells undergoing programmed cell death. For some of the genes where we do not have good enough antibodies to do immunocytochemistry and we have to physically separate stromal from glandular cells and look for enrichment, I have no feeling whatsoever that any of the genes we have identified are critical in this process. Some of these genes have been found by differential screens, looking for m-RNA's that come up rapidly following castration, for example. One thing is another caveat: In a differential library we have actually seen mitochondrial genes upregulated (e.g., cytochrome-C oxidase subunit 1). Mitochondrial genome is encoded in a polycistronic m-RNA message and then it is transluted into 8 or 10 different proteins. One of the m-RNA we can detect being upregulated within 24 hours following androgen ablation, is the cytochrome oxidase subunit 1 gene. Its' induction is transient. It comes up only for about a day or two and goes back down pretty quickly. Although we think that the mitochondria are not early targets in a programmed cell death, something is going on to effect the expression of this mitochondrial gene.

D. Livingston

Two questions: If I understand it correctly, you want to take this complex molecule and modify it such that by the esterification and other modifications, you are going to hook on a peptide which is a replica of a sequence that is present in seminal vesicles right?

J. Isaacs

Correct.

D. Livingston

How big is that peptide?

J. Isaacs

Well, the peptide is about 25 amino acids.

D. Livingston

So you are going to put 25 amino acids on. Now, if I understand it correctly, the next step is to look forward to the possibility that this hybrid molecule, this fusion protein, or fusion peptide, with the active agent on one end, is going find its way to a matrix bound copy of PSA, is that correct?

J. Isaacs

I am not so sure I would say matrix bound. It is extra-cellular.

D. Livingston

Extra-cellular positioned PSA molecule, and then I gather the next step is that the peptide gets cleaved, or at least gets bound by the enzyme?

J. Isaacs

It would cleave it "likely to go" then, due to the lypophilicity of the molecule, it is not going to go anywhere except into the cells in the local site of release.

D. Livingston

Including any normal cells in that area, is that right?

J. Isaacs

Absolutely. Normal cells would also be killed in that area.

D. Livingston

Would an antibody hooking strategy be useful here?

J. Isaacs

PSA is, unfortunately, secreted extra-cellularly.

D. Livingston

To any prostate associated protein?

J. Isaacs

There is a new one (i.e., prostate specific membrane antigen) that has just been identified and cloned. It was cloned and sequenced by the group at Memorial Sloan Kettering. This antigen does seem to be a protein which is bound on the plasma membrane, and interestingly, if anything, seems to be upregulated as the prostate cancer cells become more poorly differentiated. So there is substantial interest in this antigen for targeting therapy to prostate cancer cells.

R. Schimke

It would appear as if androgens are serving atrophic effects here and when you take them away then essentially something happens and the cells are going to die. Does this occur only in androgen sensitive tissues? What about estrogen sensitive tissues, or, in addition, are there a large number of specific trophic factors that maintain the viability of non proliferating cells? And if that is so, can you either use atrophic receptors? In other words, in the androgen dependent, not the independent state, or any number of other ways of attempting to develop specificity of delivery? You might think of other ways in which this sort of general trophic factor phenomenon may be important and you could use that in one of any number of ways to develop specificity of delivery. That is going to be an issue of importance.

J. Isaacs

Right. We and others have looked at, for example, MCF-7 estrogen dependent breast cancer cells and when you put those in animals, they are estrogen dependent and then when you remove estrogen, they activate programmed cell death. In the uterus, during menstrual cycle, removal of trophic factors also activate programmed cell death. There are many cells in the body that have androgen receptors, that respond to androgens in a trophic way but not with proliferation. So, for example, there is a muscle called the levator ani muscle, which has been used as a classic bio-assay for the anabolic effects of androgens. When you androgen ablate, the weight of the muscle decreases tremendously. In that setting, it is cellular atrophy; the cells do not undergo programmed cell death. What is interesting is that if you isolate nuclei from levator ani muscle cells and incubate them with calcium or magnesium, they do not undergo DNA fragmentation (i.e., auto-digestion), whereas the cells from the prostate, even from intact donors, when you incubate them with these ions, they will undergo DNA fragmentation. My own feeling is that part of the differentiation phenotype which controls or activates programmed cell death in each cell type is separately fixed or determined. For androgen dependent cells, whether they are seminal vesicle, coagulating gland, prostate, androgen is one of them. You can obviously activate the pathway other ways, as we have shown by elevating calcium or doing other things, but what we have been trying to do is use the normal pathway to just identify critical targets.

R. Schimke

Just let me comment. The position of a cell, either a proliferating cell or arresting cell, the default position may in fact be a cell death and unless the cell proliferates, let us say goes appropriately through mitosis, it will die. But in addition, I think you suggest that in fact the default position of a cell that is not maintained by some trophic interactions, whether that is cell-cell or hormonal, may also be to die, in which case one should spend a great deal of time trying to understand what are those interactions that are trophic. You are beginning to do that, obviously, in the prostate. Then if one can figure out how to exploit this fact, then one may well be able to get around the difficult problem that the only tumors that seem to respond are those which have a high degree of proliferating cells.

D. Papermaster

A technical question about the measurement of cell death in your tissues. You mentioned that it appeared as though the nuclei lasted only about 4 hours, and yet you were using the Gavrieli assay. Did you then multiply the number of stained cells by 6 to give you your daily cell death rates?

J. Isaacs

What we did was to castrate animals. At various times following castration, gave them back androgen, and then used the Gavrieli technique to measure the apoptotic index and to determine the half life of apoptotic bodies (i.e., approximately 4 hours). So then we adjust for the half life to determine the daily rate of cell death. If we make an estimate based on DNA content loss, we come up with the same number, which is 20% per day. That is another important point. When we do it by DNA loss, in the adult prostate, the numbers we are coming up with when we use the Gavrieli technique, suggest that you can account for basically all of the normal turnover. In other words, all of the death of the glandular prostatic cells is by terminal transferase identifiable cell loss (i.e., programmed cell death). There is a growing body of evidence that programmed cell death is not just a pathologic or a pathophysiologic activity but, in fact, is what is involved in the normal turnover of tissue.

P. Amati

When you say that the androgen dependent cells go through G0, have you followed RB phosphorylation?

J. Isaacs

We are doing this right now.

W. Giaretti

I have a short methodological question. You have shown a dual parameter flow cytometry in which the parameters were size and DNA. What is the ground to consider these signals at low DNA content and low size as apoptotic bodies?

J. Isaacs

You would call them debris?

W. Giaretti

I expect that if you call them apoptotic cell or apoptotic bodies, they would look like clouds, not just like continous signals down to zero.

J. Isaacs

You will notice that I did not attempt in that figure to quantitate the apoptotic bodies. I am always amazed that people have used flow cytometry to determine the apoptotic index based on a number of events that are sub-G0. It is clear that when a nucleus undergoes apoptosis that it breaks up into several fragments and so it is really hard to assess. I think it is a very interesting semi-quantitative method but I do not think that one can actually use it to quantitate an apoptotic index. That is why I am saying I cannot give you a direct answer. Some of that is debris, some of it is specific apoptotic bodies.

THE BCL-2 GENE : A REGULATOR OF PROGRAMMED CELL DEATH

David M. Hockenbery

Fred Hutchinson Cancer Research Center, Division of Clinical Research and Program in Molecular Medicine, Seattle, WA

INTRODUCTION

The bcl-2 gene was cloned as a novel gene located at a chromosomal translocation breakpoint in follicular B cell lymphomas by three groups in 1985 [1,2,3]. Follicular lymphomas are indolent tumors, difficult to eradicate but slow to progress, and are believed to originate from centrocytic B cells in germinal centers [4]. 85% or more of follicular lymphomas carry a t(14;18) (q32;q21) leading to the expectation that a new cellular oncogene would be activated by this translocation [5,6]. The bcl-2 gene is found on chromosome 18 at q21 and juxtaposed in a tail to head fashion to elements of the immunoglobulin heavy chain locus on chromosome 14 in the 14;18 translocation (Figure 1). As expected, the bcl-2-Ig fusion mRNA is overproduced in t(14;18) cells, leading to

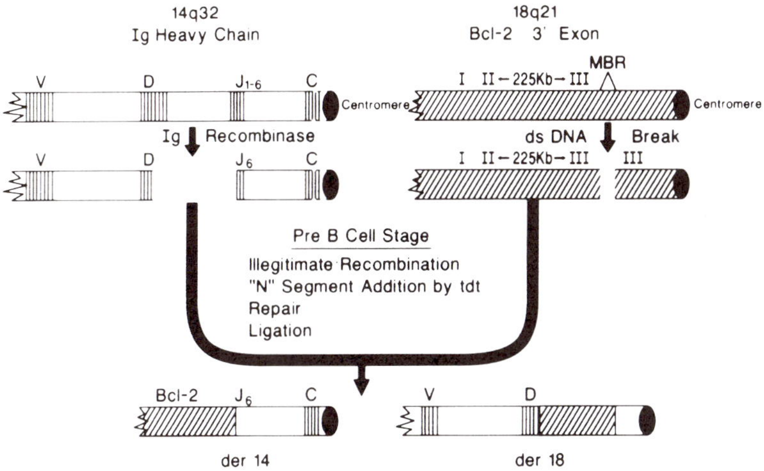

Figure 1. Structure of the t(14;18) translocation. Involvement of immunoglobulin recombinase is recognized by loss of internal D-J chromosome 14 sequences from der 14 and der 18 breakpoints as normally occurs during V(D)J recombination.

Apoptosis, Edited by E. Mihich and R.T. Schimke
Plenum Press, New York, 1994

elevated levels of the 26 kd bcl-2 protein and changes in cellular phenotype [7,8].

The breakpoints in the immunoglobulin heavy chain gene occur at the 5' end of a joining (J) region and 3' to a diversity (D) region, with loss of intervening DNA and addition of extra nucleotides characteristic of "N" segments [9]. This is notable because it dates the t(14;18) as a pre-B cell event, occurring during immunoglobulin VDJ recombination, and indicates significant normal differentiation occurs following the translocation to achieve the mature B cell phenotype of the malignant cell. Additional findings indicate the low probability of further steps culminating in full malignant transformation following an initial t(14;18). Detection of t(14;18) breakpoints by PCR analysis of DNA from tonsil samples containing reactive hyperplasia demonstrated a 54% prevalence in otherwise healthy individuals [10]. This remarkable finding has been recently confirmed [11]. In addition to other genetic events, extracellular signals may be important. Most follicular lymphomas have undergone an immunoglobulin heavy chain constant region switch recombination to the IgG isotype, implying subsequent exposure to cognate antigen may be a limiting step for progression of t(14:18)-bearing clones.

The breakpoints on chromosome 18 are also remarkably focused, suggesting sites of increased chromosomal breakage or cryptic recombination signals. The most frequent location lies in the 3' untranslated sequence of the bcl-2 gene, with 60% of breakpoints occurring within a 150bp sequence (mbr, major breakpoint region). Additional sites occur 3' to the bcl-2 gene (mcr, minor chromosomal rearrangement) and 5' to the bcl-2 gene [12,13,14]. Immunoglobulin recombination signal sequences (heptamer-space-nonamer) are not found near chromosome 18 breakpoint sequences. A putative recombination motif homologous to prokaryotic chi sequences has been found adjacent to several chromosomal breakpoints, including the mbs and mcr bcl-2 regions and within the immunoglobulin heavy chain locus [15]. Recently, a 45-Kd nuclear protein which binds chi-type sequences in the bcl-2 mbr and mcr breakpoints has been discovered [16]. Early B cell lines were found to possess a specific endonuclease activity targetting the mbr region. A plausible model involves homologous recombination between breakpoint sequences on chromosomes 14 and 18, mediated by a chi-specific recombinase and V(D)J-recombinase, with possible involvement of the 45-Kd binding protein.

RESULTS AND DISCUSSION

The initial description of the cellular function of the bcl-2 gene in 1988 examined the effect of bcl-2 overexpression on cellular viability following withdrawal of interleukin-3, a growth and survival factor from FDC.P1 promyeloid cells [17]. Bcl-2 increased viability in this context, but did not cause FDC.P1 cells to lose their dependence on IL-3 for proliferation or provide an additional proliferative stimulus. The cell death induced by IL-3 withdrawal can be shown to be typical of apoptosis or programmed cell death [18]. Characteristic internucleosomal DNA fragmentation and cellular morphological changes are prevented by bcl-2, implying that bcl-2 acts at an early step in the apoptotic pathway. These results have been confirmed in a variety of models of cellular survival factor withdrawal, although important exceptions have been noted. These cell deaths associated with IL-2, IL-6, and ciliary neurotrophic factor (CNTF) withdrawal otherwise resemble apoptosis, and the lack of bcl-2 effect suggests that multiple intracellular apoptotic pathways exist and include bcl-2-resistant and sensitive varieties [19,20].

Evidence supporting an *in vivo* role for bcl-2 in regulating cell death has been less easy to come by, in part due to the greater difficulty in quantitating cell survival *in vivo* and in part due to the need to understand the function of physiologic levels of bcl-2 within normal cells. These problems not withstanding, existing data suggests that bcl-2 functions as one, or perhaps the, important physiologic regulator of cell survival in cell lineages in which apoptosis determines cell turnover or developmental cell loss. The most compelling case can be made for B lymphocytes on which most studies to date have focussed as the cell of origin for t(14;18) bearing tumors. Transgenic mice bearing a bcl-2-Ig transgene overexpress bcl-2 in B lymphocytes and their precursors and develop polyclonal . B lymphocyte hyperplasia [21,22]. B lymphocytes are normally short-lived and the transgenic B cells are not proliferating, implying enhanced survival for their expansion. These cells have a dramatically prolonged survival in vitro, with a significant fraction surviving up to

6 months. Cells of the B lymphocyte lineage undergo cell death at early precursor, pre-B, mature B, and plasma cell stages of development reflecting requirements for correct immunoglobulin gene recombination, selection, and cell renewal. Increased numbers of plasma cells and pre-B cells are found in bcl-2 transgenic mice, suggesting that bcl-2 circumvents cell death at more than one stage of differentiation.

Immunologic memory persists in long-lived B and T cells and could plausibly be a function determined by bcl-2 regulation of cell survival. Indeed, bcl-2 transgenic mice have prolonged secondary responses to T cell-dependent antigens as measured by plaque-forming cell assay. In order to directly test for enhanced survival of memory B cells, an adoptive transfer protocol was utilized [23]. Transgenic mice and control littermates underwent primary immunization with a hapten-carrier, FITC-KLH (fluorescein isothiocyanate-keyhole limpet hemocyanin). Memory B cells were transferred 2-3 months after immunization to hapten-naive irradiated recipient mice as an enriched Percoll gradient fraction of resting B lymphocytes. In the absence of persistent cognate antigen in the new host, memory B cells rapidly decline, as assessed by immunologic response 7 days after a secondary immunization with FITC (396 indirect p.f.c. per 106 spleen cells at day of transfer to 20 p.f.c. per 106 cells at day 28-35) . However, when bcl-2 transgenics are used as the source of donor memory cells, the transferred cells remain functionally viable in the absence of antigen (433 indirect p.f.c. per 106 cells at day of transfer compared to 254 p.f.c. at day 28-35). Intracellular bcl-2 may respond to interactions with persistent antigen and stave off an opposing apoptotic tendency of memory B cells. High level, constitutive production of bcl-2 in transgenic models enables prolonged survival of memory B cells without the antigenic cues normally required.

Memory B cells and plasma cells are derived from germinal center reactions during late primary and secondary immune responses. Studies of bcl-2 expression and regulation in germinal center cell subpopulations has helped to solidify the evidence for an in vivo survival function of bcl-2. Germinal centers function to select B cells with high-affinity antibodies on their cell surfaces for differentiation into plasma cells or survival as memory B cells [24]. During a germinal center reaction, an initial 1-3 B cells are expanded by a period of rapid cell division, and undergo accelerated mutation at immunoglobulin loci to generate a large number of B cell clones with varying Ig sequence and antigen affinity. As only high-affinity clones are ultimately selected, a high rate of cell death by apoptosis occurs in the germinal center centrocyte (light) zone. The signals required for survival are believed to include interaction with limiting amounts of antigen retained on follicular dendritic cells. Immunohistochemical staining of germinal centers with monoclonal antibodies to bcl-2 demonstrates a remarkable lack of bcl-2 in the germinal center, compared with the surrounding and interfollicular zone, with the exception of scattered cells in the centrocyte area [25,26] (Figure 2). Dual staining with B cell markers reveals most of these cells are B cells. While compatible with a role for bcl-2 in regulating survival

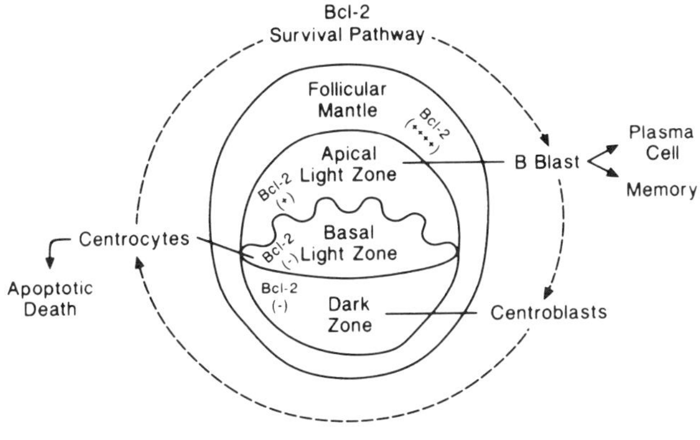

Figure 2. Organization of secondary germinal center and distribution of bcl-2 protein by immunohistochemistry. Survival function of bcl-2 mediates positive selection of centrocytes during affinity maturation of immune response.

during affinity maturation of the immune response, difficulties in recognizing cells destined to survive in histologic sections prevent testing of this hypothesis.

Elegant studies on isolated germinal center cells by Liu, et al provide the necessary additional information on bcl-2 function [27]. Centrocytic germinal center cells undergo a quicker apoptotic death *in vitro* assayed against follicular mantle zone B cells. Certain cell surface ligands are capable of rescuing these cells and promote morphologic differentiation to the expected germinal center reaction products: anti-CD40 antibody (induces memory B cell morphology) and rCD23 antigen plus rIL-1 (induces plasmablast morphology). In addition, anti-Ig-coated sheep red blood cells promote centrocyte survival, mimicking antigenic engagement of surface Ig. The reversal of the active apoptotic pathway by all these agents is associated with early induction of bcl-2 protein. An elaborate scheme for clonal selection of B cells involving activation of an apoptosis program, perhaps initiated by loss of bcl-2 expression, with subsequent rescue of those cells receiving a necessary array of signals, foremost antigenic stimulation, thus culminates in return of bcl-2 expression and blockade of cell death. An additional indicator of the complexity of this system is the recent evidence of abundant bcl-2 mRNA by in situ hybridization within the germinal center zones in the absence of protein, suggesting post-transcriptional control of gene expression as a mechanism for rapid induction of bcl-2 in positively-selected cells [28].

A still murky area is the in vivo role of bcl-2 in thymic development. The protein distribution by immunohistochemistry is highly suggestive of a role in cell survival during thymic selection. Bcl-2 is absent in most cortical thymocytes, upregulated near the cortico-medullary junction, and abundant in medullary thymocytes, cells which have successfully negotiated positive and negative selecting barriers and achieved a mature phenotype [25,18]. Transgenic mice which overexpress bcl-2 in the thymus using the lck tyrosine kinase proximal promoter have thymocytes which are dramatically resistant to several induced models of apoptotic death, such as gamma-irradiation, glucocorticoid treatment, and anti-CD3 antibodies [29]. Despite clear evidence of bcl-2 function in these examples, examination of the physiologic loss of cells associated with negative selection showed no effect. The elimination of cells by negative selection was assessed by the representation of specific V_{beta}-bearing cells normally deleted by endogenous superantigens, a model for interactions of self-peptides and major histocompatibility antigens with reactive T cell receptor bearing thymocytes. Two explanations for this result seem plausible. The first is the existence of at least two intracellular pathways to apoptosis, resulting in the same morphologic phenotype, one of which is independent of bcl-2. This fits well with the notion that morphologic responses to injury are considerably less varied than the mechanisms of damaging cells. Several examples in which bcl-2 fails to protect cells from apoptosis have been reported including withdrawal of certain survival factors like IL-2 and CNTF, cytotoxic T cell killing, and death of immature B lymphocyte cell lines following anti-IgM treatment [19,20,30].

A second possibility is one of variable activation of a single apoptotic pathway in different circumstances, rather than a simple on-off switch. In this model, the effect of bcl-2 in a particular experiment would have to be measured against the level of a quantifiable, but currently unknown, intermediate in the apoptotic pathway. Cell deaths with lower levels of intracellular apoptotic activation would be fully protected by bcl-2, but cell deaths with heightened cell death mechanisms would show only partial or no effects. There are many examples of bcl-2 protection from apoptosis in which cell death is postponed rather than fully abrogated, suggesting a partial effect. There are three published reports of transgenic overexpression of bcl-2 in the thymus which all differ in results concerning negative selection. The previously mentioned lck-bcl-2 transgene model led to no alteration in deletion of superantigen-reactive cells from mature thymocyte ($CD3^{hi}$) populations, although an increase of $CD3^{med}$ cells suggested cells were accumulating at intermediate stages of selection [29]. A E^{mu}-bcl-2 transgene model which demonstrated thymic overexpression of bcl-2 was also shown to lack superantigen-reactive cells among peripheral T cells, although some mature thymocytes did appear to escape negative selection [31]. Finally, another bcl-2/immunoglobulin enhancer transgene construct with T cell-restricted expression in some transgenic lines led to significant increases in expression of negatively selected V_{beta}-bearing cells among mature thymocytes and peripheral T cells, although less than levels found in genetic backgrounds lacking the I-E histocompatibility antigen required for the deleting function of the superantigen [32]. These variant results suggest that the ability of bcl-2 to affect cell death due to negative selection may be influenced by genetic background, transgene construction, or aspects of the superantigen-TCR interaction. A level of bcl-2 expression which is suboptimal for the intracellular apoptotic pathway activation induced by superantigen could explain these results.

The possible role of bcl-2 as an intracellular survival factor in positively-selected thymocytes remains to be assessed. The upregulation of bcl-2 coincident with the transition

Table 1 Cell deaths blocked by bcl-2

growth factor/serum deprivation	cancer drugs
heat shock	phorbol esters
ethanol	calcium ionophores
glucocorticoid	TCR-mediated T cell death
gamma-radiation	Ig-mediated B cell death
azide	Adenovirus E1A protein
tumor necrosis factor	viral cytolysis

from immature cortical thymocytes to mature medullary thymocytes has a striking resemblance to events in the germinal center, where B lymphocytes become extremely susceptible to apoptosis unless rescued by a positive signal involving the immunoglobulin antigen receptor.

The function of bcl-2 in non-lymphoid tissue is also an unaddressed issue at present. The ability of bcl-2 to function in nonlymphoid cells as an inhibitor of apoptotic death appears to be a general property and not rely on tissue-specific factors. Survival functions have been documented in fibroblasts, primary neuronal cells, and epithelial cell lines [33]. The pattern of bcl-2 expression in adult and fetal tissues is indicates that bcl-2 is present in most if not all lineages undergoing apoptotic cell death during development and normal cell turnover. This is exemplified in the adult by bcl-2 expression in glandular tissues such as breast and prostate which in adult life are capable of cellular expansion and contraction by apoptosis in response to hormonal cues, epithelia such as intestine which undergoes rapid renewal and turnover by apoptosis, and hematopoietic precursor cells which are dependent for survival on factors such as erythropoeitin which block apoptosis [26]. Expression during fetal morphogenesis correlates with sites of selective cell death and tissue remodeling such as developing nervous system, Mullerian duct-derived structures, and developing proximal nephrons [37]. Definitive evidence for the function of bcl-2 in these various contexts will await genetic manipulations of bcl-2 expression with tissue-specific transgene constructs and gene knockouts by homologous recombination. It is also possible that bcl-2 function in certain tissues will be replaced or modified by homologous proteins. Along these lines, a new gene with homology to bcl-2, MCL-1, has been cloned from a human myeloid cell line, although whether it can suppress apoptosis is unknown at present[34].

The restricted distribution of bcl-2 in certain cell lineages poses certain questions regarding the in vivo function of bcl-2. In intestinal epithelium, for example, the expression of bcl-2 is confined to the lower zones of the intestinal crypt, compatible with expression by proliferating cells and the stem cell compartment [26]. Terminally differentiated enterocytes in the crypt and villus epithelium lack bcl-2 expression. If the level of bcl-2 expression determines cellular susceptibility to apoptosis, one would expect the cells in the lower crypt to be the most resistant to apoptosis. On the contrary, when apoptosis-inducing treatments such as irradiation or cytostatic drugs are applied to the intestine, cells in the proliferating zone are profoundly sensitive to apoptosis while fully differentiated enterocytes are unaffected. One possible explanation for this dichotomy is, again, a matching of bcl-2 expression to the level of endogenous activation of an apoptotic death pathway. Cells which have higher levels of bcl-2 expression in a cell lineage may require that level to balance a cellular tendency toward apoptosis. In this way, bcl-2 may serve as a marker for cell types prone to apoptotic death. Since cells in the proliferating zone of intestinal epithelium are not known to undergo a regulated cell death, as a way to modulate intestinal growth for example, activation of an apoptotic pathway may be linked in some manner to proliferation or differentiation programs.

The variety of agents and treatments which can elicit apoptotic cell death and be counteracted by bcl-2 is astounding and still growing (Table 1). In addition to growth factor and serum withdrawal, resistance to tumor necrosis factor-mediated cytotoxicity with bcl-2 has been reported, although there are also negative reports . Physical treatments such as low-dose gamma radiation, heat shock, and 6% ethanol are counteracted by bcl-2 [35]. All classes of cancer chemotherapeutic drugs and metabolic inhibitors such as azide have been reported to be less cytotoxic with bcl-2 overexpression [36,37,31]. Many models of

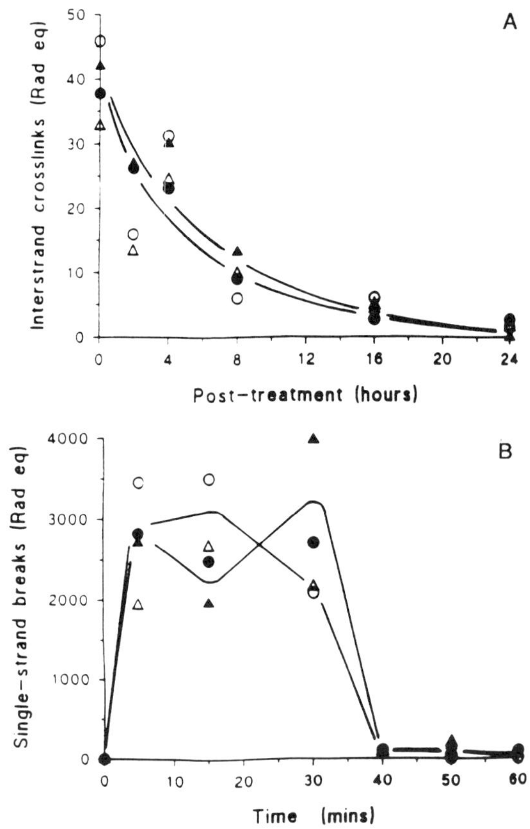

Figure 3. A. Accumulation and repair of DNA interstrand cross-links in FL5.12 cells following 30 min treatment with nitrogen mustand. B. Accumulation and repair of DNA single strand breaks in FL5.12 cells follwing 30 min treatment with camptothecin. *Open symbols* , control SFFV-neo transfectants; *Closed symbols*, SFFV-bcl-2 transfectants

apoptosis of lymphocytes, including anti-CD3-mediated death in T cells, anti-Ig-mediated death in B cells, and induction of apoptosis by calcium ionophores, phorbol esters, and glucocorticoids can be inhibited by bcl-2 [29,31]. Apoptosis triggered by adenovirus E1A protein and cytolysis normally occurring in the course of a lytic viral infection are suppressed by bcl-2 [38,39].

Knowledge of the function of bcl-2 will be central to understanding how disparate inducers of apoptosis are related, and undoubtedly, to uncovering the mechanism of apoptosis. The localization of bcl-2 within a cell may provide a clue to its function. Using a monoclonal antibody to bcl-2, the cytoplasmic immunofluorescence pattern of bcl-2 was shown to coincide with a mitochondrial fluorescent probe in a t(14;18)-bearing cell line [18]. Subcellular fractionation studies confirmed mitochondria as the only apparent localization in this cell line, and further defined the inner mitochondrial membrane as the probable membrane spanned by the C-terminal hydrophobic sequence of bcl-2. Subsequent immunofluorescent studies using different cell lines and antibodies have confirmed that mitochondria constitute a major localization for bcl-2 but have also suggested that the nuclear membrane and endoplasmic reticulum can be targets as well [27]. A possible role for bcl-2 relating to energy metabolism or ion transport may be envisioned, but no function has been proposed to date.

Many of the models of apoptotic cell death that are studied in vitro have in common a cell cycle arrest or slowing that precedes cell death. This may be elicited by a genotoxic mechanism, as in DNA damaging drugs or irradiation, or by growth factor withdrawal or interference with microtubule formation. Agents which arrest cells at different positions in the cell cycle are all capable of inducing apoptosis, and there is no evidence of a specific cell cycle position which is required for apoptosis to occur. The influence of bcl-2 on cell cycle behavior and DNA repair were examined in detail for two different DNA damaging

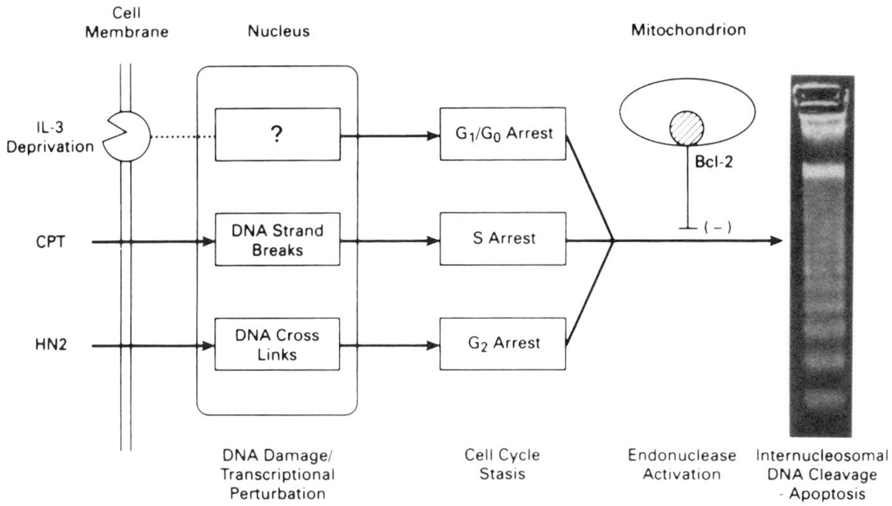

Figure 4. Model of the relationship between cell cycle arrest signals and bcl-2 function in apoptosis.

drugs, nitrogen mustard, an alkylating agent, and camptothecin, a topoisomerase I inhibitor [37]. These drugs produced G2 and S phase arrests, respectively, which were unaltered by bcl-2 overexpression sufficient to inhibit apoptotic death by 50%. The lack of effects on cell cycle behavior of bcl-2 confirms earlier reports on the G0/G1 arrest preceding cell death with IL-3 withdrawal. Examination of DNA intrastrand crosslinks after HN2 treatment and single strand breaks after camptothecin showed that formation and repair of DNA damage were not altered by bcl-2 overexpression and that repair of the DNA damage which led to cell cycle arrest precedes the onset of apoptosis (Figure 3). This suggests that apoptosis does not proceed because of irreparable DNA damage, but rather is triggered by cell cycle arrest in susceptible cells. A model can be proposed in which bcl-2, predominantly localized within mitochondria, functions distally in an apoptotic pathway which is activable by cell cycle arrest at multiple points within the cell cycle (Figure 4). Although examples of apoptotic cell death in non-cycling cells are known, e.g. following terminal differentiation in many lineages, in vitro models associated with cell arrest may be useful in searching for common intermediates in apoptotic pathways with which bcl-2 interacts.

CONCLUSIONS

The discovery of the bcl-2 gene by using the chromosomal translocation breakpoint in follicular lymphoma as a physical marker for a candidate oncogene and the deduction of its function in programmed cell death pathways has been one of the triumphs of the reverse genetic approach. More than morphologic similarities and the hallmark DNA fragmentation, the ability of bcl-2 to interdict apoptotic cell death occuring during development, responses to cellular injury, and following terminal differentiation serves to link these disparate deaths in a common mechanism. In all probability, an understanding of the function of the bcl-2 protein will reveal the essential steps in apoptotic death, and lead to interventions to alter this basic cellular process that will be useful for both conditions of excess cell growth and excess cell death, i.e. malignancies and degenerative diseases.

ACKNOWLEDGMENTS

D. M. H. is a Lucille P. Markey scholar and this work was supported by a grant from the Lucille P. Markey Charitable Trust.

REFERENCES

1. Tsujimoto, Y., Cossman, J., Jaffe, E. & Croce, C.M.,Involvement of the bcl-2 gene in human follicular lymphoma,*Science* 228: 1440-1443 (1985).
2. Cleary, M.L. & Sklar, J.,Nucleotide sequence of a t(14;18) chromosomal breakpoint in follicular lymphoma and demonstration of a breakpoint-cluster region near a transcriptionally active locus on chromosome 18, *Proc Natl Acad Sci USA* 82: 7439-7443 (1985).
3. Bakhshi, A., *et al.*,Cloning the chromosomal breakpoint of t(14;18) human lymphomas: clustering around J_H on chromosome 14 and near a transcriptional unit on chromosome 18,*Cell* 41: 899-906 (1985).
4. Horning, S.J. & Rosenberg, S.A.,The natural history of initially untreated low-grade non-Hodgkin's lymphomas,*N Engl J Med* 316: 1471 (1984).
5. Fukahara, S., Rowley, J.D., Varrakojis, D. & Golomb, H.M.,Chromosome abnormalities in poorly differentiated lymphocytic lymphoma,*Cancer Res* 39: 3119-3131 (1979).
6. Yunis, J.J., *et al.*,Multiple recurrent genomic defects in follicular lymphoma,*N Engl J Med* 316: 79-84 (1987).
7. Seto, M., *et al.*,Alternative promoters and exons, somatic mutation and transcriptional deregulation of the Bcl-2-Ig fusion gene in lymphoma,*EMBO* 7: 123-131 (1988).
8. Chen-Levy, Z., Nourse, J. & Cleary, M.L.,The bcl-2 candidate protooncogene product is a 24-kilodalton integral-membrane protein highly expressed in lymphoid cell lines and lymphomas carrying the t(14;18) translocation,*Mol Cell Biol* 9: 701-710 (1989).
9. Bakhshi, A., *et al.*,Mechanism of the t(14;18) chromosomal translocation: structural analysis of both derivative 14 and 18 reciprocal partners,*Proc Natl Acad Sci USA* 84: 2396-2400 (1987).
10. Limpens, J., *et al.*,Bcl-2/J_H rearrangements in benign lymphoid tissues with follicular hyperplasia,*Oncogene* 6: 2271-2276 (1991).
11. Aster, J.C., Kobayashi, Y., Shiota, M., Mori, S. & Sklar, J.,Detection of the t(14;18) at similar frequencies in hyperplastic lymphoid tissues from American and Japanese patients,*Am J Path* 141: 291-299 (1992).
12. Cleary, M.L., Galili, N. & Sklar, J.,Detection of a second t(14;18) breakpoint cluster region in human follicular lymphomas,*J Exp Med* 164: 315-320 (1986).
13. Tsujimoto, Y., *et al.*,DNA rearrangements in human follicular lymphoma can involve either the 5' or the 3' region of the bcl-2 gene,*Proc Natl Acad Sci USA* 84: 1329-1331 (1987).

14. Adachi, M., Cossman, J., Longo, D., Croce, C.M. & Tsujimoto, Y.,Variant translocation of the bcl-2 gene to immunoglobulin light chain gene in chronic lymphocytic leukemia,*Proc Natl Acad Sci USA* 86: 2771-2774 (1989).
15. Wyatt, R.T., Ruders, R.A., Zelenetz, A., Delellis, R.A. & Krontiris, T.G.,BCL2 oncogene translocation is mediated by a c-like consensus,*J Exp Med* 175: 1575-1588 (1992).
16. Jaeger, U., *et al.*,Mechanism of the chromosomal translocation t(14;18) in lymphoma: detection of a 45-Kd breakpoint binding protein,*Blood* 81: 1833-1840 (1993).
17. Vaux, D.L., Cory, S. & Adams, J.M.,Bcl-2 gene promotes haemopoietic cell survival and cooperates with c-myc to immortalize pre-B cells,*Nature* 335: 440-442 (1988).
18. Hockenbery, D., Nunez, G., Milliman, C., Schreiber, R.D. & Korsmeyer, S.J.,Bcl-2 is an inner mitochondrial membrane protein that blocks programmed cell death,*Nature* 348: 334-336 (1990).
19. Nunez, G., *et al.*,Deregulated Bcl-2 gene expression selectively prolongs survival of growth-factor-deprived hemopoietic cell lines,*J immunol* 144: 3602-3610 (1990).
20. Allsopp, T.E., Wyatt, S., Paterson, H.F. & Davies, A.M.,The proto-oncogene *bcl-2* can selectively rescue neurotrophic factor-dependent neurons from apoptosis,*Cell* 73: 295-307 (1993).
21. McDonnell, T.J., *et al.*,bcl-2-immunoglobulin transgenic mice demonstrate extended B cell survival and follicular lymphoproliferation,*Cell* 57: 79-88 (1989).
22. McDonnell, T.J., *et al.*,Deregulated Bcl-2-immunoglobulin transgene expands a resting but responsive immunoglobulin M and D-expressing B-cell population,*Mol Cell Biol* 10: 1901-1907 (1990).
23. Nunez, G., Hockenbery, D., McDonnell, T.J., Sorenson, C.M. & Korsmeyer, S.J.,Bcl-2 maintains B cell memory,*Nature* 353: 71-73 (1991).
24. MacLennan, I.C.M. & Gray, D.,Antigen-driven selection of virgin and memory B cells,*Immunol Rev* 91: 61-65 (1986).
25. Pezzella, F., *et al.*,Expression of the Bcl-2 oncogene protein is not specific for the 14;18 translocation,*Am J Path* 137: 225-232 (1990).
26. Hockenbery, D.M., Zutter, M., Hickey, W., Nahm, M. & Korsmeyer, S.J.,BCL2 protein is topographically restricted in tissues characterized by apoptotic death,*Proc Natl Acad Sci USA* 88: 6961-6965 (1991).
27. Liu, Y.-J., *et al.*,Germinal center cells express bcl-2 protein after activation by signals which prevent their entry into apoptosis,*Eur J Immunol* 21: 1905-1910 (1991).
28. Chleq-Deschamps, C.M., *et al.*,Topographical dissociation of BCL-2 messenger RNA and protein expression in human lymphoid tissues,*Blood* 81: 293-298 (1993).
29. Sentman, C.L., Shutter, J.R., Hockenbery, D., Kanegawa, O. & Korsmeyer, S.J.,bcl-2 inhibits multiple forms of apoptosis but not negative selection in thymocytes,*Cell* 67: 879-888 (1991).
30. Cuende, E., *et al.*,Programmed cell death by *bcl-2*-dependent and independent mechanisms in B lymphoma cells,*EMBO* 12: 1555-1560 (1993).
31. Strasser, A., Harris, A.W. & Cory, S.,bcl-2 transgene inhibits T cell death and perturbs thymic self-censorship,*Cell* 67: 889-899 (1991).
32. Siegel, R.M., *et al.*,Inhibition of thymocyte apoptosis and negative antigenic selection in *bcl*-2 transgenic mice,*Proc Natl Acad Sci USA* 89: 7003-7007 (1992).
33. Garcia, I., Martinou, I., Tsujimoto, Y. & Martinou, J.-C.,Prevention of programmed cell death of sympathetic neurons by the *bcl*-2 proto-oncogene,*Science* 258: 302-304 (1992).
34. Kozopas, K.M., Yang, T., Buchan, H.L., Zhou, P. & Craig, R.W., *MCL*1, a gene expressed in programmed myeloid differentiation, has sequence similarity to *BCL*2, *Proc Natl Acad Sci USA* **90**: 3516-3520 (1993).
35. Tsujimoto, Y.,Stress-resistance conferred by high level of bcl-2 protein in human B lymphoblastoid cell line,*Oncogene* 4: 1331-1336 (1989).
36. Miyashita, T. & Reed, J.C.,Bcl-2 oncoprotein blocks chemotherapy-induced apoptosis in a human leukemia cell line,*Blood* 81: 151-157 (1993).
37. Walton, M.I., *et al.*,Constitutive expression of human Bcl-2 modulates nitrogen mustard and camptothecin induced apoptosis,*Cancer Res.* 53: 1853-1861 (1993).

38. Rao, L., *et al.*, The adenovirus E1A proteins induce apoptosis, which is inhibited by the E1B 19-kDa and Bcl-2 proteins, *Proc Natl Acad Sci USA* 89:7742-7746 (1992).
39. Levine, B., *et al.*, Conversion of lytic to persistent alphavirus infection by the *bcl-2* cellular oncogene, *Nature* 361:739742 (1993).

Discussion

J. Ashwell

In the situation where you saw that you did not see DNA fragmentation in the *bcl*-2 double transgenic animals, did you look at the morphology for apoptotic nuclear changes?

D. Hockenbery

By EM they look classically apoptotic. So one thought is that you are actually converting something to more of a necrotic pathway since there are models in which apoptosis and necrosis perhaps form a continuum in response to certain agents and that if you block the apoptotic pathway, perhaps necrosis takes over. But these cells really look apoptotic and the time course is identical. There is really no way to distinguish cell death in this context from apoptosis that occurs associated with the DNA ladder. This echoes the work of David Ucker, looking at cells which lack an endonuclease function and you cannot distinguish the cell death in that model either.

J. Ashwell

Did you actually look at a nuclease activity in these cells?

D. Hockenbery

No, we have not.

E. Mihich

Have you tried, or has anybody tried, to alter mitochondrial function and see which kind of alteration of mitochondrial function would affect the capacity of *bcl*-2 to block apoptosis? I am reminded of a relatively old drug that LePecq at Villejuif had studied, called Ditercalinum. It was a di-basic structure, a long di-basic structure which was affecting mitochondrial function and causing death only in a very delayed fashion. In retrospect, that might be involved in some of the functions of *bcl*-2.

D. Hockenbery

We actually have not had any success but I can cite two examples from the literature in which modulation of mitochondrial function has affected cell death, and people in the audience might come up with counter-examples, but, in TNF cytotoxicity, Walter Fier's group has shown that you can modulate TNF death by blocking at different points in the electron transport chain. If you use for example, rotenone, which blocks upstream of Coenzyme-Q, complex-3, that ameliorates cell death due to TNF, and if you block downstream you actually heighten cell death. There are also reports that in camptothecin and other topoisomerase inhibitors, cytotoxicity can be modulated by oxidative phosphorylation uncouplers. These are isolated reports but suggest that mitochondrial function can modulate those pathways.

E. Mihich

I really was aiming at a functional assay, more than a toxicological assay, of the mitochondria in order to try to dissect what molecular mechanism might be involved in the function of *bcl* at the mitochondria level.

D. Hockenbery

We do not know of any functional alteration that specifically interacts with *bcl*-2 and not more generally with cell viability. We have looked, and other people would concur, that ATP does not seem to be a consistent finding in terms of the end product of mitochondrial synthesis. We have looked at oxygen consumption with various combinations of electron donor substrates and upstream electron transfer inhibitors and we do not find differences. And we have looked at calcium fluxes, and again that has not been fruitful, although there are data in the literature that John Reed has published, that there may be shifts in calcium that accompany IL-3 withdrawal that are modulated by *bcl*-2 and that is still a possibility.

D. Delia

With regard to this last cartoon and *bcl*-2 association with mitochondria, there are two important observations that need mentioning. One recent paper published in Nature shows that cells in which mitochondria are not functioning, *bcl*-2 does not alter apoptosis or drug resistance. Two ultra-structural studies show that *bcl*-2 is localized not only outside of the mitochondrial membrane but also in the perinuclear membrane. My question is: Do you have evidences, say by doing deletion studies, that the protein with the perinuclear localization may have a more important role in protecting from apoptosis?

D. Hockenbery

Addressing the first part of the question, I think that paper is a bit perplexing, that cells that lack mitochondrial DNA do undergo apoptosis and are notprotected by *bcl*-2. But one has to recognize that those cells do have mitochondria that do not function in terms of oxidative phosphorylation and so they rely on glycolysis. And *bcl*-2 does target those mitochondria because that paper did show that *bcl*-2 resided in mitochondria. Gordon Share, in collaboration with the Korsmeyer laboratory, has shown that you can *in vitro* import *bcl*-2 to mitochondria in an ATP-dependent manner. There remains the question of whether the mitochondria in those cells have a function which *bcl*-2 acts on, for example, calcium. It is not clear in that model that mitochondrial calcium stores would not function fairly normally. We know from the *bcl*-2 truncation data that *bcl*-2 can function outside mitochondria, so I suspect that nuclear localization would have a function in terms of cell death pathways. I think all you can say until we really arrive at a biochemical function for *bcl*-2 is that you have to take these results with a grain of salt. There are examples where nuclear localization has not been found; other studies have found ER localization. There may be cell type variations in these things and it may be that it functions in all the places and that different cells localize it somewhat differently. My bias is that since all of the data that I am familiar with, all the data has shown some mitochondrial localization across all the cell types that have been looked at, that mitochonddria are functionally a preferred site.

G. Bellomo

In one of your slides you mentioned *bcl*-2 localization in pancreas. Is there any cell type specific localization?

D. Hockenbery

We find it both in ductal regions and the exocrine pancreas and also within islets. We have not looked within the subtypes of islet cells.

M. Oren

You noted that the retina is a site of high *bcl*-2 expression. Has anybody looked at retinoblastoma? Is it possible that high levels of *bcl*-2 are maintained there and have something to do with protecting those cells from death?

D. Hockenbery

I do not know of any data for expression levels in retinoblastomas. It is possible, sure.

M. Fried

Has there been any progress with knockouts of *bcl*-2?

D. Hockenbery

Yes, the Korsmeyer laboratory has a knockout that is being analyzed now. These mice are viable, they are born. However, they do appear abnormal and they do die at early ages, about two, three weeks of age, and I believe the early histologic screens show abnormalities that are consistent with aberrances in cell death pathways.

M. Fried

Do the ES cells grow OK? Is there a phenotype associated with the *bcl*-2 deficient ES cells?

D. Hockenbery

The ES cells appear to grow OK and it is interesting because *bcl*-2 is expressed in ES cells.

D. Livingston

What are your thoughts about the possibility of functional redundancy for *bcl*-2? It comes to mind especially since I see the little cells that do not seem to have *bcl*-2 in them but yet are potentially very primitive.

D. Hockenbery

It is curious that some cells which lack *bcl*-2 expression do as well as, or better than, *bcl*-2 positive cells from the same lineage in survival assays. This could mean another protein with the same function is expressed, and both proteins are regulated in a complicated manner to provide a survival mechanism to all cells in that lineage. One point, though, is that if you look at developmental time points where cell death takes place as a decision process, then you really do not find places where *bcl*-2 would not be consistent with that effect. So, unless you knew there were other *bcl*-2 homologues out there, I do not think you would necessarily, based on that data, have to come up with another protein with a *bcl*-2 function. Turning that around, there are homologous *bcl*-2 proteins. One that has been reported, mcl-1, is expressed in myeloid cells and was isolated after TPA stimulation by a subtraction method. It has yet to be reported whether those have analogous functions and it is possible that *bcl*-2 may actually interact with those in some productive fashion. But I think we have to wait for those studies.

R. Horvitz

Going back to the question of mitochondrial localization, I have two questions: One, how sensitive are the assays? In other words, when you look at the mutant that is missing the C-terminus, how much might you not be able to detect if there were some *bcl*-2 still on the mitochondrion? And related to that question, has anyone done the experiment to basically dose *bcl*-2 down so you have a dose response curve for protection as a function of *bcl*-2 concentration, and, coming to the question, when you get to the point where there is no detectable *bcl*-2 on mitochondria, can you still have function even if you have the C terminus tail present?

D. Hockenbery

The sub cellular fractionation on those C-terminal deletions was quite clean. So within the limits of a Western blot to detect protein that is localized to another site, it appears that there would not be enough to be at all proportionate to the protection that is received. Now a caveat to that, I guess, is that during sub cellular fractionation you conceivably could artifactually alter localization especially for something that is not anchored in a membrane. However, the immunofluorescence is entirely consistent with a cytoplasmic localization in that setting.

R. Horvitz

The question basically is, what happens if you modulate expression so you have different cells that have different levels of *bcl*-2 expressed? Obviously, you can start at a very high level when you have a lot of product. You can go down to a zero level when you have no product. So, as you look along that curve, there is some point at which you will lose *bcl*-2 that is visibly associated with mitochondria and some point at which you will lose *bcl*-2 protective function. The question is, basically, can you show with the intact *bcl*-2 product if you can get functional *bcl*-2 at a level where you no longer can see it ? If you can, then you would question the absence of *bcl*-2 in the localization experiments as being meaningful with respect to mitochondria. Does that help?

D. Hockenbery

No, I cannot think of anyone that has really done exactly that. There is a report from John Reed, for example, that you can take a cell which, with I think what was recognized at that time as a crude antibody, had barely detectable *bcl*-2, treat it with anti-sense and decrease in survival. I think sensitivity of your detection method is an issue. *Bcl*-2 is not a terribly abundant protein within cells so, if you find a cell which lacks detectable protein on a Western blot, it is conceivable there is some *bcl*-2 that is contributing a survival function that could be in mitochondria as well as other localizations and that you are below the level of detection. No one has an RIA or other extremely sensitive assay for *bcl*-2 protein. However, the survival conferred by the cytoplasmic *bcl*-2 construct is quite out of proprotion to what might be still localized to mitochondria.

R. Horvitz

The question basically is: Is there anything that really argues against a mitochondria only function? And from what you have just said, it sounds like that is still plausible.

D. Hockenbery

Well, I think the data argue that the localization does not have to be on mitochondria for function, but the data are also consistent, I think, with the function having something to do with mitochondria.

T. Taniguchi

What is the half-life of *bcl*-2?

D. Hockenbery

In a number of cell lines it is 8 to 10 hours. I would say one caveat to that is that Mike Cleary has reported data within B-cell organs, within the germinal center, where cells are recruited to the germinal center from areas where there is a lot of *bcl*-2 protein. In the germinal center *bcl*-2 protein is not expressed but RNA is actually quite abundant in that cell, so there is a dissociation between RNA and protein. I think it is conceivable that in that physiologic context, that half-life may be quite different. There may be situations in which *bcl*-2 could be turned over quite rapidly.

M. Kastan

Just a couple of comments about the DNA damage experiments, first with the camptothecin. Despite the fact that the alkaline elution results suggest that the strand breaks are gone within a short period of time, there still could be plenty of damage left in those cells. The alkaline elution is a relatively insensitive measure of strand breaks but it is one of the best things we have unfortunately. So there still might be damage that is sensed. It would be interesting to see if a subset of those cells are undergoing apoptosis because the camptothecin should be inducing the strand breaks primarily in the cells that were in the S-phase during the pulse period. The cells in other phases of the cycle should not really be getting the strand breaks by the camptothecin. Second, with the 4HC experiments, the progenitors, the most immature progenitor cells are thought to be resistant to the 4 hydroperoxycyclophosphamide primarily because they are not cycling and because they have elevated levels of aldehyde dehydrogenase, which metabolizes the 4HC to an inactive form. So, in fact, those cells may be seeing a lot less damage.

D. Hockenbery

I guess the question would be, in a population that has *bcl*-2, those cells are also predominantly not cycling although they may be activated in terms of size. And I am not aware that anyone has really delineated differences in MDR expression between the big and small populations.

M. Kastan

Not MDR, aldehyde dehydrogenase.

D. Hockenbery

It is suggestive, from the growth properties of the cells that survive, that what is preferentially retained are the cells which do not have *bcl*-2. And what I am not aware of is if there is any data on difference in metabolism of drugs between those subsets.

G. Evan

In answer to Bob Horvitz's question, we have looked in a series of clones of fibroblasts expressing different levels of *bcl*-2, and what you see is just a progressive decline in survival potentiation. You do not see a sudden cutoff, so there are a number of things. First of all, if you analyze death rates by video microscopy, you see a longer period with high levels of *bcl*-2 before any insult will induce cell death. Once the deaths occur, they always occur with the same rate, that is, they always take 30 to 60 minutes. There seems to be no change in the speed with which cells actually die. And the other point is that it is very dependent upon the factor conditions and your base line can vary depending on just how much protection you are getting from your growth conditions. So if you are using fetal calf serum, for example, it is very much more protective than just calf serum. If you add different factors, it seems to us in the cell types we have looked at, it just adjusts your base line and so you are starting from different positions, depending on how you are culturing your cells and also on what the past history of those cells is.

D. Hockenbery

It is also true if you look at neutrophils, which really do not express *bcl*-2, that senesce by an apoptosis mechanism, that a variety of cytokines can maintain those cells but not through a *bcl*-2 dependent pathway. So it is quite possible that there are additional effects, perhaps on cell death programs, that those factors may influence.

D. Green

There are a lot of cell types in which a differentiation program that somewhat resembles apoptosis precedes, like in the differentiation of an erythrocyte or a keratinocyte. Does expression of *bcl*-2 in those settings have any influence on differentiation?

D. Hockenbery

I think that is a very interesting question. We are starting to ask that question in, for example, cells which you can differentiate, to see if *bcl*-2 can alter that program and also in ES cells and erythroid bodies and so on. I think it is interesting the parallel between some of the changes you see during differentiation and apoptosis. But there really are not, to my knowledge, any data out there that show that *bcl*-2 affects the differentiation program. Certainly in the *bcl*-2 transgenic mice you see plasma cell generation and so on. One model that is interesting from this point of view is from Naomi Rosenberg, the ability of Abelsen transformed pre-B cells to differentiate into B cells in which *bcl*-2 actually promotes that. However, in that system it looks like the differentiation program is actually already in place before *bcl*-2 effect is manifest. What *bcl*-2 does is actually to promote survival of the differentiated cells. So I think those studies need to be done and it is interesting to think that cells which are proliferating and may perhaps have a cell death program which is very active but yet cell death does not appear to occur physiologically in that population. For example, in the intestine, there is really not normal loss of cells from the proliferating zone. Cells die after they terminally differentiate. So if it is there for a purpose, then perhaps elements of an apoptotic program are tied into a differentiation program in some fashion.

P. Krammer

Could you elaborate more on blocking of anti CD-3 induced death by *bcl*-2? Now in most T-cell clones, the anti CD-3 induced death occurs after about a day, yet anti CD-3 induced signaling occurs within seconds. Now, is it known where *bcl*-2 interferes?

D. Hockenbery

No, I do not know any experiments that have tested early vs. late requirements for *bcl*-2 in that model.

P. Krammer

Does it interfere with any of the early events of anti CD-3 stimulation?

D. Hockenbery

I cannot think of any reports in which any signaling pathway, conventional signaling pathway, has been shown to be modulated by *bcl*-2.

J. Cleveland

I find it kind of unusual that there is such a rapid turnover of this protein if it is in the inner membrane of the mitochondria. I was wondering if you could comment if this is common to other proteins in this membrane?

D. Hockenbery

I think that is a very interesting question. Nothing is really known about degradation of this protein. Among inner membrane constituents, there is a variety of half lives of proteins. Some are quite long lived and some are not. I really do not know of data which explains the targeting of any of those to a degradative pathway. It would be interesting if you could show an accelerated degradation during IL-3 withdrawal, for example, to ask that question of trying to dissect how that occurs because that may have relevance to modulation of protective mechanisms in other systems which affect *bcl*-2, but I do not know of any data.

J. Cleveland

Have you identified any kind of post-translational modifications which might be important for the function of *bcl*-2?

D. Hockenbery

We have not. There is a report from John Reed's group of phosphorylation in a baculovirus system. We have looked for phosphorylation and have not found it although there are target sites for phosphorylation. I think it is still a bit of a leap to go from an insect cell with heterologous protein to make conclusions about regulation. The other thing I would say is that there really are not very many contexts in which you transfect *bcl*-2 into a cell and it does not appear to function in some apoptotic pathways. So it is hard to find a place where you would find a defect in a post-translational modification.

D. Livingston

I was wondering about a couple of ways to approach Bob Horvitz's question. I think his question is a bit apt, given the lack of quantitative data and also the fact that there are some noted examples of cases where proteins have been found in offsites and the offsite has been believed to be the correct site. One in fact, as in the case of *fms* for example, the CSF-1 receptor, the real site was in the membrane. The original observation was that the protein was associated with intermediary filaments, if you remember. So, two questions: First of all, with that as background, can you micro-inject baculovirus, *bcl*-2 into cells that do not make it, you know molecule by molecule, and ask how many molecules it takes, for example, to protect against an apoptotic insult?

D. Hockenbery

I do not see why you could not do that, but I do not know of anyone who has done it.

D. Livingston

Secondly, with the knockouts, it seems to me you can do a cytoplast binuclear re-fusion and ask if you take a *bcl*-2 cytoplast by a *bcl*-2 minus knockout nucleus, put them back together, are you protected? That does not rule out the possibility that *bcl*-2 migrates northward.

D. Hockenbery

I think those experiments would be useful to do.

D. Livingston

A third point is that the genetics, it seems to me, are not compelling. Just because it is "a membrane spanning domain", at least the computer suggests it, or one has that sense, does not mean that when you kill that region mutogenically, that the effect is to release the protein only from some membrane structure that might actually control some other interaction, for example, with some protein, which is unsuspected, and that is what you may be scoring phenotypically.

D. Hockenbery

Right, I think the only point we would make is that there is some retained function in that context where you cannot detect it in mitochondria, not that that is specifically due to lack of insertion in a membrane, and it is conceivably possible that the protein is actually dead, but by deleting that terminal segment, you have activated it so it functions in a site where it does not normally function.

D. Livingston

Absolutely. Is the T-1/2 of the mutant the same as that of the wild-type?

D. Hockenbery

For that mutant yes, although Gabriel Nunez has made some N-terminal deletions to assess localizing sequences and those degraded fairly rapidly.

D. Livingston

And those mutants are inactive?

D. Hockenbery

They are inactive, but there is really not much protein to assay because its turnover is so fast.

M. Fried

Bcl-2 appears to just delay apoptosis. But eventually that cells go through apoptotic death. Do you think that is because there is a concentration of something building up that overcomes the *bcl*-2 effect or do you think there is a second mechanism, that comes in later?

D. Hockenbery

I think you can make some predictions about that, about what type of interaction *bcl*-2 undergoes. It is unlikely that *bcl*-2, on the basis of that, acts as a kind of on-off switch for a death protein, I would imagine. So I think within the possibilities is that something accumulates and overwhelms *bcl*-2 function. One point I would make about that is that when you do these assays *in vitro* and you see that there is a delay then you say that is a partial response. That actually may be perfectly compatible with an all-or-none response *in vivo*. It may be that cells at certain points undergo transient thresholds where they either have to live or die. Once they make it to the other side which may be a short time interval, the death program may be turned off and so that level of *bcl*-2 function that you assay *in vitro*, which provides that short window of survival, may be perfectly physiologic.

D. Hockenbery

They are inactive, but there is really not much protein to assay because its turnover is so fast.

M. Fried

Bcl-2 appears to just delay apoptosis. But eventually that cells go through apoptotic death. Do you think that is because there is a concentration of something building up that overcomes the *bcl*-2 effect or do you think there is a second mechanism, that comes in later?

D. Hockenbery

I think you can make some predictions about that, about what type of interaction *bcl*-2 undergoes. It is unlikely that *bcl*-2, on the basis of that, acts as a kind of on-off switch for a death protein, I would imagine. So I think within the possibilities is that something accumulates and overwhelms *bcl*-2 function. One point I would make about that is that when you do these assays *in vitro* and you see that there is a delay then you say that is a partial response. That actually may be perfectly compatible with an all-or-none response *in vivo*. It may be that cells at certain points undergo transient thresholds where they either have to live or die. Once they make it to the other side which may be a short time interval, the death program may be turned off and so that level of *bcl*-2 function that you assay *in vitro*, which provides that short window of survival, may be perfectly physiologic.

S. Schlossman

I just wanted to bring to your attention some of the work of Paul Anderson and myself over the past couple of years, looking at perhaps some novel mechanisms by which cytolitic T-cells and NK-cells kill. These cells have an extraordinary capacity to kill their targets and they are known for the large granules that make up part of the cytoplasm and these granules are involved in the cytolitic mechanism.

For many years my laboratory has been interested in cell surface structures that contribute to both the identification of these cells and their functional programmes and we decided a few years ago to begin to make antibodies that would recognize structures that were T-cell specific or B-cell specific, that were cytoplasmic structures, and to screen such antibodies using a permeabilized cell system. Several antibodies were defined: one was TIA-1 and the other was TIA-2 where TIA stood for "T Internal Antigen". If you look at typical peripheral human T-lymphocytes, you find that the surface expresses cd-3, cd-2, cd-4, cd-8 and a variety of other antigens; if you permeabilize these cells there is very little expression of cd-3 internally. On the other hand, if you make antibodies that react with cytoplasmic antigens such as TIA-1 and TIA-2 that are not expressed on the surface of the cell, you can detect them using permeabilized cells. TIA-2 was a pan-T cell antigen. It was only found cytoplasmically and it turned out to be the zeta chain of the T-cell receptor. The reason people have never gotten a good anti-zeta chain antibody was simply that man and mouse shared the same external domain of zeta and the only differences were in the cytoplasmic domain. The other antigen which we looked at was TIA-1, which was much more restricted in its expression and it was primarily found on T-lymphocytes and was in a very small percentage of cd-4 cells, to a much larger extent in cd-8 cells. It was expressed on cytotoxic T-cell clones, NK clones and by and large not expressed on any other cells except perhaps some weak staining on myeloid cells or other cells that contain granules. So the very fact that we had an internal antigen that was restricted to T-cells, made this a rather interesting one in terms of trying to define its function. The other aspect that made it quite interesting, that it was restricted to cytolitic cells or cells that had, by limiting dilution, the capacity to become cytolitic.

The second point was that when one looked by electron microscopy at where this antigen TIA-1 was located, it decorated the granules, those large granules that were known to play a key role in the cytolitic mechanism and one could not see the expression of these antigens on the surface of the cells, so electron microscopy confirmed the observations that we had made using flow cytometry, that it was an internal antigen, but it was on the membranes of cytolitic granules and it turned out to be a 15 kilodalton protein restricted in its expression to peripheral T-cells, T-cell clones, but not some of the very early T-cells or T-cell lines that one uses in the laboratory. If one isolated cells from peripheral blood, both cd-8 cells and cd-4 cells, this 15 kilodalton protein was found predominantly in cd-8 cells, and as you know, cd-8 cells are really the vast majority of the cytolitic T-cell family and they by and large kill, in MHC restricted fashion. If you left these cells in media, the 15 kilodalton cell antigen disappeared and reappeared upon stimulation with Con A.

This protein was cloned by Paul and his colleagues in the laboratory and it turned out to a very interesting RNA binding protein. It's a member of a family of RNA binding proteins and it had three RNA binding domains, clearly a glutamine rich domain and a lysosomal targeting domain. And these 15 and 40 kilodalton structures were very closely related and it appeared, and we're still not certain, that one was derived from the other.

The important feature of this protein was the fact that if one purified the protein or made small amounts of recombinant material (unfortunately making recombinant material is very difficult, because almost every cell that you put it in dies and so one has not got still to this date, a very good expression system), one found that when one added this material to permeabilized T-lymphocytes, one saw DNA laddering, one of the hallmarks of apoptosis, and one could consider the fact that this new protein, which was distinct from other proteins known to be in cytolitic cells such as perpherin, may itself also be contributing to that particular mechanism. It took very small amounts of this to do that. This ability to cause DNA laddering and fragmentation could clearly be removed by a poli-A column and not by a poli-C column, giving credence to it being an RNA binding protein: it cannot be removed by sepharose but by sepharose anti TIA-1 column, indicating that the 15 and 40 kilodalton material that actually produced the laddering effect, was probably what we said it was; that is, this particular protein. So it seems that cytolitic T-cells had, in their granules, a material that could clearly cause DNA laddering and fragmentation and whether or not it could cause apoptosis is still not clear.

So the only point that I would like to make is that there appears to be another structure in cytolitic T-cells that is clearly contributing to killing of the target cells and may contribute to the whole process of programmed cell death. Clearly T-lymphocytes are utilizing a normal mechanism available for cell death, and exaggerating that mechanism. We would propose that TIA-1 may be contributing to the apoptotic mechanism that seems to be expressed in T-cells. We have looked for this material at the sites of graft versus host disease and found the lymphoid cells bathing those sites to be loaded with this material. And if you also look at the lymph nodes of patients with HIV infection, you find them loaded with this protein. We are right now exploring the utility of the material and its potential role in causing cell death and contributing both to the inflammatory process, graft rejection and the death perhaps, of the cd-4 population or other T-cell population in such disease as HIV.

P. Krammer

Which target cells have you used and do you have any information on the ligand on the target cell side and, second, what is the distribution of target cells you can use for such an assay?

S. Schlossman

The assays are less than perfect since we must use permeabilized cells and DNA laddering as the read out. In order to get the material into the cells, one has to either microinject it or use permeabilized cells. Presently we are using thymocytes but are trying to develop a better target system. We are simply using permeabilized human thymocytes and, although unsatisfactory, it is the best system we have.

D. Hockenbery

Can you see this within the target cells using your immuno electron microscopy?

S. Schlossman

It is very difficult to find TIA 1 in the target cells. You can convince yourself that you see little things, but compelling evidence does not exist! Similarly, other molecules, like perforin, are difficult to find in target cells.

D. Hockenbery

One quick follow-up. Have you done any mutogenesis studies to see if the glutamine rich region is important, for example?

S. Schlossman

I think we have, and that is a key region in the structure.

D. Green

I am sure you are familiar with the experiments that Arnold Greenberg and Pierre Henkart have done with the granzymes, in which DNA fragmentation can be induced directly in cells as long as you also add perforin. Will this protein induce in the presence of perforin?

S. Schlossman

I think we are trying to do a lot of those studies. It took many years to find a role for granzymes and perforin and the synergy between the two. Those studies are being done at the present time. I think that one of the difficulties is measuring effect of perforin itself in killing nucleated targets. Perhaps perforin synergizes with granzymes which then cleave on the 40kD TIA 1 moleceules into an active 15 kD molecule. Although the theory is attraction, we have no real compelling evidence to support that.

C. Riccardi

Can you block this TIA induced cell death by protein synthesis inhibitors? By treating the target cells I mean?

S. Schlossman

We are not really killing cells. The permeabilized cells that we are using are already dead. We are trying to micro-inject into ocytes and COS cells. Once we are successful and have an effect we will be in the position to use protein synthesis inhibitors.

NUCLEAR RECEPTORS AND THYMOCYTE APOPTOSIS: SHAPING THE IMMUNE REPERTOIRE

Melanie S. Vacchio, Yili Yang, and Jonathan D. Ashwell

Laboratory of Immune Cell Biology
Biological Response Modifiers Program
National Institutes of Health
Bethesda, MD 20892

INTRODUCTION

Programmed cell death (PCD) plays a critical role in the generation of the antigen-specific immune repertoire. Both the T cell receptor (TCR) ligand-binding chains and immunoglobulin are generated randomly from the rearrangement of independent gene segments. While this process ensures the diversity of the immune cell receptor repertoire, it also results in the generation of T cells that express receptors that recognize self antigens in the context of self major histocompatibility complex (MHC)-encoded molecules with high affinity. The most effective way to deal with these potentially autoreactive cells is to eliminate them, a process referred to as "negative selection". Deletion of potentially autoreactive T cells usually occurs in the thymus. Use of anti-TCR antibodies or bacterial superantigens (which mimic receptor occupancy by a natural ligand) during fetal thymic organ cultures results in PCD of immature $CD4^{+}CD8^{+}$ thymocytes[1,2], as does exposure of thymocytes from TCR-transgenic mice to the appropriate antigen[3,4]. The dying thymocytes manifest the classic apoptotic phenotype: membrane blebbing, cytoplasmic condensation, and chromatin clumping, accompanied by endonuclease-mediated cleavage of internucleosomal DNA[1]. Injection of cyclosporin A, which blocks activation-induced T cell hybridoma and thymocyte apoptosis[5,6], prevents the deletion of a population of autoreactive T cells when the thymus regenerates after sublethal irradiation[7,8].

While some thymocytes express anti-self TCRs and are actively killed, other thymocytes have aberrantly rearranged TCR genes resulting in no cell surface receptor, or express a TCR that has little or no avidity for self. These thymocytes die within the thymus, presumably by apoptosis. In the middle of these two extremes (high avidity anti-self and no recognition of self) are the thymocytes that express a TCR with so-called "low to moderate" avidity for self antigen/MHC. These thymocytes do not die, but mature and populate the periphery as antigen-specific effector T cells (positive selection). Fig. 1 schematically illustrates the hypothetical relationship between TCR avidity for self antigen/MHC and thymocyte fate.

Apoptosis, Edited by E. Mihich and R.T. Schimke
Plenum Press, New York, 1994

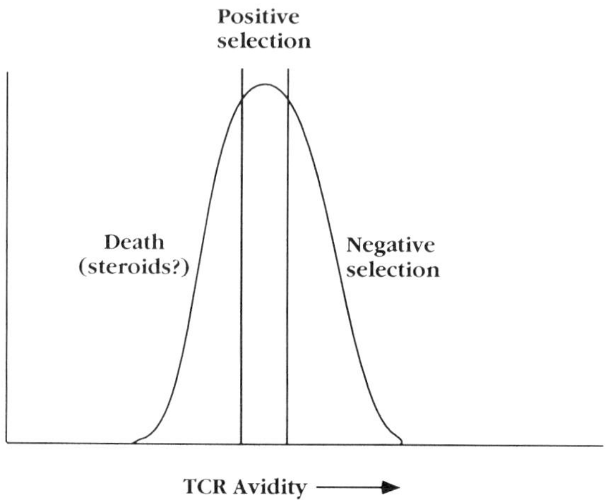

Figure 1. Schematic representation of the relationship between TCR avidity and thymocyte selection. The reason that thymocytes with low avidity TCRs die is unknown, but may due, at least in part, to glucocorticoids.

Favoring this model are reports that increasing the overall strength of the TCR/ligand interaction can cause negative selection in a situation that otherwise results in positive selection. For example, TCR transgenic mice expressing a receptor that normally results in positive selection in H-2^b mice were crossed with transgenic mice that overexpress the CD8 co-receptor, one function of which is to increase the avidity of the TCR/ligand interaction. Rather than being positively selected, in H-2^b mice the thymocytes that simultaneously expressed the transgenic TCR and overexpressed CD8 in H-2^b mice resulted in negative selection[9]. How activation through the TCR can lead to two such different cellular outcomes (death versus survival) is unresolved.

EFFECT OF THYMIC STEROIDS ON SELECTION

In addition to TCR occupancy, glucocorticoids are a potent means of inducing PCD in thymocytes[10,11]. While mature peripheral T cells and the more mature population of thymocytes that express either CD4 or CD8, but not both, are resistant, immature $CD4^+CD8^+$ thymocytes are extremely sensitive to glucocorticoid-induced PCD[12,13]. The thymus is actually sensitive to physiologic levels of glucocorticoids: increases in circulating glucocorticoids due to stress cause the thymus to involute[14,15], and decreases in glucocorticoid levels after adrenalectomy cause thymic hypertrophy[16]. The biological significance of this relationship is unclear.

Although TCR occupancy and glucocorticoids independently cause T cell hybridoma and thymocyte PCD, simultaneous *in vitro* treatment with both results in cell survival (mutual antagonism). This finding prompted the speculation that exposure to both of these stimuli during thymocyte differentiation may be a basis for positive selection[17,18,19]. In this model, high avidity TCR/ligand interactions induce signals too potent to be antagonized, while low to moderate avidity TCR/ligand interactions transduce weaker signals that can be antagonized by ambient glucocorticoids, allowing the cell to survive and be positively selected. This type of model, involving both TCR-mediated and non-TCR mediated signaling, would provide a mechanistic basis for the hypothesis that the strength of the TCR/ligand interaction determines positive versus negative selection. For this model to be viable, thymocytes would need to be exposed to glucocorticoids early in ontogeny, when the T cell antigen-specific repertoire begins to form. Thymocyte differentiation is ongoing during fetal, neonatal, and post-natal life, but tapers off as animals approach puberty. Circulating levels of glucocorticoids, however, are extremely low during fetal and neonatal life[20,21]. Because the thymus has been reported to have some endocrine properties[22,23,24], and because extra-adrenal sites of steroid synthesis have been reported,[25,26] we asked whether the thymus can produce steroids, and if so, would they influence antigen-specific thymocyte selection.

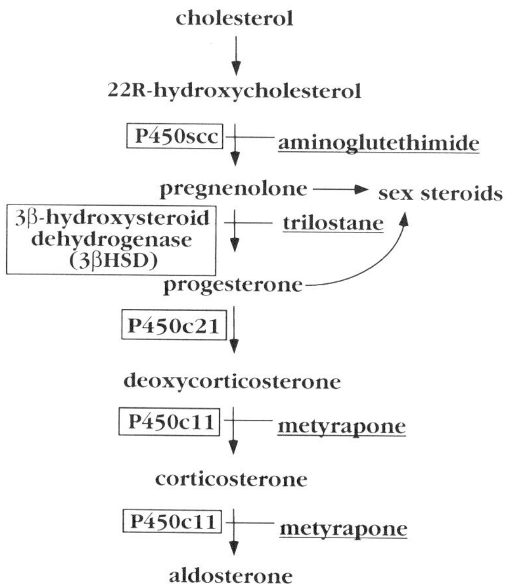

Figure 2. Simplified pathway of steroid biosynthesis. Note that the major circulating glucocorticoid in rodents is corticosterone. Enzymes are boxed. Inhibitors of steroid synthesis are underlined.

Expression of steroidogenic enzymes in the thymus

P450scc, the initial enzyme in the synthesis of steroids, cleaves the side chain of cholesterol to yield pregnenolone (Fig. 2). Presence of this enzyme is the determining factor in whether a tissue is capable of steroidogenesis; its expression has only been found in steroidogenic tissues, such as the adrenal gland, placenta, sex organs, and the brain[25,27,28,29]. Sections of thymus were analyzed for the expression of P450scc. An antiserum to P450scc specifically stained a small subset of thymic cells whose morphology (large size and abundant cytoplasm) indicated they were not thymocytes. Because large numbers of thymocytes made it difficult to visualize other cells in the thymus, mice were sublethally irradiated to eliminate thymocytes without disturbing the epithelium. Analyses of thymuses 4 days post-irradiation dramatically enhanced the visibility of radiation-resistant stromal and epithelial elements (Fig. 3). Cells that stained positively with the anti-P450scc serum were distributed throughout the thymus, with the most prominent staining occurring in the cortex (Fig. 3b). Similar staining patterns were observed with an anti-P450c11 serum, which

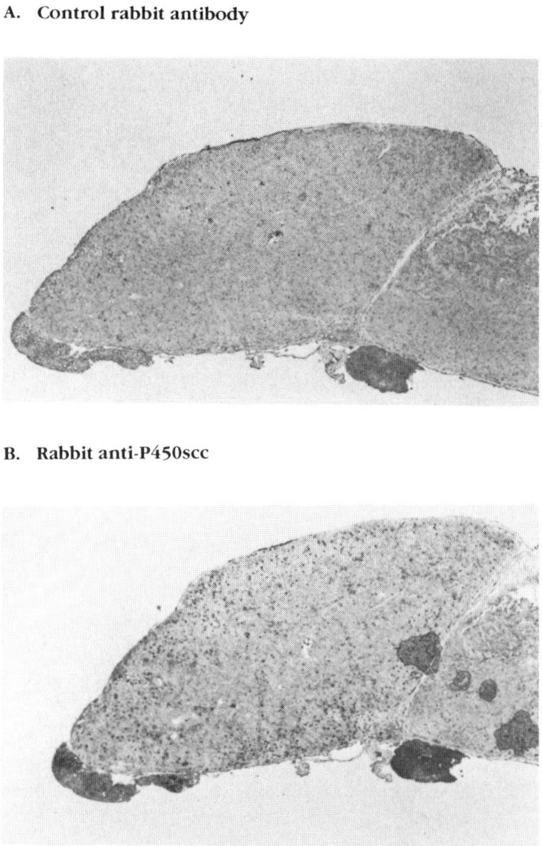

Figure 3. P450scc is expressed in the thymus. Control antibody or anti-P450scc serum was used to stain sections of mouse thymus obtained from mice that had been sublethally irradiated (600 cGy) 4 days previously.

recognizes an enzyme responsible for the conversion of deoxycorticosterone to the biologically active corticosterone, but no staining was observed with an antiserum against P450aromatase, an enzyme involved in sex steroid but not glucocorticoid biosynthesis (data not shown). Tissue sections were also stained with a monoclonal antibody specific for a macrophage antigen, Mac-1. Very few Mac-1^+ cells were detected, far fewer than then number of cells staining positively for P450scc (data not shown). Staining of thymus sections with anti-cytokeratin demonstrated that the predominant cell type remaining in the thymus was epithelial, and that P450scc-expressing cells were a relatively small subset of the epithelial cells present in the thymus (data not shown). Therefore, enzymes necessary for steroid synthesis are present in the thymus, and appear to be expressed in a subpopulation of radiation-resistant cells. The majority of the P450scc$^+$ cells are epithelial, although a small population of P450scc$^+$ non-epithelial cells cannot be excluded.

Synthesis of pregnenolone in the thymus

To determine if steroids are synthesized in the thymus, different cell populations were cultured *in vitro* for 4 d and the supernatants assayed for pregnenolone. Although thymocytes produced no detectable pregnenolone under any conditions, this steroid precursor could be measured in cultures of collagenase-treated thymus cells (Fig. 4). Pregnenolone was not detected in the supernatants when collagenase-treated thymus cells were cultured alone, but was found when the cells were cultured with the precursor of pregnenolone, 22R-hydroxycholesterol (Fig. 4A). ACTH, which increases mobilization of cholesterol from intracellular stores to the inner mitochondrial membrane and causes upregulated expression of P450 enzymes[30], further enhanced the synthesis of pregnenolone, the optimal concentration being 30 nM (Fig. 4A). Collagenase-treated thymus cells from mice 4 days after sublethal irradiation were cultured in the presence of ACTH and 22R-hydroxycholesterol, and the amount of pregnenolone in the medium compared to levels seen with collagenase-treated normal thymus. Consistent with the enrichment of radioresistant P450scc-expressing cells observed by immunocytochemistry, ~100-fold fewer cells were required to generate levels of pregnenolone synthesis equivalent to that seen with normal collagenase-treated thymus (Fig. 4B). Two conclusions can be drawn. First, P450scc is present and functional in the thymus, resulting in pregnenolone synthesis by radiation-resistant thymic cells. Second, the steroidogenic machinery in the thymus, like that in the adrenal, is regulated by ACTH.

Thymic nurse cells (TNC) are subcapsular epithelial cells that are distinguishable from other epithelial cells because they surround (engulf) large numbers of thymocytes[31,32]. Populations enriched for TNC and thymic rosettes (T-ros, composed of thymocytes bound to macrophages and dendritic cells) were cultured for 6 days (Fig. 4C). Significant levels of pregnenolone were detected in TNC cultures, while cultures of T-ros had detectable, but much lower, levels of pregnenolone. This could either be due to contamination of T-ros with TNC during purification, but may also reflect the possibility that macrophages and/or dendritic cells might produce relatively small amounts of pregnenolone.

Ontogeny of thymic steroid synthesis

If thymic steroids play a role in thymocyte selection, P450scc activity would be expected to be detectable earlier in the thymus than in the adrenal. Immunocytochemistry revealed that fetal thymus expresses abundant P450scc in radiation-resistant cells (data not shown). Steroidogenic activity was measured in T-depleted thymus cells from mice at 7 and 48 days of age (Fig 4D). After 5 days of culture in the presence of ACTH and 22R-hydroxycholesterol, production of pregnenolone was measured. Cells from the one week old mice produced 104 pg pregnenolone/ml, compared to 41.5 pg/ml by the seven week old mice (a decrease of 55%). Thus, the ontogeny of thymus-synthesized steroid production is the inverse to that of the adrenal, being highest in early life and declining thereafter.

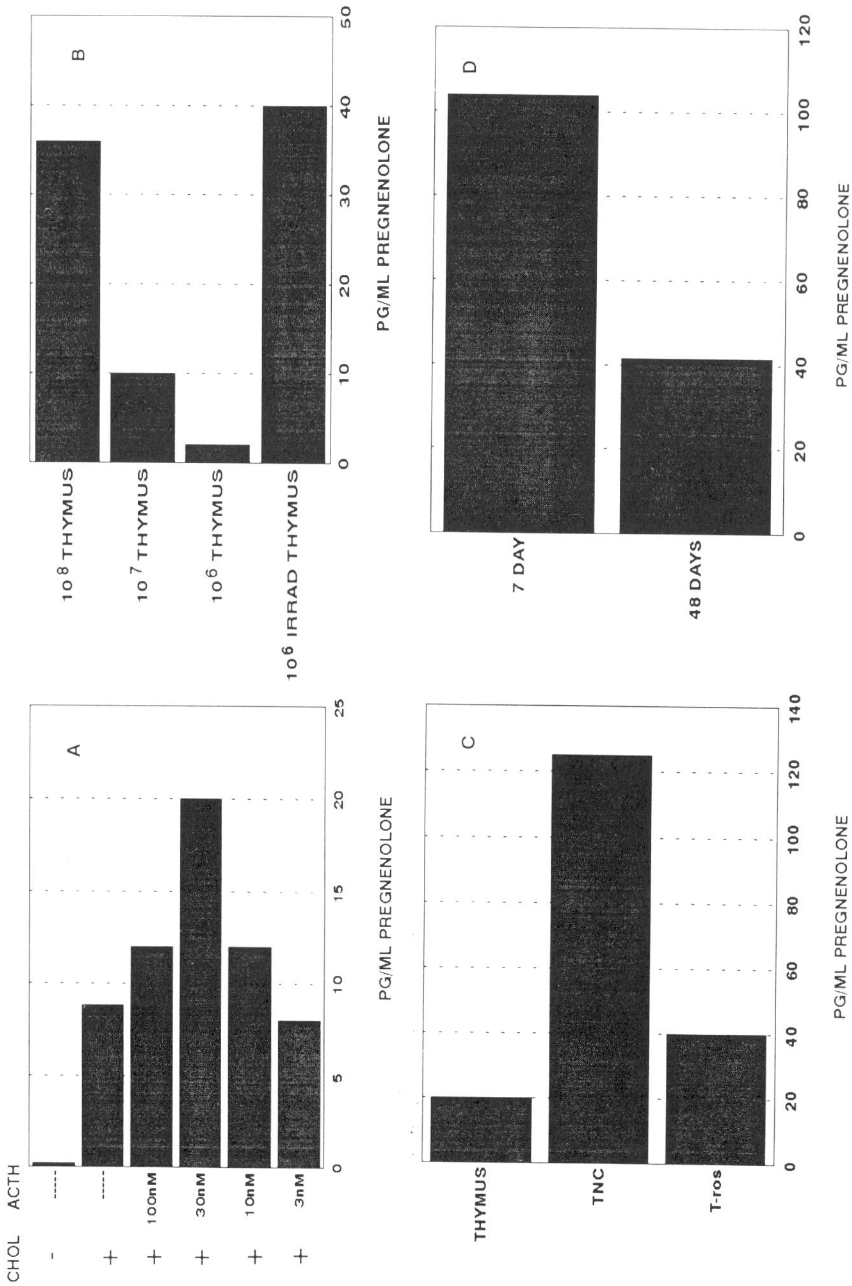
A
CHOL
ACTH
100nM
30nM
10nM
3nM
PG/ML PREGNENOLONE
B
10^8 THYMUS
10^7 THYMUS
10^6 THYMUS
10^6 IRRAD THYMUS
PG/ML PREGNENOLONE
C
THYMUS
TNC
T-ros
PG/ML PREGNENOLONE
D
7 DAY
48 DAYS
PG/ML PREGNENOLONE

Figure 4. Pregnenolone is synthesized in the thymus. A) Pregnenolone synthesis is responsive to ACTH. 10^8 collagenase-treated thymus cells were cultured for 5 days in the presence or absence of 6.25 µM 22R-hydroxycholesterol, the indicated concentrations of ACTH and 5 µM trilostane (Fig. 2). After 4 d, pregnenolone in the supernatants was measured by RIA. B) T-depletion of the thymus enriches cells capable of synthesizing pregnenolone. Titrated numbers of collagenase-treated normal thymus or 10^6 collagenase-treated thymus cells from mice sublethally irradiated (600 cGy) were cultured for 5 days in the presence of 6.25 µM 22R-hydroxycholesterol, 10 nM ACTH, and 5 µM trilostane. C) Thymic nurse cells synthesize pregnenolone. 10^6 thymic nurse cells (TNC), T-rosettes (T-ros), or collagenase-treated whole thymus cells were cultured for 5 days in the presence of 6.25 µM 22R-hydroxycholesterol,, 10nM ACTH, and 5 µM trilostane. D) Effect of age on thymic steroid synthesis. Collagenase-treated thymus cells were T-depleted by anti-Thy-1.2 plus complement treatment. 10^6 T-depleted cells from either 7 d or 48 d old mice were cultured in the presence of 6.25 µM 22R-hydroxycholesterol, 10 nM ACTH, and 5 µM trilostane.

Glucocorticoids and thymocyte selection

To evaluate the influence of endogenously synthesized steroids on TCR-mediated selection, ligation of the TCR with monoclonal antibodies was used to mimic receptor occupancy in fetal thymic organ culture[1]. Fetal thymuses were cultured in serum-free medium with H57-597 (H57), a TCR Cb-specific antibody[33], in the presence or absence of metyrapone, a specific inhibitor of P450c11 (Fig. 2). Organ cultures with H57 alone resulted in a 28% decrease in $CD4^+CD8^+$ thymocytes (83.8%➔60.1%) (Table 1, Expt. 1). Although cultures with metyrapone alone had little effect, addition of both H57 and metyrapone decreased the number of $CD4^+CD8^+$ thymocytes by 71% (84%➔24.1%). Importantly, the synergy between anti-TCR killing and corticosterone production blockade (61.5%➔17.5%) was reversed by addition of exogenous corticosterone (57.6%➔52.6%) (Table 1, Expt. 2). Additional experiments were performed with glucocorticoid antagonist RU-486. Culture of thymocytes with H57 alone resulted in a 17% decrease in $CD4^+CD8^+$ thymocytes (82.4%➔ 65.5%) (Table 1, Expt. 3). In this experiment, RU-486 alone had a small effect on $CD4^+CD8^+$ cell numbers, although this was not regularly seen in other experiments. In combination with H57, RU-486 synergistically enhanced loss of $CD4^+CD8^+$ thymocytes (65.5%➔11.6%).

Table 1
Increased negative selection occurs in the absence of thymus-derived steroids

		% $CD4^+CD8^+$ Medium alone	% $CD4^+CD8^+$ Anti-TCR
Experiment 1			
	Medium	83.8	60.1
	Metyrapone	84.1	24.1
Experiment 2			
	Medium	81.7	61.5
	Metyrapone	ND	17.5
	Medium + Corticosterone	76.6	57.6
	Metyrapone + Corticosterone	ND	52.6
Experiment 3			
	Medium	82.4	65.5
	RU-486	72.1	11.6

Individual d17 thymuses were split into 3-4 pieces and cultured under the conditions above. Results in each experiment were obtained from individual animals and are representative of 4-8 animals tested in each experiment. ND means not determined.

RETINOIC ACID AND THYMOCYTE DEVELOPMENT

Like glucocorticoids, retinoic acid (RA) binds an intracellular (intranuclear) receptor that in turn binds DNA and regulates gene transcription. Retinoids are vitamin A (retinol) derivatives that affect growth and differentiation of many cells, including those of the immune system[34,35,36]. One of the major retinol metabolites is retinoic acid, for which two receptor families exist: retinoic acid receptors (RARs) and retinoid X receptors (RXRs)[37,38,39]. These receptors belong to a superfamily that includes the glucocorticoid, thyroid hormone, and vitamin D receptors. All-*trans* RA binds RARs, while 9-*cis* RA, a stereoisomer of all-*trans* RA, can bind both RARs and RXRs[40,41,42]. Experiments were performed to evaluate whether signals delivered through these nuclear receptors could also affect TCR-mediated signaling for apoptosis.

RA effects on T cell hybridoma activation and death

The T cell hybridoma 2B4.11 was cultured in the presence of the immobilized anti-CD3-specific monoclonal antibody, 2C11, with or without RA. Activation with anti-CD3 induced 40% specific DNA fragmentation. While all-*trans* RA alone had no effect on cell viability (data not shown), the presence of 10^{-5} M all-*trans* RA eliminated 2C11-induced DNA fragmentation (Fig. 5A). Studies have shown that while all-*trans* RA and 9-*cis* RA are equivalent in potency with regard to RAR activation, 9-*cis* RA is 10-30 fold better than all-*trans* RA at activating RXRs[39,41]. Comparison of the ability of 9-*cis* RA and all-*trans* RA to inhibit activation-induced PCD revealed that 9-*cis* RA was also 10-fold more potent than all-*trans* RA at inhibiting TCR-mediated PCD (Fig. 5A), consistent with the involvement of RXRs in the signaling pathway. 9-*Cis* RA also blocked DNA fragmentation induced by the specific antigen, pigeon cytochrome *c*. (Fig. 5B). On the other hand, PCD induced by dexamethasone was not blocked by even high concentrations of 9-*cis* RA (Fig. 5C). To address whether RA blocks all manifestations of T cell hybridoma activation, effects of 9-*cis* RA on T cell proliferation and IL-2 production were evaluated. T cell hybridomas undergo activation-induced growth inhibition that is distinct from activation-induced cell death. Activation-induced growth inhibition is the result of a G_1/S cell cycle block, which occurs more quickly than PCD and, unlike PCD, is not blocked by cyclosporin A. Anti-CD3-indued growth inhibition of 2B4.11 cells was not prevented by 9-*cis* RA[43]. RA partially inhibited activation-induced IL-2 production by 2B4.11 cells (~ 50%) (Fig. 5D), although it had very little effect on IL-2 production by normal splenic T cells stimulated with anti-CD3 antibodies[43]. RA, therefore, has different effects on different manifestations of T cell hybridoma activation, the most profound being its ability to prevent PCD.

RA inhibits antigen-specific deletion of thymocytes

Based upon its effects on T cell hybridoma viability, experiments were performed to evaluate whether antigen-specific thymocyte death (negative selection) could be inhibited by RA. Swat et al. have developed an *in vitro* model system of negative selection that utilizes TCR aβ transgenic mice in which greater than 90% of the thymocytes express a transgenic TCR specific for the male antigen H-Y[4]. Culture of female thymocytes on monolayers of thymic APCs expressing H-Y/H-$2D^b$ results in the death of normal $CD4^+CD8^+$ ($CD4^{hi}CD8^{hi}$) thymocytes, yielding a population of cells with reduced levels of both CD4 and CD8 ($CD4^{lo}CD8^{lo}$). Thymocytes from female TCR transgenic mice were cultured with either female or male H-$2D^b$ thymic APCs (Fig. 6). After 36 hr of culture with female (H-Y^-) APC,

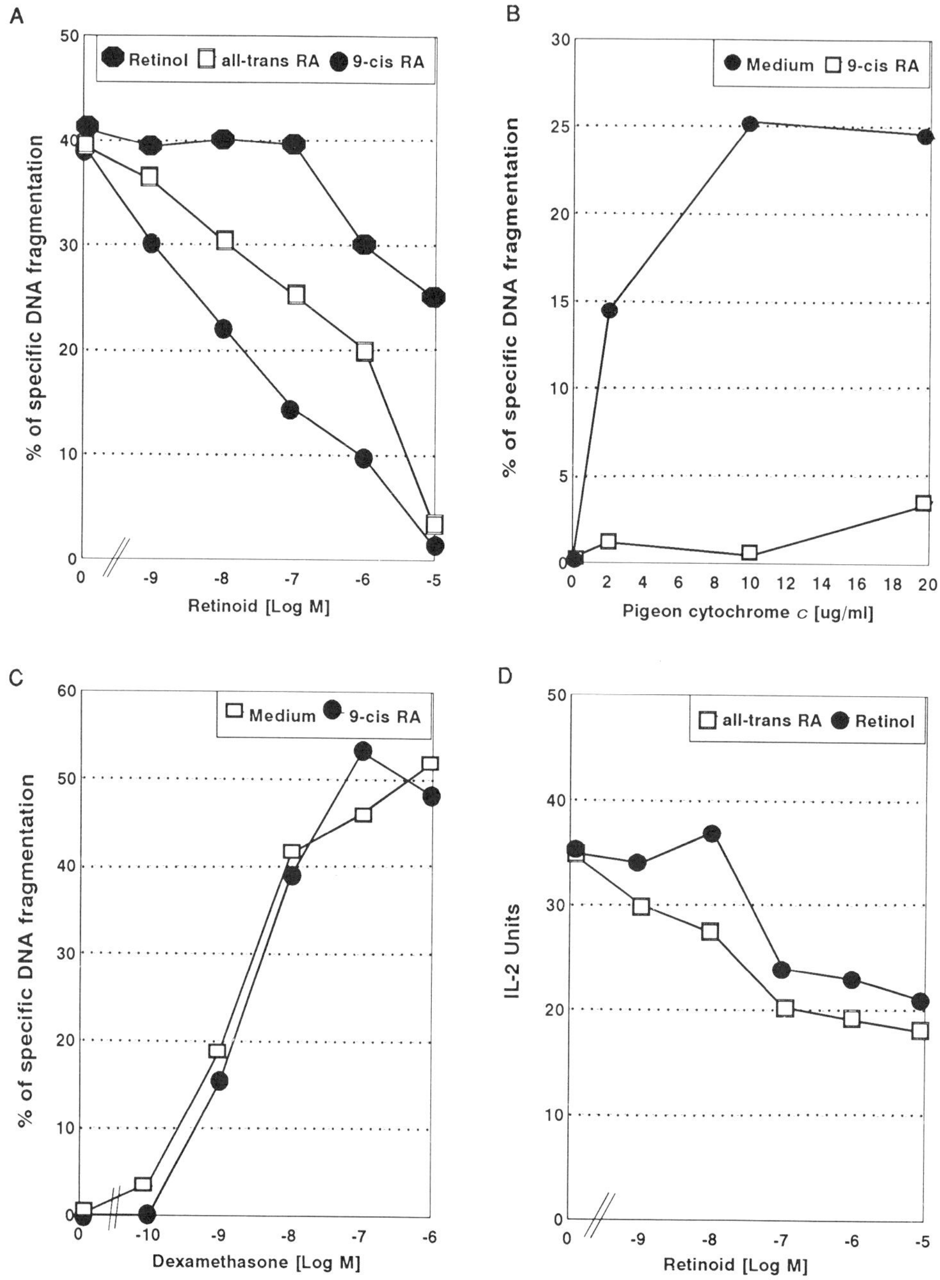

Figure 5. Effect of RA on T cell hybridoma TCR-mediated activation. A) 9-*cis* RA is a more potent inhibitor of PCD than all-*trans* RA or retinol. [^{3}H]-thymidine-labeled 2B4.11 T hybridoma cells were cultured with anti-TCR antibodies in the presence of the indicated concentrations of retinoid. Incubation of the retinoids with 2B4.11 cells alone induced no DNA fragmentation. Specific DNA fragmentation was measured after 16 hr. B) RA inhibits antigen-driven PCD. [^{3}H]-thymidine-labeled 2B4.11 T hybridoma cells were cultured with 10^5 irradiated antigen presenting cells (APCs) (LK 35.2[44]) and the indicated concentrations of pigeon cytochrome *c* with or without 9-*cis* RA (10^{-6} M). C) RA has no effect on glucocorticoid-induced PCD. [^{3}H]-thymidine-labeled 2B4.11 T hybridoma cells were cultured with anti-TCR antibodies in the presence of the indicated concentrations of 9-*cis* RA. D) RA partially inhibits IL-2 production by 2B4.11. 2B4.11 T hybridoma cells were cultured in the presence of 2C11 (1 μg/well) and the indicated concentrations of retinoids. After 24 hr, supernatant was removed and assayed for IL-2.

29.1% of the cells were $CD4^{hi}CD8^{hi}$. Culture of the same thymocytes with $H\text{-}Y^{+}$ male thymic APCs, however, resulted in a dramatic decrease in the number of $CD4^{hi}CD8^{hi}$ cells (to 9.1%). Culture with 10^{-6} M all-*trans* RA, or 10^{-7} M 9-*cis* RA, resulted in total recovery of $CD4^{+}CD8^{+}$ cells. This difference in potency was maintained over a full dose-response curve[43], and indicates that RXRs mediate the inhibitory effects of RA on TCR-mediated thymocyte as well as T cell hybridoma PCD.

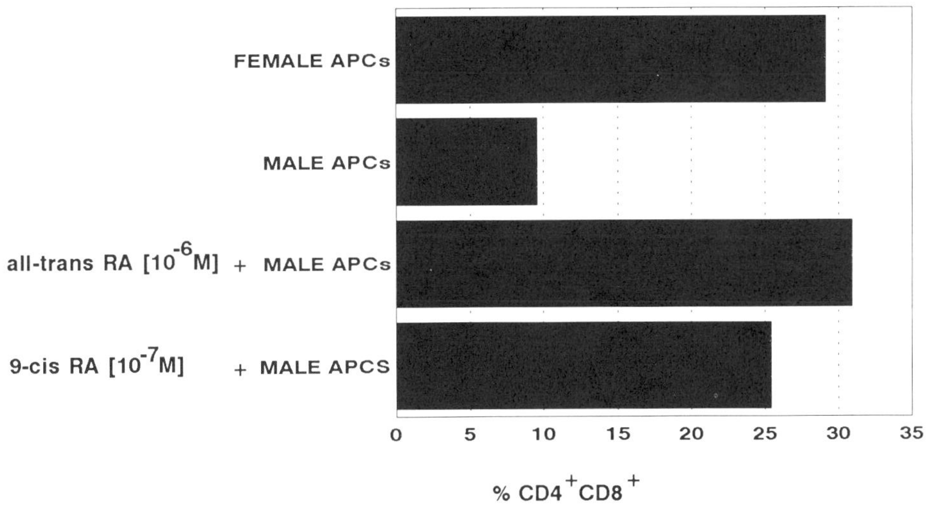

Figure 6. Inhibition of antigen-specific thymocyte deletion by RA. Thymocytes from female transgenic mice expressing TCR ab genes specific for the male H-Y antigen in association with $H\text{-}2D^{b}$ were isolated and cultured on either male or female thymic APC monolayers for 36 hr in the presence or absence of RA. Cells were harvested and analyzed for the expression of $CD4^{+}CD8^{+}$ cells by flow cytometry.

RA depletion during thymocyte development

Since RA inhibits activation-induced T cell hybridoma and thymocyte death *in vitro*, we asked if RA might participate in normal thymocyte selection. If this were the case, one might expect that in its absence there would be enhanced antigen-driven deletion of thymocytes. To address this, BALB/c mice were placed on vitamin A-deficient diets or vitamin A-deficient diets supplemented with all-*trans* RA. At 14 weeks and 20 weeks of age, thymuses were harvested and CD4/CD8 expression analyzed by flow cytometry (Table 2). At 14 weeks of age vitamin A-deficient mice had approximately 40% the number of thymocytes as those on the all-*trans* RA-supplemented diet. This number dropped to 7% at 20 weeks. The most significant drop at 20 weeks of age was in $CD4^{+}CD8^{+}$ cells, the population most sensitive to antigen selection ($26 \times 10^{-6} \rightarrow 0.5 \times 10^{-6}$). This result is supported by other studies in which thymic atrophy was noted in vitamin A-deprived mice[45]. A trivial explanation could be that this effect on the immune system is indirect, potentially resulting from, for example, elevated levels of glucocorticoids in the nutritionally-deprived mice. However, the data are also compatible with the more intriguing possibility that the presence of RA protects thymocytes with TCRs that have a low to moderate avidity for self antigens, and that in the absence of RA, this low avidity interaction is capable of inducing PCD.

Table 2
Thymus phenotype of vitamin A-deficient mice fed with or without all-*trans* RA

Age (weeks)	Mice	Thymocyte Number ($x10^{-6}$)	$CD4^-CD8^-$ x 10^{-6} (%)	$CD4^+CD8^+$ x 10^{-6} (%)	$CD4^+CD8^-$ x 10^{-6} (%)	$CD4^-CD8^+$ x 10^{-6} (%)
14	RA^+	35.6	2.3 (6.4)	27.8 (78.4)	4.4 (12.3)	1.0 (2.9)
	RA^-	14.8	1.2 (7.9)	9.6 (64.9)	2.7 (18.8)	1.2 (8.5)
20	RA^+	32.0	1.7 (5.4)	26.0 (81.4)	3.7 (11.6)	0.4 (1.3)
	RA^-	2.5	1.2 (46.0)	0.5 (20.1)	0.6 (23.4)	0.3 (10.6)

Pregnant BALB/c mice were fed a vitamin A-deficient diet (Teklad), either supplemented (RA^+) or not (RA^-) with all-*trans* RA (3 µg/g of food), which was maintained during suckling of the litters. Offspring were weaned at three weeks and maintained on the same diets thereafter. At the ages indicated, mice were sacrificed and the thymuses analyzed for total cell recovery and distribution of CD4 and CD8 cell surface markers. One RA^+ and two RA^- mice were tested at each time point.

IMPLICATIONS

These studies explore two independent hormone/nuclear transcription factor systems that influence thymocyte selection *in vitro* and, possibly, *in vivo*. The results suggest that TCR avidity for self antigens is not the only parameter that determines the fate of immature thymocytes. Rather, the interplay between TCR-mediated activation and other, modifying, signals influences the maturation and selection of developing T cells. The mutual antagonism of steroid and TCR-mediated cell death *in vitro* led to speculation that simultaneous exposure of thymocytes to steroids and TCR-ligands during development might influence thymocyte selection. The results in this study support this possibility. First, steroids are produced *in situ* by radioresistant epithelial cells. Second, deprivation of steroids resulted in more efficient TCR-mediated thymocyte deletion in fetal thymic organ culture. The data also indicate that the biologically relevant steroids are glucocorticoids. P450c11, the enzyme responsible for conversion of deoxycorticosterone to corticosterone, is present and functional. Exogenous corticosteroids reversed the synergistic effect of the steroid synthesis blockade on TCR-mediated thymocyte loss, and addition of a glucocorticoid antagonist in fetal thymic organ cultures resulted in more efficient TCR-mediated deletion. This is consistent with the hypothesis outlined in Fig. 1, and with the notion that steroids allow $CD4^+CD8^+$ thymocytes to survive despite TCR occupancy; removal of steroids thus "converts" positive to negative selection. This raises a number of interesting questions, such as whether deficiencies of thymic steroidogenesis might result in an aberrant T cell antigen-specific repertoire, and could too much steroid production rescue thymocytes with relatively high reactivity with self antigens, allowing autoreactive T cells to escape? The observation that enzymes required for steroid synthesis are present in the thymus and result in the synthesis of functionally active steroids may explain, at least in part, several previously reported findings. For example, adrenalectomy of adult, but not young, rats results in thymic hypertrophy[46]. Local production by thymuses in young animals may provide sufficient levels of steroids to regulate thymus cellularity. Another finding has been that mRNA for ACTH is present in the thymus, and under certain circumstances peripheral lymphocytes produce ACTH[47,48]. As shown in Fig. 4, ACTH enhances thymic steroid production *in vitro*, and local production of ACTH may regulate thymic steroid production *in vivo*. Finally, thymic synthesis of steroids may help explain the finding that bone marrow-derived dendritic cells in the thymus are primarily responsible for negative selection, while thymic epithelial cells are primarily responsible for positive selection[49,50]. The TCR/glucocorticoid antagonism model proposes that glucocorticoids are necessary to prevent moderate avidity TCR/ligand interactions from causing thymocyte death. Since thymic epithelial cells appear to be the primary source of local steroid production, they would be expected to be especially effective in this regard. Bone marrow-derived APCs would mainly mediate negative selection, because they either constitute a small fraction of the steroid-producing cells and/or produce relatively small amounts of the steroids to countermand TCR-mediated deletion signals. This may be particularly true in the case of thymocytes enveloped by TNCs, which forms a unique microenvironment that could allow directed delivery of steroids to the thymocytes. TNC are thought to have a special role in thymocyte development, since a thymocyte population sensitive to selection — glucocorticoid-sensitive, $CD4^+CD8^+$, relatively high levels of TCR — are sequestered within TNCs[51]. Steroids produced by TNCs may act as paracrine factors that readily diffuse between contacting cells, reaching significantly higher intracellular glucocorticoid levels than if released into the circulation.

RA, most probably by activation RXRs, also inhibits activation-induced thymocyte PCD. Previous studies have shown that vitamin A deficiency is associated with immune function abnormalities[36,45,52]. This might be due, at least in part, to over-efficient negative selection, resulting in the disappearance of $CD4^+CD8^+$ cells and potentially useful portions of the antigen-specific T cell repertoire. As with thymic steroids, RA may play a role in thymus selection by antagonizing TCR-mediated negative selection, allowing more thymocytes to survive and be selected than would otherwise have been if selected by TCR avidity alone. Although high doses of RA have been found to be toxic, doses of RA slightly higher than in normal diets result in thymic hypertrophy[53], and it is an interesting speculation that T cells that would normally be deleted may have survived in these animals.

The hypotheses put forward here require further testing. Experiments that would help to determine what role, if any, glucocorticoids and retinoic acid have in *antigen-specific* T cell development include the study of animals with altered levels of the appropriate receptors, and nutritional manipulation of transgenic mice bearing TCRs of defined specificity, with analysis of thymocyte development in the presence or absence of the appropriate antigen.

REFERENCES

1. C.A. Smith, G.T. Williams, R. Kingston, E.J. Jenkins, and J.J.T. Owen, Antibodies to CD3/T-cell receptor complex induce death by apoptosis in immature T cells in thymic cultures, *Nature*, 337:181 (1989).
2. E.J. Jenkinson, R. Kingston, C.A. Smith, G.T. Williams, and J.J. Owen, Antigen-induced apoptosis in developing T cells: a mechanism for negative selection of the T cell receptor repertoire, *Eur. J. Immunol.*, 19:2175 (1989).
3. K.M. Murphy, A.B. Heimberger, and D.Y. Loh, Induction by antigen of intrathymic apoptosis of $CD4^+CD8^+TCR^{lo}$ thymocytes in vivo, *Science*, 250:1720 (1990).
4. W. Swat, L. Ignatowicz, H. von Boehmer, and P. Kisielow, Clonal deletion of immature $CD4^+8^+$ thymocytes in suspension culture by extrathymic antigen-presenting cells, *Nature*, 351:150 (1991).

5. M. Merćep, P.D. Noguchi, and J.D. Ashwell, The cell cycle block and lysis of an activated T cell hybridoma are distinct processes with different Ca^{2+} requirements and sensitivity to cyclosporine A, *J. Immunol.*, 142:4085 (1989).
6. Y. Shi, B.M. Sahai, and D.R. Green, Cyclosporin A inhibits activation-induced cell death in T-cell hybridomas and thymocytes, *Nature*, 339:625 (1989).
7. M.K. Jenkins, R.H. Schwartz, and D.M. Pardoll, Effects of cyclosporin A on T cell development and clonal deletion, *Science*, 241:1655 (1988).
8. E.K. Gao, D. Lo, R. Cheney, O. Kanagawa, and J. Sprent, Abnormal differentiation of thymocytes in mice treated with cyclosporine A, *Nature*, 336:176 (1988).
9. E.A. Robey, F. Ramsdell, D. Kioussis, W. Sha, D. Loh, R. Axel, and B.J. Fowlkes, The level of CD8 expression can determine the outcome of thymic selection, *Cell*, 69:1089 (1992).
10. A.H. Wyllie, Glucocorticoid-induced thymocyte apoptosis is associated with endogenous endonuclease activation, *Nature*, 284:555 (1980).
11. J.J. Cohen, and R.C. Duke, Glucocorticoid activation of a calcium-dependent endonuclease in thymocyte nuclei leads to cell death, *J. Immunol.*, 132:38 (1984).
12. G.E. Roelants, J. London, K.S. Mayor-Whithey, and B. Serrano, Peanut agglutinin. II. Characterization of the Thy-1, Tla and Ig phenotype of peanut agglutinin-positive cells in adult, embryonic and nude mice using double immunofluorescence, *Eur. J. Immunol.*, 9:132 (1979).
13. P.D. Boyer, and E.V. Rothenberg, IL-2 receptor inducibility is blocked in cortical-type thymocytes, *J. Immunol.*, 140:2886 (1988).
14. H. Selye, Thymus and adrenals in the response of the organism to injuries and intoxication, *Brit. J. Exp. Path.*, 17:234 (1936).
15. T.F. Dougherty, Effect of hormones on lymphatic tissue, *Phys. Rev.*, 32:379 (1952).
16. H.L. Jaffe, The influence of the suprarenal gland on the thymus. III. Stimulation of the growth of the thymus gland following double suprarenalectomy in young rats, *J. Exp. Med.*, 40:753 (1924).
17. C.M. Zacharchuk, M. Merćep, P. Chakraborti, S.S. Simons, Jr., and J.D. Ashwell, Programmed T lymphocyte death: cell activation- and steroid-induced pathways are mutually antagonistic, *J. Immunol.*, 145:4037 (1990).
18. C.M. Zacharchuk, M. Merćep, and J.D. Ashwell, Thymocyte activation and death: a mechanism for molding the T cell repertoire, *Antigen and clone-specific immunoregulation*, 636:52 (1991).
19. M. Iwata, S. Hanaoka, and K. Sato, Rescue of thymocytes and T cell hybridomas from glucocorticoid-induced apoptosis by stimulation via the T cell receptor/CD3 complex: a possible *in vitro* model for positive selection of the T cell repertoire, *Eur. J. Immunol.*, 21:643 (1991).
20. S.J. Henning, Plasma concentrations of total and free corticosterone during development in the rat, *Am. J. Physiol.*, 235:E451 (1978).
21. L. Savu, H. Zouaghi, and E.A. Nunez, Serum inflammatory responses of transcortin binding activities and of total and free corticosterone and progesterone levels in developing rats: a kinetic approach, *Int. J. Tiss. Reac.*, 7:443 (1985).
22. V. Geenen, J.J. Legros, P. Franchimont, M. Baudrihaye, M.P. Defresne, and J. Boniver, The neuroendocrine thymus: coexistence of oxytocin and neurophysin in the human thymus, *Science*, 232:508 (1986).
23. V.K.M. Han, A.J. D'Ercole, and P.K. Lund, Cellular localization of somatomedin (insulin-like growth factor) messenger RNA in the human fetus, *Science*, 286:193 (1987).

24. V.K.M. Han, D.J. Hill, A.J. Strain, A.C. Towie, J.M. Lauder, L.E. Underwood, and J. D'Ercole, Identification of somatomedin/insulin-like growth factor immunoreactive cells in the human fetus, *Pediatr. Res.*, 22:245 (1987).
25. C. LeGoascogne, P. Robel, M. Gouezou, N. Sananes, E.E. Baulieu, and M. Waterman, Neurosteroids: cytochrome P-450scc in rat brain, *Science*, 237:1212 (1987).
26. I. Jung-Testas, Z.Y. Hu, E.E. Baulieu, and P. Robel, Neurosteroids: biosynthesis of pregnenolone and progesterone in primary cultures of rat glial cells, *Endocrinology*, 125:2083 (1989).
27. F. Mintami, T. Shimizu, R. Ueno, Y. Ishimura, S. Izumi, N. Komatsu, and K. Watanabe, Cytochrome P450-11β and P450scc in adrenal cortex: zonal distribution and intramitochondrial localization by the horseradish peroxidase-labeled antibody method, *J. Histochem. Cytochem.*, 30:1066 (1982).
28. H.J. Geuze, J.W. Slot, K. Yanagibashi, J.A. McCracken, A.L. Schwartz, and P.F. Hall, Immunogold cytochemistry of cytochromes P-450 in porcine adrenal cortex, *Histochem.*, 86:551 (1987).
29. A.H. Payne, Hormonal regulation of cytochrome P450 enzymes, cholesterol side-chain cleavage and 17α-hydroxylase/C17-20 lyase in Leydig cells, *Biol. Reprod.*, 42:399 (1990).
30. I. Hanukoglu, R. Feuchtwanger, and A. Hanukoglu, Mechanism of ACTH and cAMP induction of mitochondrial cytochrome P450 system enzymes in adrenal cortex cells, *J. Biol. Chem.*, 265:20602 (1990).
31. H. Wekerle, and U.-P. Ketelsen, Thymic nurse cells—Ia-bearing epithelium involved in T-lymphocyte differentiation?, *Nature*, 283:402 (1980).
32. H. Wekerle, U.-P. Ketelsen, and M. Ernst, Thymic nurse cells. Lymphoepithelial cell complexes in murine thymuses: morphological and serological characterization, *J. Exp. Med.*, 151:925 (1980).
33. R.T. Kubo, W. Born, J.W. Kappler, P. Marrack, and M. Pigeon, Characterization of a monoclonal antibody which detects all murine α/β T cell receptors, *J. Immunol.*, 142:2736 (1989).
34. C.W. Ragsdale, Jr., and J.P. Brockes, Retinoids and their targets in vertebrate development, *Curr. Opin. Cell Biol.*, 3:928 (1991).
35. B.L.M. Hogan, C. Thaller, and G. Eichele, Evidence that Hensen's node is a site of retinoic acid synthesis, *Nature*, 359:237 (1992).
36. A.C. Ross, Vitamin A status: relationship to immunity and the antibody response, *Proc. Soc. Exp. Biol. Med.*, 200:303 (1992).
37. A. Krust, P.H. Kastner, M. Petkovich, A. Zelent, and P. Chambon, A third human retinoic acid receptor, hRARγ, *Proc. Natl. Acad. Sci. USA*, 86:5310 (1989).
38. D.J. Mangelsdorf, E.S. Ong, J.A. Dyck, and R.M. Evans, Nuclear receptor that identifies a novel retinoic acid response pathway, *Nature*, 345:224 (1990).
39. D. Mangelsdorf, U. Borgmeyer, R.A. Heyman, J.Y. Zhou, E.S. Ong, A.E. Oro, A. Kakizuka , and R.M. Evans, Characterization of three RXR genes that mediate the action of 9-*cis* retinoic acid, *Genes Dev.*, 6:329 (1992).
40. A.A. Levin, L.J. Sturzenbecker, S. Kazmer, T. Bosalkowski, C. Huselton, G. Allenby, G. Speck , C.I. Kratzeisen, M. Rosenberger, A. Lovey, and J. Grippo, 9-*Cis* retinoic acid stereoisomer binds and activates the nuclear receptor RXRα, *Nature*, 355:359 (1992).
41. R.A. Heyman, D.J. Mangelsdorf, J.A. Dyck, R.B. Stein, G. Eichele, M. Evans, and C. Thaller, 9-*Cis* retinoic acid is a high affinity ligand for the retinoid X receptor, *Cell*, 68:397 (1992).

42. X.-K. Zhang, J. Lehmann, B. Hoffmann, M.I. Dawson, J. Cameron, G. Graupner, T. Hermann, P. Tran, and M. Pfahl, Homodimer formation of retinoid X receptor induced by 9-*cis* retinoic acid, *Nature*, 358:587 (1992).
43. Y. Yang, M.S. Vacchio, and J.D. Ashwell, 9-*Cis* retinoic acid inhibits activation-driven T cell apoptosis: implications for retinoid X receptor involvement in thymocyte development, *Proc. Natl. Acad. Sci. USA*, in press(
44. J. Kappler, J. White, D. Wegmann, E. Mustain, and P. Marrack, Antigen presentation by Ia^+ B cell hybridomas to H-2-restricted T cell hybridomas, *Proc. Natl. Acad. Sci. USA*, 79:3604 (1982).
45. K.P. West, Jr., G.R. Howard, and A. Sommer, Vitamin A and infection: public health implications, *Annu. Rev. Nutr.*, 9:63 (1989).
46. T. Inomata, and T. Nakamura, Influence of adrenalectomy on the development of the neonatal thymus in the rat, *Biol. Neonate*, 55:238 (1989).
47. E.M. Smith, W.J. Meyer, and J.E. Blalock, Virus-induced corticosterone in hypophysectomized mice: a possible lymphoid adrenal axis, *Science*, 218:1311 (1982).
48. E. Batanero, F.E. De Leeuw, G.H. Jansen, D.F. Van Wichen, J. Huber, and H.J. Shuurman, The neural and neuro-endocrine component of the human thymus, *Brain Behav. Immunol.*, 6:249 (1992).
49. J. Sprent, D. Lo, E.-K. Gao, and Y. Ron, T cell selection in the thymus, *Immunol. Rev.*, 101:172 (1988).
50. F. Ramsdell, T. Lantz, and B.J. Fowlkes, A nondeletional mechanism of thymic self tolerance, *Science*, 246:1038 (1989).
51. B.A. Kyewski, Thymic nurse cells: possible sites of T-cell selection, *Immunol. Today*, 7:374 (1986).
52. M.A. Smith, D.R. Parkinson, B.D. Cheson, and M.A. Friedman, Retinoids in cancer therapy, *J. Clin. Oncol.*, 10:839 (1992).
53. E. Seifter, G. Rettura, and S.M. Levenson, Decreased resistance of C3H/HeHa mice to C3HBA tumor transplants: increased resistance due to supplemental vitamin A, *J. Natl. Cancer Inst.*, 67:467 (1981).

Discussion

M. Oren

As the epithelial cells in the thymus contact both the cells that are negatively selected and the ones that are not, have you looked whether steroid receptor expression is altered in any way, or diverges in any way, in the thymocytes that contact those epithelial cells?

J. Ashwell

We have looked a bit at steroid receptor expression in the thymus. It is expressed in double positive cells, of course, and it is also expressed in single positive cells. You want to know if we looked at epithelial cells?

M. Oren

No, thymocytes. When you present the T-cell hybridoma antigen, is there any effect on steroid receptor expression?

J. Ashwell

We have not looked in T cell hybridomas, but it is certainly known that activated T-cells upregulate the steroid receptor. In fact, that is a classic response to treatment with ConA or PHA. The receptor is upregulated, making the cells more sensitive to some steroid effects. But the paradox is that you would think activated cells would be more sensitive to killing by steroids but, in reality, it is exactly the opposite: It makes the cells less sensitive to killing by steroids.

M. Oren

Second question: Given that TPA induces apoptosis in T-cells and steroid receptors act as antagonists of the AP-1 dependent pathway, have you looked at whether that protection may be mediated by blocking AP-1 dependent gene expression?

J. Ashwell

That is a very good question. Our working hypothesis is that the mutual antagonism is mediated by inhibitory interactions, physical or at the DNA level, between AP-1 and the glucocorticoid receptor. We have reproduced the work of others with our cells, in that glucocorticoids inhibit transcription of a TRE-regulated reporter gene and, if you choose the right GRE, T-cell receptor activation of PKC inhibits transcription of a GRE regulated reporter gene. But that does not tell you that these interactions are biologically important. We have been trying to use trans-dominant negative Jun to see if we can block or modulate programmed cell death, either directly or at the level of antagonism. So far, we have failed to get high enough levels of expression in stable transfectants to ask the question, but it is a very good question.

J. Gruber

Did you see differences in subpopulations after treatment with exogenous e.g., dexamethasone, or endogenously elevated steroids?

J. Ashwell

We have not looked at exogenous treatment with dexamethasone very much. Basically, when people treat with exogenous steroids they use high doses that get rid of all double positive cells. It does not matter what the T-cell receptor specificity is, you lose all the double positives and preserve the single positives. I do not see why, *a priori*, one would expect a difference between steroids made in the thymus and those made in the periphery, except possibly for two reasons. One is quantitative, that there is more per cell. There is obviously more steroid production in the adrenal than in the thymus. I would guess that only about 1% of the thymic epithelial cells were positive for the steroidogenic enzymes, whereas if you stain the adrenal, virtually all of the cells are positive; every cell there is making high levels of steroids. The other reason, and one of the areas of speculation in our model, is that since the epithelial cells surround the thymocytes, you could have directed delivery of steroids. So although the overall quantity of steroids in the thymus is small, the local concentrations can be quite high. A corollary to that, one that most immunologists often do not consider, is that the large majority of circulating steroids is bound to protein. Ninety to 95% or more, is bound to corticoid binding globulin or albumin, and so the effective level of steroids in the serum is much lower than the total steroid concentration. In the case where one has local secretion and uptake, or perhaps cell-cell contact and diffusion of steroids, the problem is avoided, and whatever steroids are produced locally are close to 100% biologically active, as opposed to steroids made in the adrenals which circulate in the serum.

J. Gruber

Did you look at T-cells expressing the $\gamma\delta$ receptors?

J. Ashwell

No, we have not looked at $\gamma\delta$ at all.

D. Hockenbery

Do retinoids have effects on T-cell viability independent of TCR interactions, and have you looked in your hybridomas to see where retinoids affect the cascade of events? Do they act downstream of cell cycle block, for example?

J. Ashwell

Actually, in hybridomas, the retinoids are completely benign. They do not affect growth even up to 10^{-5}M, which is the highest we have gone. They had no effect on steroid-induced death. We did not look specifically at thymocytes, but with 36 hour cultures quite a few cells die nonspecifically. Retinoic acid does not prevent that--you do not see better background viability.

S. Schlossman

How do you explain differential sensitivity of immature T-cells versus mature T-cells?

J. Ashwell

Immature T-cells are exquisitely sensitive to steroids' cytotoxic effects, and peripheral T-cells are pretty resistant. There is also species variability--human cells are more resistant than murine cells to steroids. The answer to your question is, to my knowledge, not known. There are obviously a number of different variables one has to consider. Steroid receptor levels are the first, and as I mentioned earlier, steroid receptor levels do go up during activation. I would not not be surprised if thymocytes that are recognizing antigen have more receptors than mature T cells. That is a speculation. Steroid receptors are also quite complicated, and other proteins, such as heat schock proteins and FKBP, interact with glucocorticoid receptors and can modulate ligand binding or receptor function. We have considered a number of these possibilities, and people in the laboratory are working on them.

D. Green

This is an exquisite model because not only does it explain positive and negative selection, it also explains some paradoxes associated with it, explains the loss of cells that fail to express T-cell receptor, it revives the stochastic model of positive selection. But let us try to make a prediction from this model and then ask Dave Hockenbery if it holds. Since we know that *bcl*-2 expression in the thymus of transgenic mice will block glucocorticoid induced apoptosis, one prediction is that there is an increase in positive selection in those animals and Dave has already shown us that. The other prediction is that you will see an accumulation of cells that have aberrantly rearranged T-cell receptors, that are failing to express T-cells receptors because now the glucocorticoids are not inducing death of those cells.

J. Ashwell

That is a good question. I want to point out that we do not hang our hats on the possibility that steroids cause death of TCR-negative cells, although it is a possibility. But the prediction from that would be that, in the absence of steroid killing, you would get a hypercellular thymus, right? David, do you find any change in cell number or viability?

D. Hockenbery

Well, the *bcl*-2 transgenic mice have normal sized thymuses. However, if you take adrenals out of an adult mouse where the majority of cortical steroid is being produced in the adult mouse, the thymus increases in size dramatically.

J. Ashwell

One of the points I did not go over was the observation that if you take adrenals out of an older animal, in this case three weeks of age, the thymus hypertrophies. If you take the thymus out of a young animal, less than three weeks of age, it does not. That correlates nicely with the ontogeny data I showed for steroid production, suggesting that thymic local steroid production is all you need to maintain cellularity in young mice, and as the animal gets older, the adrenal takes over that function.

D. Hockenbery

There does not appear to be an overabundance of TCR negative cells in the periphery in the positive selection models, but specific details of TCR rearrangement have not really been looked at.

D. Livingston

What do you predict the phenotype would be of a knockout mouse missing the enzyme just proximal to the pregnenolone point in the pathway which was fed exogenous cortisol from the time it was conceived through birth?

J. Ashwell

So you have knocked out endogenous production but you have supplied....

D. Livingston

You have supplied cortisone exogenously plus mineral corticoid?

J. Ashwell

That is a complicated biological scenario you have painted. I do not think there is anything special about the biochemical product of the thymic steroid pathway versus the adrenal pathway--I think they are identical. The question really becomes: Does local concentration matter? And does directed delivery, if you will, or paracrin delivery of steroids make a difference compared to systemic delivery of steroids? I would like to think that the microenvironment does matter. There is a very critical balance here. As I mentioned, thymocytes are very sensitive to the lethal effects of steroids, so I think there is a relatively narrow window that you have between steroid effects that result in positive selection and those that result in non-antigen-specific deletion of thymocytes. That could be extremely difficult to titer. Actually, we have done experiments trying to titer steroids by supplying them parenterally, but so far have not found any antigen-specific effects.

D. Livingston

Is it not more than 6-5 pickem? In short, if you are right, that mouse is effectively scid but otherwise OK? After all, you have argued that you need to have local production of glucocorticoid to bind the glucocorticoid receptor and that, plus antigen, leads to positive selection. So, if you cannot substitute exogenous steroid for that, do not you end up scid, if your model is correct?

J. Ashwell

If you can knockout the steroid receptor in thymocytes, a tissue-specific knockout of the steroid receptor, what would happen? If that is your question, I will predict that you would lose the double positive thymocytes.

D. Livingston

No. I could not concede that. But would you not lose positive cells this way too?

J. Ashwell

Ultimately, if this were the only determining factor. It this were the only criterion, every T cell with a receptor that had any affinity for self at all would be deleted.

D. Livingston

So you would effectively have a scid?

J. Ashwell

If you like, yes.

D. Livingston

So, is such an experiment in progress anywhere?

J. Ashwell

Yes, we have transgenic animals now. We are doing experiments designed actually to try to knock out thymic glucocorticoid receptor.

M. Oren

You could envisage that part of the morphogenic effect of retinoic acid is by keeping particular cells alive. Have you tried to extend your observations to prevention of cell death in other cell types or developmental processes?

J. Ashwell

We have certainly thought of that, but the answer is no, we have not. It is an intriguing possibility, but we are not really developmental biologists. We are getting into this because of our interest in T-cell ontogeny.

J. Isaacs

Right, but the clinical observation: There are a variety of individuals who are attempting to use retinoic acid to sensitize cells to chemotherapy to enhance the cell death and there are many examples, in culture, of just the opposite of what you have described. That retinoic acid sensitizes cells to undergo programmed cell death.

J. Ashwell

That is right. First of all, it is a very complicated system. There are two different families of retinoic acid receptor, and each family has at least three isoforms. The effects of retinoic acid are pleiotropic, and they can be quite tissue-specific. Retinoic acid is being used to treat APL and Kaposi's Sarcoma. In some cases retinoic acid causes terminal differentiation, in some cases apoptosis, and in some cases enhanced viability. We do not know enough about the receptor or how it works as a transcription factor to understand its different effects.

J. Cleveland

This is a very nice model, but I do not understand how cells with low avidity die. How do you explain that? It seems like with this system that you are getting a balance of mitogenic signals in a certain window. That is how that works for positive selection, but how do you explain the lower avidity receptor, and how that negative selection works?

J. Ashwell

When you say low avidity, you mean low to zero? Well, there are several possibilties here. You could explain it all in this model if you wanted to. Remember I showed with hybridomas that unopposed steroids kill. You need T cell receptor occupancy to counter the steroid death signal. You could argue that thymocytes with very low to zero affinity receptors die because of unopposed steroid effects, that there is enough circulating or locally-produced steroids to result in thymocyte death.

J. Cleveland

Yes, but the key here, it seems to me, is the amount of signaling from the receptor, is that correct?

J. Ashwell

Yes, which is at the moment unquantifiable.

J. Cleveland

Yes, I was wondering how you would read that out.

J. Ashwell

For many years people have been trying to understand the relationship between receptor affinity, receptor avidity, and functional responses. There is no easy way of reading out receptor affinity yet. There are some numbers in the literature, but the determination of avidity is complicated by 20,000 to 40,000 receptors per cell that contact a ligand bound to another molecule on a second cell membrane. There is, as yet, not even a mathematical model to explore what avidity means under these circumstances. All one can discuss is relative not absolute values.

REGULATION OF CELL GROWTH BY TRANSCRIPTION FACTORS, IRF-1 AND IRF-2

Tadatsugu Taniguchi, Hisashi Harada, Motoo Kitagawa, Nobuyuki Tanaka, Nobumasa Watanabe, and Tohru Kimura

Institute for Molecular and Cellular Biology,Osaka University Yamadaoka 1-3, Suita-shi, Osaka 565, Japan

INTRODUCTION

Cytokine transmit signals to cell interior, resulting in the activation/repression of the genes critical for cell growth and differentiation. One class of cytokines is the interferons (IFNs), which represent a heterogeneous family of multifunctional cytokines that typically manifest antiviral activity. In fact, viral infections commonly trigger induction of the type-I IFN (IFN-αs and IFN-β) genes in a variety of cells, and these newly produced IFNs can, in turn, activate a set of IFN-inducible genes. IFNs also exhibit anti-proliferative effects on many normal and transformed cells[1,2]. In addition, it has been reported that one of the IFN species, IFN-γ, plays a critical role in induced cell death of effector T lymphocyte[3].

Although the molecular mechanisms underlying the cytokine-mediated cellular response still remains obscure, recent studies have revealed the importance of transcription factors as the critical targets of cytokines; those factors, once induced or activated, regulate the expression of a variety of genes required for the cytokine-induced cellular responses[4,5,6].

Interferon Regulatory Factor-1 (IRF-1) and IRF-2, were originally discovered as transcriptional regulators for the type-I interferon (IFN) genes[7,8,9]. DNA sequences recognized by IRFs have been also found in the regulatory regions, termed IFN-stimulated regulatory elements (ISREs), in a number of IFN-inducible genes[8,9,10]. IRF-1 and IRF-2 bind to the same DNA motifs, termed IRF-E [G(A)AAA$^{G}/_{C}$$^{T}/_{C}$GAAA$^{G}/_{C}$$^{T}/_{C}$] through highly conserved DNA binding domains[10]. cDNA transfection studies have shown that IRF-1 functions as a transcriptional activator, and IRF-2 represses IRF-1 action[9,11,12,13,14]. In addition, the genes for IRF-1 and IRF-2 are both virus- and IFN-inducible[8,9,15]. In IFN-stimulated cells the IRF-2 gene induction occurs only after induction of IRF-1[9]. In IFN-treated or

Apoptosis, Edited by E. Mihich and R.T. Schimke
Plenum Press, New York, 1994

virus-infected cells, the IRF-2 protein is more stable than the IRF-1 protein (half-lives of 8 hours and 30 min, respectively)[16]. Thus in growing cells, IRF-2 is more abundant than IRF-1, but after stimulation by IFN or viruses, the amount of IRF-1 increases relative to IRF-2[16] .

Taken together these observations, a transient increase in the IRF-1/IRF-2 ratio may be a critical event in the regulation of cell growth by IFNs. Consistent with this notion are the findings that transgenic mice carrying the human IRF-1 gene linked to the human immunoglobulin gene enhancer are deficient in developing B lymphocytes[17] and that the expression of IRF-1 at high levels is inhibitory for *in vitro* growth of some cell lines (N. T., unpublished data). As an approach to further assess the role of IRF-1 and IRF-2 in cell growth regulation, we examined the effect of perturbation of the IRF-1/IRF-2 ratio in NIH 3T3 cells.

RESULTS AND DISCUSSION

Oscillation of the IRF-1/IRF-2 ratio during cell cycle

Expression of IRF-1 and IRF-2 mRNAs in mouse NIH 3T3 cells were analysed throughout the cell cycle. Cells were growth-arrested by serum starvation for 24 hours (G1 arrest), and were then induced to transit the cell cycle by serum restoration. A [^{3}H]-thymidine uptake assay and flow cytometric analysis revealed that DNA synthesis began 8 to 12 hours after serum stimulation (H.H., unpublishe data). S1 analysis indicated that IRF-1 mRNA expression reached a peak (about five transcripts per cell) in growth-arrested cells, declined sharply after serum stimulation and then increased gradually, prior to the onset of DNA synthesis[18] . At 2 hours after stimulation, IRF-1 mRNA expression was five times lower than in growth-arrested cells. In contrast, IRF-2 mRNA expression remained essentially constant throughout the cell cycle. Western (protein) blotting analysis with antibodies to IRF-1 indicated that expression of the protein also oscillated throughout the cell cycle[18].

Overexpression of IRF-2 in NIH 3T3 cells

To examine the effect of perturbing the IRF-1:IRF-2 ratio on cell growth, we first generated NIH 3T3 cell clones that overexpressed IRF-2. The plasmid pAct-2[12] which contains mouse IRF-2 complementary DNA (cDNA) cloned downstream of the chicken β-actin gene promoter, was transfected into NIH 3T3 cells. Control transfections were done with pAct-C, a plasmid devoid of the cDNA insert. After selection for a *neo*-resistance, we obtained several clones that expressed high levels of IRF-2 mRNA and we selected three (R21, R25, and R27) for further analysis. The expression of IRF-2 mRNA in these cells was more than 10 times higher than the basal expression observed in the pAct-C-transfected cells (C2, C3). On the other hand, the IRF-2 binding activities (monitored by gel-shift analysis) were only four to nine times higher than the control cells[18].

Although cells overexpressing IRF-2 did not exhibit any obvious morphological changes, they displayed marked differences in other growth properties. The IRF-2-overexpressing cells grew at a rate similar to that of control cells, but reached a cell density that was about three times higher (Table 1).

Furthermore, all of the IRF-2 overexpressing cells formed colonies in methylcellulose gel with an efficiency ranging from 6 to 19 %, whereas no colony formation was seen with the control cells (Table 1)[18]. As another test of the oncogenic potential of IRF-2, we constructed a recombinant retrovirus vector, pGDIRF2, that directed expression of the mouse IRF-2 cDNA[18]. NIH 3T3 cells were infected with the pGDIRF2 retrovirus at a high

Table 1. Growth properties of cell lines overexpressing IRF-2 and controls.

Cell line	Growth in monolayers *		Growth in methyl cellulose gel	Tumorigenicity ‡	
	Doubling time (hours)	Saturation density (10^6 cells)	Efficiency (%) †	Tumors/ injection	Latency (weeks)
C2	30.6 ± 0.0	1.87 0.06	0, 0	0 / 7	—
C3	28.7 ± 0.8	1.87 0.05	0, 0	0 / 5	—
R21	25.7 ± 1.1	4.93 0.14	7, 12	6 / 6	2-3
R25	28.8 ± 0.3	4.30 0.04	6, 6	6 / 6	2-3
R27	27.5 ± 0.6	4.40 0.01	10, 19	6 / 6	2-3

* Cells were seeded at 2x104 cells per 35 mm plate and grown in DMEM supplemented with 10 % FCS and 700 mg/ml G418. Saturation density was the number of the cells in culture after reaching confluency. The cell size of those clones remained essentially the same as that of the parental cells. Numbers are the means of duplicate assays + the standard deviation.

† The colony-forming assay was performed essentially as follows. Briefly, cells (105) were suspended with 1.3% methyl cellulose gel dissolved in culture medium and overlayed on an agarose bed composed of 0.53% agarose and culture medium. Colonies were scored 3 weeks after plating.

‡ Nude mice (Balb/c nu/nu; Clea Japan, Inc.; 4-6 week old) were injected subcutaneously on both flanks with 2x106 cells resuspended in 200 ml of serum-free DMEM. Cells were scored as tumorigenic if a visible nodule appeared at the site of injection and increased thereafter.

multiplicity of infection (m.o.i.) and the cells were directly tested in a colony formation assay on methylcellulose gel. As shown in Table 2, the cells infected by the IRF-2-expressing virus, but not by the control pGD virus, formed colonies at high efficiency. Assuming that all the cells were infected by the virus, the colony-forming efficiency was similar to that of the R21, R25, and R27 clones (see Table 1).

Table 2 . Efficiency of colony formation in methyl cellulose gel following retroviral introduction of IRF-2 gene.

	Efficiency of colony formation in methyl cellulose gel (%)	
	pGD	pGDIRF2
experiment 1	<1 , <1	17 , 15
experiment 2	<1 , <1	12 , 16

NIH3T3 cells were infected with recombinant retrovirus pGDIRF2 or control virus pGD (m.o.i.=10)[18]. After 48 hours, cells were split into methyl cellulose gel as described in Table 1 footnote. Colonies were scored 3 weeks after plating. Stable integration of proviral genome was virtually 100% as revealed by neo-resistance of an aliquot of the cells.

Enhanced tumorigenicity of the clones overexpressing IRF-2

In order to test whether or not the observed cell growth properties of the R21, R25, and R27 clones are accompanied with the enhancement of tumorigenicity, cells (2 x 10^6) from the above three clones were injected subcutaneously into nude mice. As summarized in Table 1, tumors developed within 2 to 3 weeks and continued to grow unrestricted, although they showed no signs of metastasis. No tumors developed in nude mice injected with cells from the control C2 and C3 clones during the same time period. Collectively, these results suggest that IRF-2 has an oncogenic potential.

Reversion of the transformed cells by the expression of IRF-1

The above results imply that restrained cell growth depend upon a balance bewteen IRF-1 and IRF-2 expression. When this balance is perturbed by the overexpression of IRF-1 cell proliferation may be inhibited[17] whereas the overexpression of IRF-2 may promote cell transformation. Frow this view point, we next addressed whether the transformed phenotype displayed by NIH 3T3 cells overexpressing IRF-2 could be reverted to the original phenotype by increasing the expression of IRF-1. To test this possibility, we introduced into the IRF-2-transformed cells a 19-kb DNA segment containing all exons, as well as the promoter region [455 bp upstream of the major cap site, (H.H., unpublished data)], of the human IRF-1 gene. The R21, R25, and R27 cells were cotransfected with a plasmid (pUCHIRF1B) carrying the IRF-1 gene and a hygromycin (*hgr*) resistance gene. Clones showing *hgr*-resistance were selected and subsequently screened for stable integration of the IRF-1 gene. The transfectants R21-1, R21-2, R25-2, R27-3, and R27-4 were derived from parental clones R21, R25, and R27. S1 analysis revealed steady state expression of human IRF-1 mRNA in most of these clones, albeit at the different levels (Table 3). The different levels of the IRF-1 mRNA expression correlated with the IRF-1 binding activity. The transfected IRF-1 gene was also virus-inducible in all clones (Table 3) and, in a separate set of experiments, it was shown that the promoter sequence within the cloned gene was also IFN-inducible (H.H., unpublished data). Expression of IRF-2 mRNA and protein, assessed by Northern blotting and immunoprecipitation analyses, respectively, was similar to that in the parental cells. Interestingly, however, the tumorigenicity of the transfected cells was suppressed (Table 3). Moreover, extent of suppression was correlated with ectopic IRF-1 mRNA expression; clones R25-2 and R27-3, both of which expressed high amounts of human IRF-1 mRNA showed strong suppression; clones R27-4 and R21-1, which expressed lower amounts of IRF-1 mRNA, showed somewhat weaker suppression, and clone R21-2, in which steady state expression of human IRF-1 mRNA was not detectable, showed no suppression (Table 3).

Table 3. Growth properties of cell lines expressing human IRF-1.

Cell line	Human IRF-1 mRNA expression (copies/cell)		Growth in monolayers *		Growth in methyl cellulose gel	Tumorigenicity ‡	
	(NDV) —	(NDV) +	Doubling time (hours)	Saturation density (10^6 cells)	Efficiency (%) †	Tumors/ injection	Latency (weeks)
21-1	1	6	28.0 ± 0.1	2.66 ± 0.11	4, 4	2 / 6	2
R21-2	<1	7	28.1 ± 0.2	3.72 ± 0.38	6, 8	6 / 6	2-3
R25-2	24	463	28.9 ± 0.3	2.04 ± 0.37	0, 0	0 / 5	—
R23-3	16	542	28.8 ± 1.5	1.67 ± 0.07	0, 1	0 / 6	—
R24-4	3	65	29.8 ± 1.1	2.52 ± 0.02	5, 5	3 / 6	3-3.5

The mRNA copy number was determined by S1 mapping analysis and calculated by densitometric analysis as described[18]. All the other assays were performed as described in Table 1 footnote.

Concomitant with the loss or reduction of the transformed phenotype, the R21-1, R25-2, R27-3, and R27-4 cell clones exhibited a loss or reduction of other transformation-associated traits, such as increased cell saturation density and anchorage-independent growth (Table 3). In addition, there was a marked reduction in the colony-forming potential of R27 cells infected with a retrovirus (pGDIRF1) that directs the expression of mouse IRF-1 (Table 4)[18]. Thus, the IRF-2-induced transformation of NIH 3T3 cells appears to be reversible by the introduction and increased expression of IRF-1 gene.

Table 4. Colony formation of R27 cells following retroviral induction of IRF-1 gene

	Number of colonies in methyl cellulose gel per 5000 cells*		
	m.o.i.	pGD	pGDIRF1
Experiment 1	0.3	417, 389	308, 287
Experiment 2	1	423, 408	187, 225
Experiment 3	10	415, 432	196, 124

R27 cells were infected with recombinant retrovirus pGDIRF1 or control virus pGD at different multiplicities of infection (m.o.i.=0.3, 1, 10)[18]. Colonies were counted per 5000 cells. All experiments were done in duplicate. *For each assay, a total of 5000 cells, judged viable by visual inspection under a microscope, were counted and the number of colonies formed among them was scored.

CONCLUSIONS AND PERSPECTIVE

The work described here shows that transcription factors that participate in the IFN system regulation also play a role in regulating cell growth. We infer that the ability of IFNs to inhibit cell proliferation may be due, at least in part, to a transient induction of IRF-1; however, this possibility has not yet been directly tested. Presumably, overexpression of IRF-2 induced cell transformation by suppressing the cell growth-restraining function of IRF-1. Although the mechanism by which IRF-1 and IRF-2 affect cell growth is unknown, we think it is likely that IRF-1 functions in a manner analogous to the tumor suppressor p53[19,20] -that is- it activates a set of genes whose products are required for the negative regulation of cell growth, whereas IRF-2 represses those genes. Many, if not all, of the IFN-inducible genes contain promoter elements that bind IRFs[6,21] and some of the genes such as the 2'-5' oligoadenylate synthetase and double-stranded RNA-dependent protein kinase (dsRNA-PK) appear to be involved in the inhibition of cell proliferation[22] . Interestingly, the activity of the 2'-5' oligoadenylate synthetase oscillates throughout the cell cycle[23] and its expression is upregulated by IRF-1[14,24]. The dsRNA-PK itself has recently been shown to have anti-oncogenic potential[25,26].

One of the most frequent cytogenetic abnormalities in human leukemia and myelodysplasia (MDS) is a loss of chromosome 5 or an interstitial deletion of the long arm of chromosome 5 [del(5q) or "5q-"]. A tumor suppressor gene has been hypothesized to lie in 5q31, the most commonly deleted "critical" region. The IRF-1 gene has been mapped to 5q31.1, and it has been shown that one or both IRF-1 alleles are deleted in each of 13 representative cases of MDS and leukemia with del(5q) or translocation 5q31[27]. Furthermore, inactivating structural rearrangements of the IRF-1 gene, accompanied by deletion of the residual allele, were found in a case of de novo acute leukemia. Thus, IRF-1 may be a critically deleted tumor suppressor gene in the del(5q); loss of one or both IRF-1 alleles may contribute to unrestrained cellular proliferation thereby promoting the development of human leukemia and MDS[27].

Work is in progress to obtain mice lacking either IRF-1 or IRF-2 or both. The availability of such mice will certainly help clarifying further the role of these factors.

ACKNOWLEDGEMENTS

We thank Ms. Y. Kasakawa for her assistance. This work was supported in part by a grant-in-aid for Special Project Research, Cancer Bioscience, from Ministry of Education, Science and Culture of Japan, and Takeda Science Foundation.

REFERENCES

1. De Maeyer, E., and De Maeyer-Guignard, J. , 1988, Interferons and Other Regulatory Cytokines (New York: John Wiley and Sons).
2. Vilcek, J. ,1990, Interferons. *In* Peptide Growth Factors and Their Receptors, Handbook of Experimental Pharmacology, M. A. Sporn and A.B.Roberts, eds. (Berlin: Springer-Verlag), 3.
3. Liu, Y., and Janeway, Jr., A., 1990, Interferon γ plays a critical role in induced cell death of effector T cell : A possible third mechanism of self-tolerance. *J. Exp. Med.* 172: 1735-1739.
4. Lewin, B., 1991, Oncogenic conversion by regulatory changes in transcription factors. *Cell* 64: 303.
5. Karin, M., and Smeal, T., 1992, Control of transcription factors by signal transduction pathways: The beginning of the end. *TIBS* 17:418.
6. Tanaka, N. and Taniguchi, T., 1992, Cytokine gene regulation: regulatory *cis*-elements and DNA binding factors involved in the interferon system. *Advances in Immunol.* 52: 263.
7. Fujita, T., Sakakibara, J., Sudo, Y., Miyamoto, M., Kimura, Y., and Taniguchi, T., 1988, Evidence for a nuclear factor(s), IRF-1, mediating induction and silencing properties to human IFN-β gene regulatory elements. *EMBO J.* 7: 3397.
8. Miyamoto, M., Fujita, T., Kimura, Y., Maruyama, M., Harada, H., Sudo, Y., Miyata, T., and Taniguchi, T. ,1988, Regulated expression of a gene encoding a nuclear factor, IRF-1, that specifically binds to IFN-β gene regulatory elements. *Cell* 54: 903.
9. Harada, H., Fujita, T., Miyamoto, M., Kimura, Y., Maruyama, M., Furia, A., Miyata, T., and Taniguchi, T. ,1989, Structurally similar but functionally distinct factors, IRF-1 and IRF-2, bind to the same regulatory elements of IFN and IFN-inducible genes. *Cell* 58: 729.
10. Tanaka, N., Kawakami, T., and Taniguchi, T., 1993, Recognition sequences of IRF-1 and IRF-2; Regulators of cell growth and the interferon system. *Mol. Cell. Biol.* in press.
11. Fujita, T., Kimura, Y., Miyamoto, M., Barsoumian, E. L., and Taniguchi, T. , 1989, Induction of endogenous IFN-α and IFN-β genes by a regulatory transcription factor, IRF-1. *Nature* 337: 270.
12. Harada, H., Willison, K., Sakakibara, J., Miyamoto, M., Fujita, T., and Taniguchi, T. 1990, Absence of the type I IFN system in EC cells: Transcriptional activator (IRF-1) and repressor (IRF-2) genes are developmentally regulated. *Cell* 63: 303.
13. Näf, D., Hardin, S. E., and Weissmann, C. ,1991, Multimerization of AAGTGA and GAAAGT generates sequences that mediate virus inducibility by mimicking an interferon promoter element. *Proc. Natl. Acad. Sci. USA* 88: 1369.
14. Reis, L. F. L., Harada, H., Wolchok, J. D., Taniguchi, T., and Vilcek, J., 1992, Critical

role of a common transcription factor, IRF-1, in the regulation of IFN-β and IFN-inducible genes. *EMBO J.* 11: 185.
15.Pine R., Decker, T., Kessler, D. S., Levy, D. E., and Darnell, J. E., Jr., 1990, Purification and cloning of interferon-stimulated gene factor 2 (ISGF2): ISGF2 (IRF-1) can bind to the promoters of both beta interferon- and interferon-stimulated genes but is not a primary transcriptional activator of either. *Mol. Cell. Biol.* 10: 2448.
16.Watanabe, N., Sakakibara, J., Hovanessian, A. G., Taniguchi, T., and Fujita, T., 1991, Activation of IFN-β element by IRF-1 requires a post-translational event in addition to IRF-1 synthesis. *Nucl. Acids Res.* 19: 4421.
17.Yamada, G., Ogawa, M., Akagi, K., Miyamoto, H., Nakano, N., Itoh, S., Miyazaki, J., Nishikawa, S., Yamamura, K., and Taniguchi, T., 1991, Specific depletion of the B-cell population induced by aberrant expression of human interferon regulatory factor 1 gene in transgenic mice. *Proc. Natl. Acad. Sci. USA* 88: 532.
18.Harada, H., Kiatagawa, M., Tanaka, N., Yamamoto, H., Harada, K., Ishihara, M., and Taniguchi, T. ,1993, Anti-oncogenic and oncogenic potentials of Interferon Regulatory Factors-1 and -2. *Science* 259: 971.
19.Levine, A. J., Momand, J., and Finlay, C. A., 1991, The p53 tumour suppressor gene. *Nature* 351: 453.
20.Kern, S. E., Pietenpol, J. A., Thiagalingam, S., Seymour, A., Kinzler, K. W., and Vogelstein, B. ,1992, Oncogenic forms of p53 inhibit p53-regulated gene expression. *Science* 256: 827.
21.Stark, G. R., and Kerr, I. M., 1992, Interferon-dependent signalling pathways: DNA elements, transcription factors, mutations, and effects of viral proteins. *J. Interferon R.* 12: 147.
22.Revel, M., and Chebath, J., 1986, Interferon-activated genes. *Trends Biol. Sci.* 11: 166.
23.Jacobsen, H., Krause, D., Friedman, R. M., and Silverman, R. H., 1983, Induction of ppp(A2'p)*n*A-dependent RNase in murine JLS-V9R cells during growth inhibition. *Proc. Natl. Acad. Sci. USA* 80: 4954.
24.Au, W.-C., Raj, N. B. K., Pine, R., and Pitha, P. M., 1992, Distinct activation of murine interferon-α promoter region by IRF-1/ISGF-2 and virus infection. *Nucl. Acids Res.* 20: 2877.
25.Koromilas, E. A., Roy, S., Barber, N. Glen., Katze, G. Michael., and Sonenberg, N., 1992, Malignant transformation by a mutant of the IFN-inducible dsRNA-dependent protein kinase. *Science* 257: 1685.
26.Meurs, E. F., Galabru, J., Barber, G. N., Katze, M., G., and Hovanessian, A., 1993, Tumor suppressor function of the interferon-induced double-stranded RNA-activated protein kinase. *Proc. Natl. Acad. Sci. USA*, 90:232.
27.Willman C. L., Sever, C. E., Pallavicini, M. G., Harada, H., Tanaka, N., Slovak, M. L., Yamamoto, H., Harada, K., Meeker, T. C., List, A. F., and Taniguchi, T., 1993, Deletion of IRF-1, mapping to chromosome 5q31.1, in human leukemia and preleukemic myelodysplasia. *Science* 259:968.

Discussion

R. Schimke

That was a very nice paper. I have two questions of you. Is the IRF-1 in high levels in a number of tissues? Basically it is a question of whether the active suppression, high levels of this protein might be involved in active maintenance of a non proliferative state. Have you looked in non proliferating cells, not just 3T3 cells in the arrested state?

T. Taniguchi

That is an interesting possibility. We have not examined yet in the non proliferating state whether this IRF-1 is very high.

R. Schimke

Because you might say that in the serum starved state, if that is analogous to a G-0 state, maybe this protein at high levels is involved in maintaining that in an inactive G-0 state. The other question I had is, since you can titrate IFR-1 and 2 and control proliferation/non proliferation, have you looked for, or is there any indication that there might be, overproduction, perhaps even amplification, of the IRF-2, as the other side of the coin, one being a deletion, and the other an overproduction?

T. Taniguchi

We have not examined this very extensively at this moment, but in those leukemia patients, in all the cases, there is neither deletion nor amplification of IRF-2. The IRF-2 gene remains intact in those patients, and other cells we have not examined so far but it is an interesting possibility.

D. Livingston

I wonder if, before you get through answering questions from people here, you could remind people about the *myc* story for just a minute.

T. Taniguchi

Yes, now the *myc* story has been published some time ago but I did mean to mention a little bit about how this works in haematopoietic cell cycles. Of course *myc* is important for cell proliferation and we have been studying the haematopoietic cell cycle progression induced by IL-2. One of the IL-3 dependent haematopoietic cells we have been working with is called BAF-BO-3, and this has been kindly provided by Mary Collins in London. IL-2 has three chains: alpha, beta and gamma. In BAF-BO3, the beta chain, but not the alpha and gamma chains, is missing but, if you introduce the beta chain, this cell becomes responsive to IL-2 for proliferation. On the other hand, if you express another growth factor receptor, like an EGF receptor, when you stimulate these cells with EGF or TGF alpha, DNA synthesis starts and the cell goes to S-phase but the cell cycle cannot progress further. And we have measured all the cell cycle controlling genes so far available and the only defect we found here was the lack of C-myc gene expression. So C-myc gene expression is not induced by EGF receptor, at least very efficiently, and therefore DNA synthesis occurred, but further progression of the cell cycle did not occur, suggesting the possible involvement of C-myc after the DNA synthesis stage. Therefore, we introduced C-myc gene constitutively. The cells still remain factor dependent. Interestingly, when you treat them under a viral promoter with EGF, these cells go through the cell cycle without affecting DNA synthesis. So this is a sort of indication that C-myc also plays a critical role in the late phase of the cell cycle. For further details, please see our paper (*Cell* 70:57, 1992).

M. Fried

In the NIH cells and tumor cells overproducing IRF-2, did you look to see if any of the genes that have the consensus site are repressed at the transcriptional level? Can you revert the tumor cells by transfecting in IRF-1?

T. Taniguchi

The question is whether we have identified any possible target genes critical for tumor suppression. The answer is unfortunately no, not yet.

M. Fried

Is there any difference in the transcription of those genes that have the consensus site?

T. Taniguchi

We have not examined all of them but what we have done was to try to revert the phenotype of the IRF-2-transformed cells by introducing one of the interferon inducible genes which has been shown to have potentially anti-oncogenic activity, that is, double strand RNA-dependent protein kinase. So far we do not have any positive results.

M. Fried

Have you put IRF-1 back into the human tumor cells?

T. Taniguchi

That is also an ongoing experiment because, so far, it has been very difficult to obtain MDS-derived cells in culture.

M. Fried

What is known about p53 in these cells?

T. Taniguchi

We have not examined that. It would be interesting to see whether p53 is affected by IRF-1 in this system.

C. Croce

Did you look at the abnormalities in other alleles, particularly in acute leukemia?

T. Taniguchi

In one case of acute leukemia, that I mentioned to you, one gene is deleted, another one is rearranged. So this is one example.

C. Croce

Is that consistent? Do you see it in every case?

T. Taniguchi

Well, we have only examined 13 cases, including older patients, so the number of patients analyzed may not be enough, statistically speaking. And whether protein is produced, or messenger protein is produced, has not been examined in this experiment and we are now analyzing that.

C. Croce

Do you see rearrangement of IRF-2 in any case of acute myelogenous leukemia?

T. Taniguchi

No, not as far as we have examined.

J. Cleveland

As you know, I am really interested in your EGF model. At the end you said that if you put in *bcl*-2 that you rescued proliferation in EGF?

T. Taniguchi

That is right. However, I would like to emphasize that this is still preliminary.

J. Cleveland

And that is long term growth or is that just cycling?

T. Taniguchi

It is just what I heard from a colleague when I left my laboratory a few days ago, and he said it is proliferating and he wanted to make sure that it is long term proliferation.

J. Cleveland

And what is the status of *myc* expression when you rescue with *bcl*-2?

T. Taniguchi

That has not been examined.

J. Cleveland

The other thing I would just mention in passing is the other model for this kind of mapping function of *myc* and entry into S-phase. With Martine Roussell's experiments, it is clear in the CSF-1R cells that they fail to enter S-phase so effects appear to depend on the cell type and you can get a different kind of arrest.

T. Taniguchi

You are talking about Martine's experiment in 3T3 cells? Yes, that is different.

J. Cleveland

These cells failed to incorporate BUDR into their nuclei.

T. Taniguchi

This is a 3T3 cell expressing MCSF (CSF-1) receptor with the mutation at tyrosine 809. And this receptor cannot be induced for cell proliferation unless you ectopically express the C-myc gene. In this case you are saying there is no DNA synthesis induced by this mutated form of CSF-1 receptor.

M. Kastan

I just wanted to answer the question about p53 status in myelodysplastic syndrome. p53 is rarely mutated in primary leukemias. There are a couple of exceptions in mature B-cell leukemia lymphomas like Burkitts that have a significant percentage of cases that are mutated and the other situation is in AML's that arose from myeloid dysplasias. There is a measurable percentage, as opposed to primary AML's, that did not arise from myeloid dysplasias, so it might be playing a role in some of the transition phases here.

T. Taniguchi

In some cases, p53 mutations have been reported in AML's. N-ras is also rather frequently mutated, so it is clear that there is a number of events happening for the development of these leukemia.

M. Fried

Have you tried to make knockout mice without IRF-1?

T. Taniguchi

Yes, the mice are viable and it looks like certain growth abnormality can be seen in embryonal fibroblasts but it is too premature to make any solid conclusions. Mice seem to be viable and, as far as I can tell, at 3 or 4 months of age they do not develop leukemia as one might have expected. And of course the question is what happens if you treat those mice with some other carcinogens or oncogene activations. What we are doing is to introduce some of the proto-oncogenes and whether there is any cooperations with other oncogenes. Now in terms of interferon production, it is interesting that, again these are preliminary results, when you induce the embryonic fibroblast of IRF-1 knockout mice by double stranded RNA, the interferon production for both alpha and beta goes down dramatically, several to ten-fold, suggesting that IRF-1 is important, as we thought. But, when you treat the same cells with viruses, the production level remains the same. So clearly there seems to be alternative pathways which are turned on in virus infected cells.

MYC, FAS, APOPTOSIS, AND IMMUNE TOLERANCE

Douglas R. Green, Yufang Shi, Jaqueline Glynn, Artin Mahboubi, and Rona J. Mogil

La Jolla Institute for Allergy and Immunology, La Jolla, CA USA

INTRODUCTION

The problem of immunological tolerance, the apparent ability of the immune system to discriminate between self and "nonself" is almost as old as the field of immunology. Arguably, Ehrlich's description of the phenomenon and the ensuing belief that it is a fundamental attribute of the immune system delayed the recognition of the existence of autoimmune disease until the 1950's (1). However, once it was realized that the immune system could respond to self molecules, the idea that self/nonself discrimination was "learned" (and therefore potentially fallible) came into vogue. It is therefore likely that this recognition prompted the appearance of models of immune tolerance, and served as the impetus for the development of the clonal selection theory (2,3). The application of this theory to explain tolerance through the deletion of potentially autoreactive cells as proposed by Burnet (3,4) persist in almost unchanged form today.

From the perspective of this volume, the most important feature of these models of immune tolerance is that they rely upon the deletion of specific clones of lymphocytes with the potential for self-reactivity. This deletion translates into apoptotic cell death, as we will see. In the model proposed by Burnet (3), lymphocytes each express one of a large number of possible receptors for antigen, and it is the expansion of a given clone in response to its specific antigen that allows the generation of an immune response. He further proposed that during development any clone with the potential to respond to self antigens is deleted from the system and therefore only nonself responses can occur. He likened this removal to the death of nonlymphoid cells in response to bacterial toxins (4).

In the late 1980's, direct evidence was obtained for the process of clonal deletion in developing T cells (5,6). Burnet's model was only slightly modified in that we now know that T cell development proceeds well beyond embryonic life (and probably throughout life) and therefore it is the stage of lymphocyte maturation rather than that of the individual that determines when deletion (now called "negative selection") rather than proliferation occurs in response to antigenic stimulation. Therefore, when a foreign antigen enters the system, developing T cells capable of responding to that antigen undergo negative selection (just as they would if it were a self antigen). However, T cells that had developed prior to the appearance of the foreign antigen respond to it and a specific immune response ensues. Thus, rather than discriminating between self and nonself, per se, the immune system is selected to discriminate between antigens that are always present versus those that are only sometimes (or never) present. The result has the appearance of self/nonself discrimination.

The development of tolerance in response to persistent stimulation of immature lymphocytes is called "central tolerance", and in T cells this occurs predominantly in the thymus. Another form of clonal deletion which occurs outside of the thymus and affects mature T cells has more recently been described and has been called "peripheral deletion".

Apoptosis, Edited by E. Mihich and R.T. Schimke
Plenum Press, New York, 1994

This occurs in peripheral T cells confronted with relatively high doses of antigen, resulting in an expansion of the specific T cells followed by a decline in numbers, often to below the starting levels (7,8,9). Whether or not this deletion is vital for the maintenance of unresponsiveness is not clear, but it is nevertheless likely that this well established phenomenon plays at least some roles in immune homeostasis.

Here, we consider the role of apoptosis in both central and peripheral deletion in T lymphocytes and examine the function of two molecules in these processes: c-Myc and Fas/Apo-1.

RESULTS AND DISCUSSION

Activation-Induced Apoptosis and Negative Selection

Several years ago, two groups tested the hypothesis that negative selection could be explained as a consequence of activation-induced cell death, i.e., that activation through the T cell receptor (TCR) results in death rather than proliferation of immature cells. Jenkinson and colleagues (10) examined this idea by exposing thymocytes in fetal thymus organ culture to an antibody against CD3ε, a component of the TCR complex. While this antibody usually induces proliferation of mature T cells, they found that thymocytes underwent an activation-induced apoptosis. Similarly, we demonstrated (11,12) that administration of anti-CD3ε in vivo rapidly resulted in apoptotic death of immature thymocytes but not peripheral T cells. Further, we found that cyclosporin A, an immunosuppressive drug that had been shown to inhibit negative selection, blocked this activation-induced apoptosis (11). Later, a similar experiment was performed using mice transgenic for a single TCR; administration of the antigenic peptide for which this TCR is specific resulted in activation-induced apoptosis in immature thymocytes (13).

Another mechanism of negative selection that does not appear to involve apoptosis has recently been described (14). Thymocytes that are at an early $CD4^-$, $CD8^-$, TCR^+ stage, or $CD4^-$, $CD8^{low}$, TCR^+ stage fail to mature into $CD4^+$, $CD8^+$, TCR^+ cells if the TCR is engaged (i.e., specific ligand is present). This activation-induced developmental arrest cannot explain all negative selection, however, since at least some $CD4^+$, $CD8^+$ thymocytes that bear self-reactive TCR are not deleted until a later stage of the maturation pathway, at or just prior to the shift to single-positive cells (5,15). Thus, while activation-induced developmental arrest of $CD4^-$, $CD8^-$ cells occurs, its importance is not yet established, and activation-induced apoptosis of $CD4^+$, $CD8^+$, TCR^+ cells remains likely to be a major mechanism of negative selection.

The phenomenon of activation-induced apoptosis in immature thymocytes, though well established, is not well understood. This is partly due to difficulties in working with these fragile cells in vitro. An alternative, then, is the use of T cell hybridomas. We (11,16) and others (17,18) observed that cell lines generated by the fusion of mature T cells with an immature thymoma rapidly undergo apoptosis following activation. This has proven to be a useful model for the study of activation-induced apoptosis.

c-Myc and Activation-Induced Apoptosis

Our studies on activation-induced apoptosis in T cell hybridomas revealed a role for the c-Myc protein in this phenomenon. We employed antisense oligonucleotides to down-regulate c-Myc protein levels as had been done by others (19,20). While activation of antisense-treated hybridomas resulted in lymphokine production, apoptotic death following activation was profoundly inhibited (21). Thus, c-Myc appears to be required for activation-induced apoptosis in these cells. Activation of thymocytes induces an up-regulation in c-*myc* mRNA (22) and recently the ability of a calcium ionophore to induce apoptosis in immature thymocytes was correlated with expression of c-*myc* (23), suggesting that a requirement for c-Myc might exist for activation-induced apoptosis in thymocytes as well.

Ashwell and colleagues (24) demonstrated that activation of T cell hybridomas results in a cell cycle arrest, and this was shown to be independent of cell death. We therefore asked whether antisense oligonucleotides corresponding to c-*myc* influence this phenomenon. To separate the phenomena of activation-induced apoptosis and cell cycle arrest, cyclosporin A (CsA) was added to some of the cultures (24). As shown in Figure 1, the ability of c-*myc*-antisense oligonucleotides to block cell death did not correspond to a reversal of the block to cell proliferation. This was especially clear in B, where CsA inhibited cell death and thus the effects of the oligonucleotides on proliferation could be seen independly of their effects on cell death. Therefore, inhibition of c-Myc protein expression by antisense oligonucleotides did not reverse activation-induced inhibition of ^{3}H-TdR incorporation in the T hybridoma cells. Thus, inhibition of c-Myc expression prevents activation-induced apoptosis but not activation-induced cell cycle arrest.

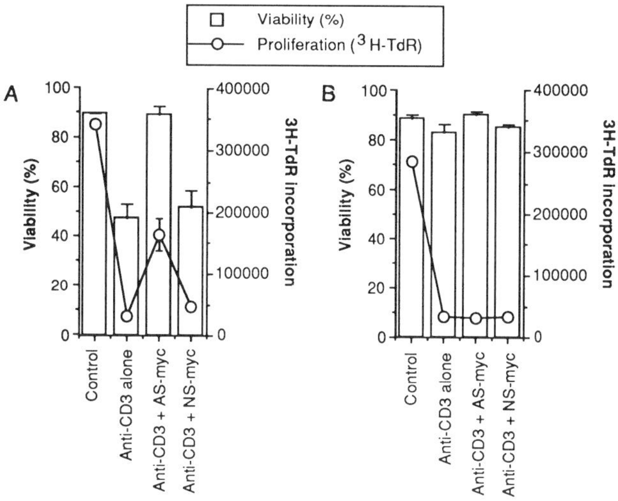

Figure 1. Antisense oligonucleotides corresponding to c-*myc* prevent activation-induced apoptosis but not cell cycle arrest in T hybridoma cells. A. T hybridoma cells were cultured alone (control) or on anti-CD3-coated plastic (anti-CD3). Antisense (AS) or nonsense (NS) oligonucleotides corresponding to the translation start of c-*myc* were added at time 0, as described (21). After overnight culture, viability was assessed by trypan blue exclusion or proliferation was assessed by a 1 hr pulse with ^{3}H-TdR. B. As in A, but with the addition of 150 nM cyclosporin A to all cultures.

This observation is in keeping with other observations on the functions of c-Myc in apoptosis, and provides support for an emerging model (25,26). The model is based on observations in nonlymphoid systems, in which expression of c-*myc* has been shown to drive apoptosis when growth factors are limiting (25-27). In this model shown in Figure 2, c-Myc is seen as a stimulus for progression towards <u>both</u> cell death and cell proliferation, and the outcome of expression of this protein is determined by additional factors. For example, growth factors (25-27) or other oncogene products, such as Bcl-2 (28,29) block the progression towards apoptosis, and under these influences the cells proceed to proliferation. In contrast, ligation of the TCR complex in T cell hybridomas blocks the pathway towards proliferation, and cells proceed to c-Myc-induced apoptosis.

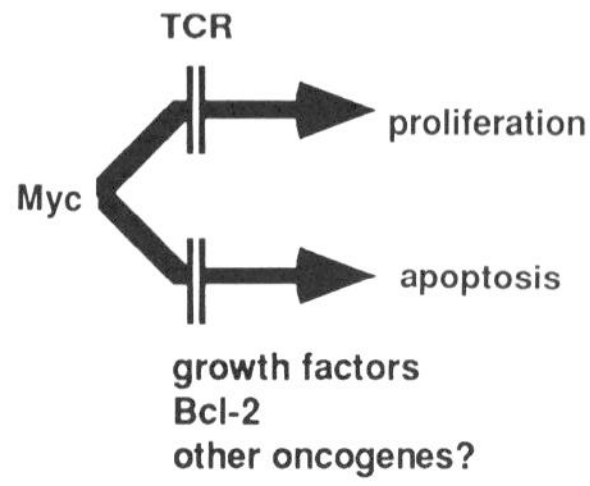

Figure 2. A model of c-Myc-induced apoptosis. Expression of c-Myc protein in a cell opens two opposing pathways, one leading to cell proliferation and the other to apoptotic cell death. The choice of pathway is made on the basis of other influences affecting the cell. Growth factors and other oncogene products (e.g., Bcl-2) can block the apoptotic pathway, leading the cell to undergo Myc-driven proliferation. Inhibition of proliferation, e.g. by TCR ligation on a T cell hybridoma, leads the cell towards c-Myc-driven apoptosis.

The role of c-Myc in activation-induced apoptosis is further supported by recent observations on the effects of transforming growth factor β (TGFβ) in this system. TGFβ inhibits expression of c-*myc* in several systems (30) and we have observed an inhibition of c-*myc* mRNA in T cell hybridomas treated with TGFβ (result not shown). This correlated with an inhibition in activation-induced apoptosis in these cells (not shown). While these observations are consistent with our findings on the effects of c-*myc* antisense and are in line with the above model, TGFβ has been shown to induce apoptosis in some cell types (e.g., 31). It is likely that in such situations c-Myc does not perform the same function that we have proposed here.

Fas/Apo-1 in Central and Peripheral Deletion

Another gene product known to be involved in some forms of apoptosis is the Fas/Apo-1 protein. Unlike c-Myc, this is a cell surface protein which upon crosslinking with specific antibodies transduces a signal resulting in apoptosis (32,33). Fas/Apo-1 is expressed on murine thymocytes and on activated lymphocytes (33,34), and thus it might play some role in apoptosis associated with tolerance (35). This idea was made even more attractive when it was discovered that the lpr gene defect in mice is due to a defect in the Fas/Apo-1 gene (36). This defect results in a massive lymphoproliferative disorder and acceleration of autoantibody production.

Negative selection in lpr/lpr mice is normal (37,38) although recovery from neonatal tolerance induction appears to be accelerated (39). Nevertheless, the presence of negative selection suggests one of two possibilities: either activation-induced apoptosis proceeds normally in the absence of Fas/Apo-1, or else additional mechanisms compensate for defects in this process.

To address these possibilities we employed young MRL lpr/lpr mice (4-6 weeks old) which had not yet displayed any indication of disease. We reasoned that any defects we might observe would be directly due to the defect in Fas/Apo-1 rather than a secondary effect resulting from the disease process. We injected these animals with staphylococcal enterotoxin B (SEB) which has been shown to stimulate T cells bearing $V\beta8^+$ T cell receptors (40). This stimulation therefore results in activation-induced apoptosis and deletion in $V\beta8^+$ thymocytes. As shown in Figure 3, no differences were observed in SEB-induced deletion of thymocytes between MRL lpr/lpr mice and age-matched wild type (+/+) controls. Thus, Fas/Apo-1 does not appear to be an essential molecule for the phenomenon of activation-induced cell death in immature thymocytes.

Administration of SEB into mice triggers another deletion event which is also associated with apoptosis, a phenomenon called "peripheral deletion" (8,9). Following injection, numbers of mature, $V\beta8^+$ T cells increase in the periphery, then rapidly decline. This decline is associated with onset of apoptosis in the $V\beta8^+$ cells (9). We therefore examined this phenomenon in young MRL lpr/lpr versus +/+ mice. Animals were injected with SEB and the percentages of Vβ8+ cells in peripheral blood were determined daily. As shown in Figure 4, the percentage of Vβ8+ T cells quickly rose in both types of animals. However, the subsequent decline in numbers that was observed in MRL +/+ mice was significantly delayed in the MRL lpr/lpr animals. Even 3-4 weeks after a single injection of SEB, the percentages of Vβ8+ cells in lpr/lpr mice were significantly higher than those seen in SEB-treated +/+ animals (not shown). This difference was probably not due to hyper-responsiveness of the young lpr/lpr T cells, since proliferation of these cells in response to

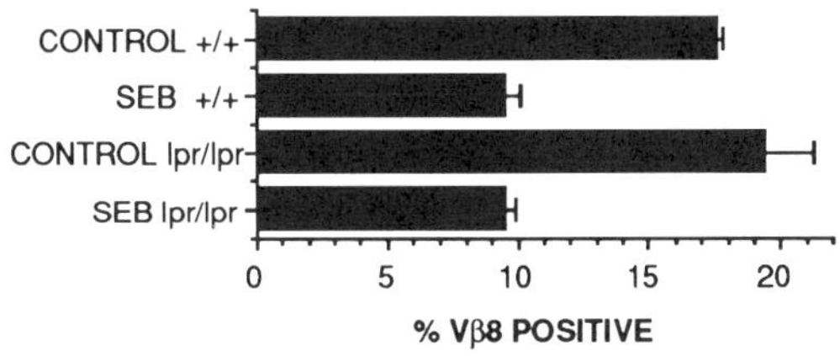

Figure 3. SEB-induced deletion of Vβ8+ thymocytes in lpr/lpr mice. 5 week old MRL lpr/lpr or +/+ mice (Jackson Laboratories) were injected i.v. with 250 μg of SEB (Sigma) and thymocytes were quantitated 18 hrs later for Vβ8+ cells using FITC-anti-Vβ8 (Pharmingen) as described (10). Means ± s.d. from 3 animals/group are shown.

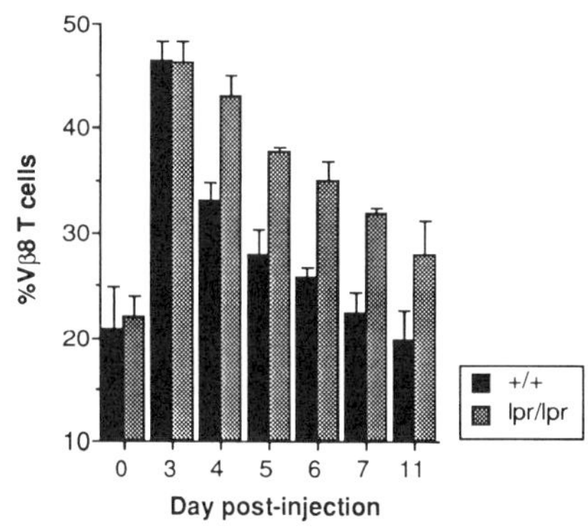

Figure 4. Defect in SEB-induced peripheral deletion in mature T cells of young lpr/lpr mice. 6 week old MRL lpr/lpr or +/+ mice (Jackson Laboratories) were injected i.v. with 250 μg of SEB (Sigma) and peripheral blood lymphocytes were examined on the days shown for %Vβ8+ cells, using FITC-anti-Vβ8 as described (10). Means ± s.d. from 3 animals/group are shown. Day 0 values are the averages of measurements from 3 age-matched uninjected mice/group determined on 4 separate days. Very similar results were obtained in other experiments using animals 4-5 weeks old (not shown).

SEB in vitro was not distinguishable from that of cells from +/+ mice (not shown). Further, we examined the incidence of apoptotic death in Vβ8+ cells taken at the peak of expansion (day 3) and using methods described by others (9) we found that apoptosis was prominent among these cells from +/+ but not lpr/lpr animals (not shown). Thus, the differences in numbers of Vβ8+ cells we observed in vivo are probably due to differences in rate of apoptosis in the expanded population. Therefore, the apoptosis associated with peripheral deletion appears to depend upon expression of the Fas/Apo-1 molecule.

It is, of course, tempting to ascribe the massive lymphoproliferative disorder seen in older *lpr/lpr* mice to the defect in peripheral deletion we have observed in the young animals. However, the relationship between these phenomena is not clear. The expanded cell population present in the spleens and lymph nodes of these mice are $TCR\alpha\beta^{+}$, $CD4^{-}$, $CD8^{-}$, and $B220^{+}$ (41), a phenotype not associated with normal, mature T cells. Although it is possible that this phenotype arises from T cells with a conventional phenotype, however we did not observe any loss of CD4 or CD8 in the cell population that expanded in response to SEB (not shown). Alternatively, these cells might arise due to defects in negative selection that are not obvious on a gross level (42), defects in positive selection (43), or via other processes.

While the ability of the lpr defect to accelerate autoantibody production might be an important consequence of a failure in peripheral deletion in T cells, other mechanisms might be at least as important. For example, Zhou, et al (39) found that recovery from negative selection following neonatal tolerance induction is more rapid in lpr/lpr mice than in +/+ controls, and the resulting defect in neonatal tolerance may contribute to autoimmunity. Further, activated B cells express Fas/Apo-1 (33) and therefore the role of this molecule in B cell deletion mechanisms must also be taken into account if we are to understand autoimmunity in these mice.

Regardless, our findings represent direct links between Fas/Apo-1, apoptosis induction in T cells, and the phenomenon of peripheral deletion in vivo, and thus one role for this interesting molecule in maintenance of immunological homeostasis is established.

CONCLUSION

Apoptosis and Immune Tolerance

The concept of clonal deletion provided a framework for understanding at least one form of immune tolerance, and while other mechanisms of tolerance undoubtedly exist (see below) deletion is at least one of the most critical. The conceptual leap from "deletion" to "cell death" is not a large one, and Burnet saw this connection, as noted above. However, it was not until recently that it was realized that the activation-induced cell death which underlies clonal deletion is apoptotic in nature. Thus, the elucidation of the molecular mechanisms of apoptosis has become highly significant for the understanding of important forms of immune tolerance.

As we have seen, an understanding of the roles of c-Myc and Fas/Apo-1 in triggering apoptosis in some systems has begun to provide insights into mechanisms of central and peripheral tolerance. Of course, there remains much to do. Nevertheless, these represent the first inroads into an understanding of these phenomena at a molecular level.

Deletion mechanisms are not the only route to immune nonresponsiveness. Rather than undergo apoptotic death cells may become anergic, failing to respond to activation signals (44). Such clonal anergy almost certainly plays roles in immune regulation. Similarly, active inhibition of specific immune responses by cytokines (45) or possibly by other mechanisms is critically important under some circumstances. Thus, there are many different mechanisms for the establishment and maintenance of immune tolerance, and this should not be surprising, given the extreme consequences of a failure to avoid antiself responses.

From a different perspective, though, it may be easy to see why not all tolerance mechanisms are deletional in nature. Any specific immune response is determined by both the concentration of antigen and the affinity of receptors that can recognize the antigen. If the potential for generating all possible anti-self responses (even low affinity recognition events) were deleted from the system, relatively few potential responses would remain. Parasites might quickly exploit the resulting "holes" in the immune repertoire, with an insuing loss of effective defense. Thus, by deleting only the most flagrant self-reactive clones the system retains the potential to generate some anti-self responses if required for the elimination of a pathogen. Responsiveness can be recovered in anergic cells (44) and inhibitory mechanisms can be countered (45). The autoimmunity that accompanies immune responses to many types of infection probably represent examples of this effect.

Thus, while apoptosis is not the key to all forms of immunological tolerance, its study sheds light on some of the major mechanisms for the acquisition of self/nonself discrimination and homeostasis by the immune system. As we gain a deeper understanding of the process of apoptosis, we find new ways to explore how this remarkable form of cell death maintains the normal function of the immune system.

ACKNOWLEDGEMENTS

The authors thank their colleagues who have contributed to the research reviewed here, including Reid Bissonnette, Tom Cotter, Fernando Echeverri, and Susan Oja. This work was supported by grant AI 31591 from the U.S. National Institutes of Health.

REFERENCES

1. R.S. Schwarz and S.K. Datta. Autoimmunity and autoimmune diseases, *in:* "Fundamental Immunology, Second Edition," W.E. Paul, ed., Raven Press Ltd, New York (1989).
2. N.K. Jerne, The natural-selection theory of antibody formation, *Proc. Natl. Acad Sci. USA* 41: 849 (1955).
3. F.M. Burnet. A modification of Jerne's theory of antibody production using the concept of clonal selection. *Aust. J. Sci.* 20: 67 (1957).
4. F.M. Burnet. "The Clonal Selection Theory of Aquired Immunity" Cambridge University Press, Cambridge (1959).
5. A. Herman, J. W. Kappler, P. Marrack, A. M. Pullen, Superantigens: Mechanism of T-cell stimulation and role in immune responses, *Annu. Rev. Immunol.* 9: 745 (1991).
6. Kisielow, P., W. Swat, R. Rocha, H. v. Boehmer, Induction of immunological unresponsiveness in vivo and in vitro by conventional and super-antigens in developing and mature T cells, *Immunol. Rev.* 122: 69 (1992)
7. S. Webb, C. Morris, and J. Sprent, Extrathymic tolerance of mature T cells: clonal elimination as a consequence of immunity,*Cell* 63, 1249 (1990)
8. L.A. Jones, L.T. Chin, D.L. Longo, and A.M. Kruisbeek, Peripheral clonal elimination of functional T cells,*Science* 250, 1726 (1990).
9. Y. Kawabe, and A. Ochi, Programmed cell death and extrathymic reduction of $V\beta8^+$ $CD4^+$ T cells in mice tolerant to Satphylococcus aureus enterotoxin B,*Nature* 349, 245-248 (1991).
10. C.A. Smith CA, G.T. Williams, R. Kingston, E.J. Jenkinson, and J.T.T. Owen, Antibodies to CD3/T-cell receptor complex induce death by apoptosis in immature T cells in thymic organ cultures, *Nature* 337: 181 (1989).
11. Y. Shi, B.M. Sahai, and D.R. Green, Cyclosporin A inhibits activation-induced cell death in T cell hybridomas and in thymocytes, *Nature*, 339: 625 (1989).
12. Y. Shi, R.P. Bissonnette, N. Parfrey, M. Szalay, R.T. Kubo, and D.R. Green, In vivo administration of antibodies to the CD3-T cell receptor complex induces cell death (apoptosis) in immature thymocytes. *J. Immunol.* 146: 3340 (1991).
13. K.M. Murphy, A.B. Heimberger, and D.Y. Loh, Induction by antigen of intrathymic apoptosis of $CD4^+CD8^+TCR^{lo}$ thymocytes in vivo, Science 250: 1720 (1990).
14. Y. Takahama, E.W. Shores EW, and A. Singer, Negative selection of precursor thymocytes before their differentiation into CD4+CD8+ cells, *Science* 258: 653 (1992).
15. C.J. Guidos, J.S. Danska, C.G. Fathman, and I.L. Weissman, T cell receptor-mediated negative selection of autoreactive T lymphocyte precursors occurs after commitment to the CD4 or CD8 lineages, *J. Exp. Med.* 172: 835 (1990).
16. Y. Shi, M.G. Szalay, L. Paskar, M. Boyer, B. Singh, and D.R. Green, Activation-

induced cell death in T cell hybridomas is due to apoptosis: Morphological aspects and DNA fragmentation, *J. Immunol.* 144: 3326 (1990).

17. D.S. Ucker, J.D. Ashwell, and G. Nickas, Activation-driven T cell death. I Requirements for de novo transcription and translation and association with genome fragmentation, *J. Immunol.* 143:3461 (1989).
18. C. Odaka, H. Kizaki, and T. Tadakuma, T cell receptor-mediated DNA fragmentation and cell death in T cell hybridomas, *J. Immunol*. 144: 2096 (1990).
19. R. Heikkila, G. Schwab, E. Wickstrom, S. L. Loke, D. H. Pluznik, R. Watt, and L. M. Neckers, A c-myc antisense oligodeoxynucleotide inhibits entry into S phase but not progress from G0 to G1, *Nature* 328: 445 (1987).
20. A. Harel-Bellan, D.K. Ferris, M. Vinocour, J.T. Holt, W.L. Farrar, Specific inhibition of c-myc protein biosynthesis using an antisense synthetic deoxy-oligonucleotide in human T lymphocytes, *J. Immunol.*, 140: 2431 (1988)
21. Y. Shi, R. P. Bissonnette, J. M. Glynn, L. J. Guilbert, T. G. Cotter, and D. R. Green, Inhibition of activation-induced apoptosis in T cells by antisense oligodeoxynucleotides corresponding to *c-myc. Science* 257, 212-214 (1992).
22. J.S. Riegel, E.R. Richie, and J.P. Allison, Nuclear events after activation of $CD4^+CD8^+$ thymocytes, *J. Immunol.* 144: 3611 (1990).
23. J. Nikolic-Zugic, G. Stella, and S. Andjelic, Reciprocal expression of bcl-2 and c-myc defines susceptibility of immature thymocytes to Ca^{++}-induced apoptosis, *J. Immunol.* 150: 152A (1993).
24. M. Mercep, P.D. Noguchi, and J.D. Ashwell, The cell cycle block and lysis of an activated T cell hybridoma are distinct processes with different Ca^{2+} requirements and sensitivity to cyclosporine A, *J Immunol* 142:4085 (1989).
25. G.I. Evan, A. H. Wyllie, C. S. Gilbert, T. D. Littlewood, H. Land, M. Brooks, C. M. Waters, L. Z. Penn, and D. C. Hancock. Induction of apoptosis in fibroblasts by c-myc protein, *Cell* 69: 119-128 (1992).
26. R.P. Bissonnette, Y. Shi, A. Mahboubi, J.M. Glynn, and D.R. Green, c-Myc and apoptosis, *in*: "Apoptosis, The Molecular Basis of Cell Death, Volume 2," L.D. Tomei and F.O. Cope, eds., Cold Spring Harbor Lab. Press, Cold Spring Harbor, in press.
27. D.S. Askew, R. A. Ashmun, B. C. Simmons, and J. L. Cleveland. Constitutive *c-myc* expression in an IL-3-dependent myeloid cell line suppresses cycle arrest and accelerates apoptosis, *Oncogene* 6: 1915 (1991).
28. R.P. Bissonnette, F. Echeverri, A. Mahboubi, and D.R. Green, Apoptotic cell death induced by c-myc is inhibited by bcl-2, *Nature* 359: 552 (1992).
29. A. Fanidi, E.A. Harrington,, and G.I. Evan, Cooperative interaction between c-myc and bcl-2 proto-oncogenes, *Nature* 359: 554 (1992).
30. J. A. Pietenpol, J.T. Holt, R.W. Stein, and H.L. Moses, Transforming growth factor $\beta 1$ suppression of c-myc gene transcription: Role in inhibition of keratinocyte proliferation, *Proc. Natl. Acad. Sci. USA* 87: 3758 (1990).
31. W. Bursch, F. Oberhammer, R.L. Jirtle, M. Askari M, et al., Transforming growth factor-beta 1 as a signal for induction of cell death by apoptosis, *Br J Cancer* 67:531-6 (1993)
32. S. Yonehara, A. Ishii, M. Yonehara, A cell-killing monoclonal antibody (anti-Fas) to a cell surface antigen co-downregulated with the receptor for tumor necrosis factor, *J.Exp.Med.* 169: 1747 (1989).
33. B.C. Trauth, C. Klas, A.M.J. Peters, S. Matzku, P. Moller, W. Falk, K.M. Debatin, and P.H. Krammer, Monoclonal antibody-mediated tumor regression by induction of apoptosis, *Science* 245: 301 (1989).
34. T. Miyawaki, T. Uehara, R. Nibu, T. Tsuji, A. Tachie, S. Yonehara, and N. Taniguchi, Differential expression of apoptosis-related Fas antigen on lymphocyte subpopulations in human peripheral blood,*J. Immunol* 149: 3753 (1992).
35. Cohen, P.L., and R.A. Eisenberg, The lpr and gld genes in systemic autoimmunity: life and death in the Fas lane, *Immunol. Today* 13, 427-428 (1992).
36. Watanabe-Fukunaga, R., C. I. Brannan, N. G. Copeland, N. A. Jenkins, and S. Nagata, Lymphoproliferation disorder in mice explained by defects in Fas antigen that mediates apoptosis,*Nature* 356: 314 (1992).

37. P.A. Singer, R.S. Balderas, R.J. McEvilly, M. Bobardt, and A.N. Theophilopoulos, Tolerance-related V beta clonal deletions in normal $CD4^-8^-$, TCR-alpha/beta$^+$ and abnormal lpr and gld cell populations, *J. Exp. Med.* 170: 1869 (1989).
38. C.L. Sidman, J.D. Marshall, and H. von Boehmer, Transgenic T cell receptor interactions in the lymphoproliferative and autoimmune syndromes of lpr and gld mutant mice, *Eur. J. Immunol.* 22: 499 (1992).
39. T. Zhou, H. Bluethmann, J. Zhang, C.K. Edwards, and J.D. Mountz, Defective maintenance of T cell tolerance to a superantigen in MRL-lpr/lpr mice, *J. Exp. Med.* 176, 1063-1072 (1992).
40. E.J. Jenkinson, R. Kingston, C.A. Smith, G.T. Williams, and J.T.T. Owen, Antigen-induced apoptosis in developing T cells: a mechanism for negative selection of the T cell receptor repertoire, *Eur. J. Immunol.* 19, 2175-2177 (1989).
41. T. Giese, and W.F. Davidson, Evidence for early onset, polyclonal activation of T cell subsets in mice homozygous for lpr, *J. Immunol.* 149:3097-3106 (1992).
42. Mountz JD; Edwards CK, Murine models of autoimmunity: T-cell and B-cell defects,*Curr Opin Rheumatol* 4:612 (1992).
43. P.A. Singer, and A.N. Theofilopoulos, T-cell receptor Vβ repertoire expression in murine models of SLE, *Immunol. Rev.* 118: 103-127 (1990).
44. Jenkins MK, The role of cell division in the induction of clonal anergy, *Immunol Today* 13:69 (1992).
45. T.R. Mosmann and R.L. Coffman, TH1 and TH2 cells: Different patterns of lymphokine secretion lead to different functional properties, *Annu. Rev. Immunol.* 7: 145 (1989).

Discussion

M. Oren

Could you tell us anything about the involvement of *myc* in radiation induced T-cell apoptosis?

D. Green

No, I do not have any information about that. Of course, as you presented, there is a role for p53 in radiation induced apoptosis in lymphocytes. We cannot yet find any role for p53 in this phenomenon. Expression of mutant p53 or deletion of p53 with HPV-E6 have had no effect on this cell death. With radiation we just have not looked yet, so we do not know.

D. Hockenbery

When you modulate *myc* expression in your T-cell hybridomas, do you affect the cell cycle behavior after engagement of CD3?

D. Green

Very good question. Actually, no, not during the time course of the study, so there does not seem to be any inhibition of DNA synthesis or expansion of the cells. Now if we look after a couple of days in the antisense we do not accumulate as many cells, we are starting to get a block. As we might expect, there are high levels of *myc* message and there is a reasonable amount of protein here. I do not think we are eliminating all of it.

D. Hockenbery

So they arrest in the same place at the same time?

D. Green

Yes, this is not due to arresting the cells prior to their ability to undergo apoptosis.

J. Ashwell

Do you think if you could get the potentially autoreactive thymocytes out of the thymus and into the periphery, that you would see clonal expansion of anti-self T-cells? Let me propose a possible experiment. If prior to receptor expression you were to give them cyclosporin A, they do not delete in the thymus. If they positively select, and there is some data to think they might, some percent would then get out. I ask that because you showed not only do T-cells not die when you superantigen to lpr mice, they actually overexpand.

D. Green

If we did this in lpr/hpr mice in other words? We have not done it. It is a great question. As you know, there is a phenomenon where cyclosporin administration over a 6 week period leads to a cyclosporin-A induced auto-immune syndrome, which has been associated with the appearance of potentially auto-reactive cells in the periphery. The involvement of those cells in auto-immunity has been suggested in the cyclosporin-A induced abnormalities. It would be very interesting to look in an LPR mouse and see whether that is dramatically enhanced. The trick would be to get it very early before there is any LPR syndrome appearing, but that should still be possible. It is a good experiment.

INDUCTION OF APOPTOSIS BY CELL-CYCLE PHASE SPECIFIC DRUGS

Steven W. Sherwood and Robert T. Schimke

Department of Biological Sciences
Stanford University
Stanford, CA 94305

INTRODUCTION

The events of apoptosis represent the outcome of processes defining a common physiological pathway of dying cells, and the inputs to this pathway are complex and highly varied[1-5]. It is clear that the actions of cytotoxic agents in mammalian cells frequently feed into the apoptosis pathway, and an extremely wide variety of agents/treatments have been shown to induce the events of apoptosis in cultured mammalian cells[6,7]. It is important to note, however, that cells vary widely in responses to specific cytotoxic agents and the induction of apoptosis by a given agent can vary between cell lines.

We have been interested in the mechanisms of cytotoxicity and genotoxicity of antiproliferative agents which act in a cell cycle specific manner[8-10]. The "system" we have used consists of several common cultured mammalian cells treated with cell cycle phase specific antiproliferative drugs. In this and similar systems, the cytotoxic outcome of antiprolifertive treatment commonly occurs via apoptosis[6]. Our focus, however, has not been on drugs specifically as inducers of apoptosis, but rather on the impact of cell cycle phase specific drugs on the integration of cell cycle processes and the operation of cell cycle checkpoints. Apoptosis, in our "system" occurs in the context of different cell cycle perturbations and we have examined drug-induced apoptosis from the perspective of the cell cycle disturbances induced by drug treatment. Our goal has been to determine whether the cell cycle correlates of apoptosis provide insight into the mechanism(s) of induction or progression of the process.

MATERIALS AND METHODS

Cell lines: HeLa S3 cells were obtained form the ATCC and maintained as monolayer cultures in alpha-MEM medium supplimented with 10% fetal calf serum, glutamine and penicillin-streptomycin. Cells were synchronized by the mitotic-shakeoff method as described in[11]. Cell viability was measured by vital-dye exclusion using propidium iodide.

Flow Cytometry: Cell cycle analysis was done using ethanol-fixed cells stained with propidium iodide by standard procedures. Cell size was measured using forward angle and orthogonal light scatter (FALS and 90LS, respectively,[14]). Flow cytometry was performed on a Coulter EPICS 753 flow cytometer.

Apoptosis, Edited by E. Mihich and R.T. Schimke
Plenum Press, New York, 1994

DNA preparation and agarose gel electrophoresis: Total cellular DNA was prepared by standard detergent-lysis, phenol extraction methods. Purified DNA was treated with RNase and electrophoresed in 1.2% agarose gels to determine the pattern of DNA degradation.

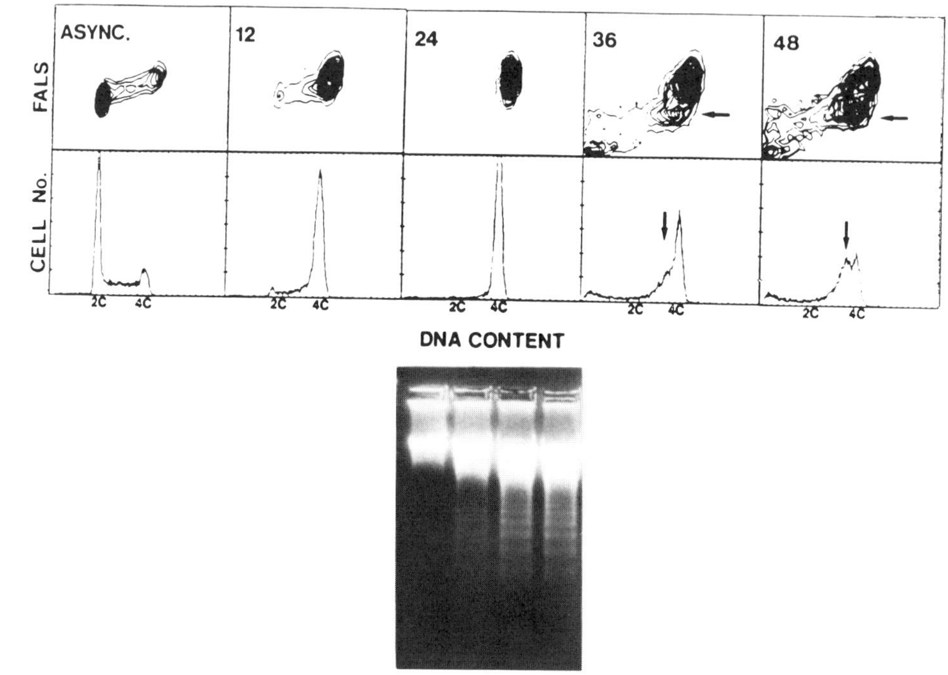

Figure 1. Induction of apoptosis in asynchronous HeLa S3 cells treated with high concentrations of colcemid. Asynchronous cells were exposed to 70 ng/ml colcemid, fixed in ethanol at times indicated and analysed for DNA content and cell size by flow cytometry. Apoptotic cells are first apparent as a supopulation of cells with reduced light scatter (size) and DNA content at 36 hrs (arrows). Total genomic DNA isolated from treated cells show time-dependent increase in DNA degradation.

RESULTS

Induction of Apoptosis in HeLa Cells by Anti-Mitotic Drugs

Mitosis represents an important target for antiproliferative drug action and anti-mitotic drugs are widely used in cancer chemotherapy. Such agents inhibit spindle formation and toxicity is generally attributed to cell cycle arrest in mitosis. We have examined cell line differences in checkpoint controls, assayed as the capacity of cells to proceed through mitosis into a tetraploid cell cycle in the absence of a mitotic spindle and found that HeLa cells, and many other human cell lines display "stringent" mitotic checkpoint control. "Stringent" cells differ from cells with a so-called "relaxed" control (best exemplified by Chinese hamster ovary cells) in that they are unable to proceed into a tetraploid cell cycle in the absence of a mitotic spindle[10]. These cells remain arrested in mitosis in the presence of concentrations of anti-tubulin drug sufficient to block spindle formation, and the phenotype of mitotically-arrested stringent cells includes persistent elevation of intracellular cyclin B protein levels and histone H1 kinase activity[10]. This checkpoint appears to sense the state of the spindle and may be analogous, if not homologous with the mitotic checkpoint defined in yeast by MAD or bub mutants[12,13].

HeLa cells held at this checkpoint by continuous exposure to colcemid will undergo apoptosis *if* the mitotic arrest is of sufficient duration. Figure 1 shows the accumulation of asynchronously growing cells in mitosis when exposed continuously to a high concentration of colcemid (>50 ng/ml). Cells arrest in a pseudoprometaphase state with condensed chromosomes and undergo apoptosis in this arrested state. Apoptotic cells are first evident in the population 32-36 hrs after the addition of colcemid to the medium. Apoptosis is defined in these experiments by cell shrinkage, loss of DNA staining with DNA-specific fluorochromes, apoptotic chromatin morphology and DNA degradation visible in agarose gels. Apoptotic cells are visible in the histograms shown as cells with reduced light scatter (size) and DNA content (Figure 1). Using cells synchronized by mitotic shakeoff, we have determined that cells are arrested in mitosis for approximately 1 cell cycle (18hr) before the onset of apoptosis. Thus, HeLa cells exposed to spindle-inhibiting concentrations of colcemid fail to traverse a mitotic checkpoint which monitors spindle function, and if arrested for a length of time equivalent to the time it would normally take to complete one cell cycle, undergo apoptosis.

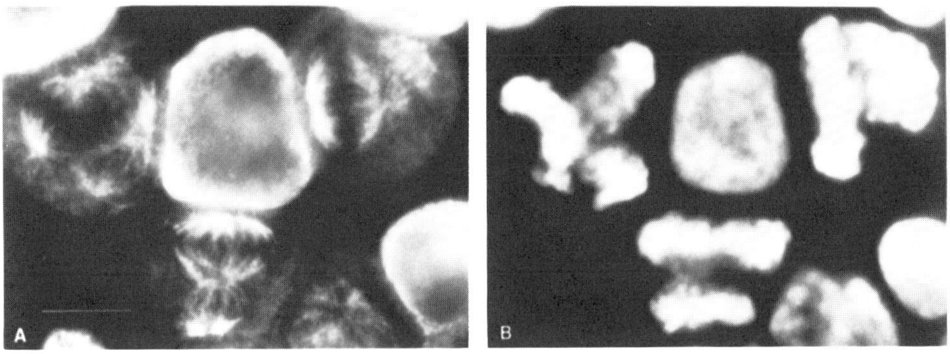

Figure 2. Multipolar mitosis resulting from continuous exposure to low concentrations of colcemid. Asynchronously growing HeLa S3 cells were treated with 5 ng/ml colcemid and fixed for for staining with anti-tubulin antibody. Panel A shows the three telophase daughter cells of a tripolar mitosis (white arrow shows residual midbody of the anaphase spindle); Panel B shows Hoechst staining of the nuclei, and Panel C the phase-contrast image of the three cells.

The cytotoxicity of low concentrations of colcemid (5-10 ng/ml), or alternatively, pulse treatment with high concentrations (eg. 12 hr exposure to 70 ng/ml) also involves apoptosis, but in a different cell cycle context. Under these conditions multipolar spindles are induced which result in multipolar divisions (Figure 2). The resulting daughter cells (which we term microcells), appear normal morphologically (plasma membrane intact, normal appearing interphase nucleus and chromatin) and physiologically (cells exclude vital-dyes and have normal non-specific cytoplasmic esterase activity measured by carboxyfluorescein diacetate hydrolysis), but are variably hypodiploid. These cells subsequently undergo apoptosis in interphase, presumably because of genetic imbalances. In Figure 3 we show the result of an experiment in which synchronized HeLa cells were blocked for 12 hrs in mitosis with a high concentration of colcemid and then washed free of drug. This procedure generates a synchronous cohort of microcells for analysis and from such studies we conclude that microcells persist intact for approximately 18 hrs after they are "born" before apoptosis begins. In contrast to the situation in which cells are actually blocked in mitosis, apoptosis in microcells occurs in an interphase state which we describe as "pseudo-G1" ("pseudo-G1"

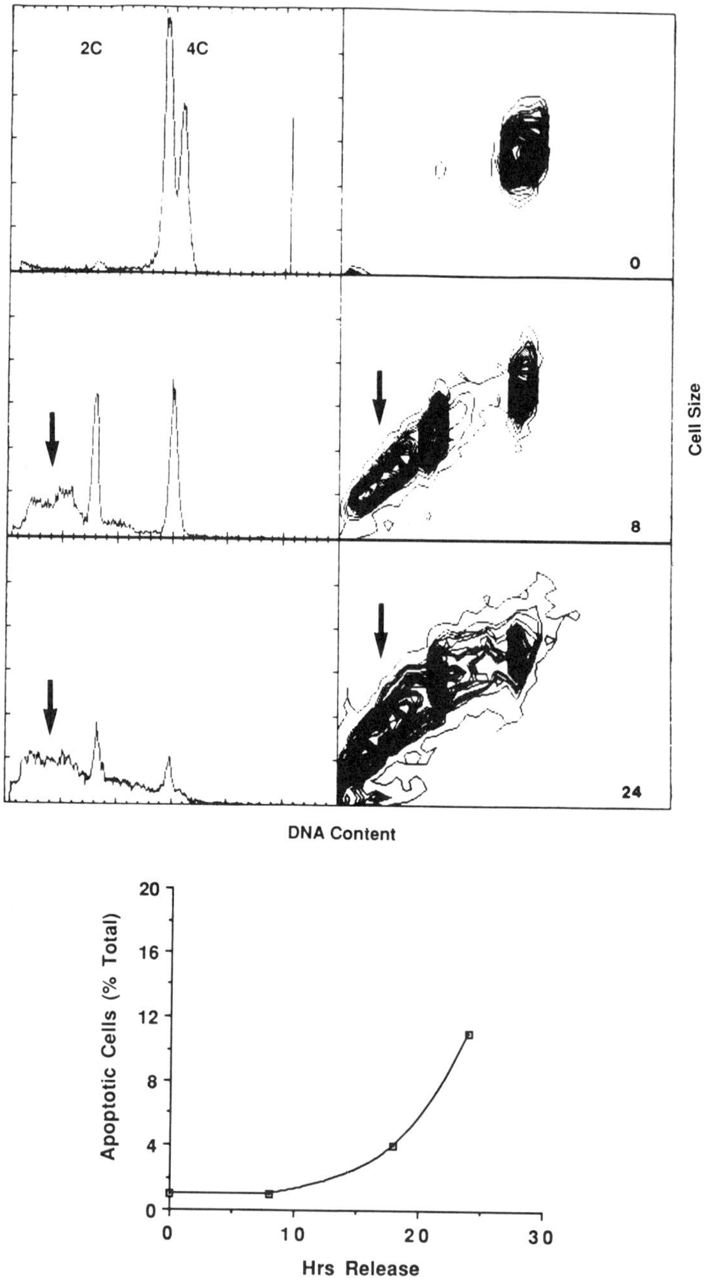

Figure 3. Induction of multipolar mitosis in synchronized HeLa S3 cells transiently arrested in mitosis with colcemid. Cells were synchronized by mitotic shakeoff and blocked in mitosis for 12 hrs with a high concentration of colcemid (70 ng/ml). Cells released from the block undergo multipolar mitosis generating a synchronous population of "pseudo-G1" microcells evident in DNA histograms (arrows) at the hours shown. The frequency of apoptotic cells (determined by fluorescence microscopy of nuclear morphology and by DNA degradation) does not begin to rease signficantly for 12-18 hrs indicating that microcells are not apopotic at the time of their origin.

meaning that there is no real way to define a cell cycle position although the cells are similar to normal G1's insofar as they represent post-mitotic cells which re-enter interphase). This experiment also demonstrates that the "stringent" mitotic checkpoint revealed in HeLa S3 cells by the inability of cells to proceed through mitotic arrest in the presence of colcemid operates in an all-or-none fashion. Cells with an abnormal (i.e. multipolar) spindle will proceed with division but when spindle formation is completely blocked cells fail to exit mitosis.

The duration of cell cycle "arrest" induced by colcemid appears to be critical in the induction of apoptosis. This is true whether the cells are in mitosis or in the interphase following an aberrant mitosis. Mitotic HeLa S3 cells must be held at the checkpoint for approximately 16-18 hrs; cells released from mitotic arrest prior to this critical interval undergo aberrant multipolar mitosis mechanistically identical to that seen in cells treated continuously with low concentrations of colcemid (Figure 3). The genetically unbalanced daughter cells of multipolar mitoses can be a argued to be in an "arrested" cell cycle state (pseudo-G1) for a critical period prior to undergoing apoptosis. Few/none of these cells appear to progress into S-phase (by immunofluorescence staining of BrdU-substituted DNA) and so, while the cells complete mitosis in a viable state, they are unable to continue cycling and unable to enter a stable quiescent or differentiated state. It is important to note that apoptosis occurs in these cells in the absence of the drug and thus cannot be ascribed to a direct apoptosis-inducing effect of the drug. Figure 4 summarizes the relationship of apoptosis to the mitotic checkpoint in colcemid-treated HeLa S3 cells.

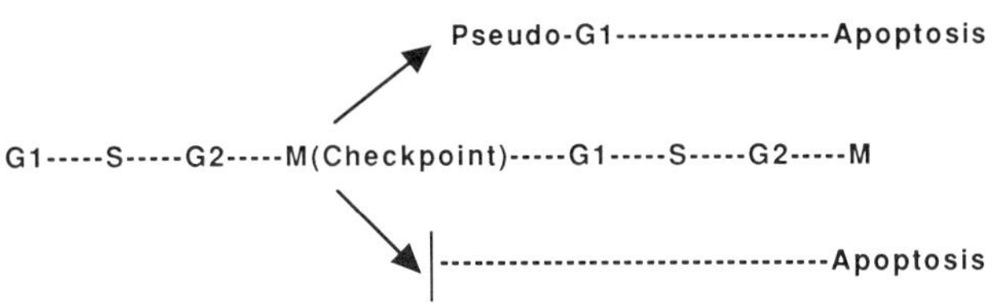

Figure 4. Summary of the cell cycle pathways leading to apoptosis in colcemid-treated HeLaS3 cells. The duration of cell cycle stasis prior to the onset of apoptosis is similar (18hrs) in cells arrested in mitosis with high doses of colcemid as well as in the length of time cells reside in "pseudo-G1" following multipolar mitosis.

The Induction of Apoptosis by Inhibitors of DNA Synthesis

The general importance of the interaction of antiproliferative drugs with cell cycle control mechanisms and the concequences of these interactions on the time and point in the cell cycle when apoptotic cells first appear is further emphasized by results obtained with inhibitors of DNA synthesis. We have used as a model of this class of agent, the DNA polymerase inhibitor aphidicolin. When synchronized HeLa S3 cells are exposed to aphidicolin after they have been in G1 for 2-4 hrs (S-phase begins 8-10 hrs), we find that the time (hrs post-mitosis) as well as cell cycle position (based on cellular DNA content) when cells begin to undergo apoptosis varies as a function of aphidicolin concentration. High concentrations of APC (>1.0 ug/ml) block cells at the G1/S boundary where they begin to undergo apoptosis (seen in Figure 5 as a loss of DNA staining and reduced cell size) beginning 36 hrs post-mitosis. These concentrations represent the levels of drug which completely inhibit DNA synthesis and while cells may carry-out very early S-phase processes (SWS, unpubl.) they fail to proceed through S phase.

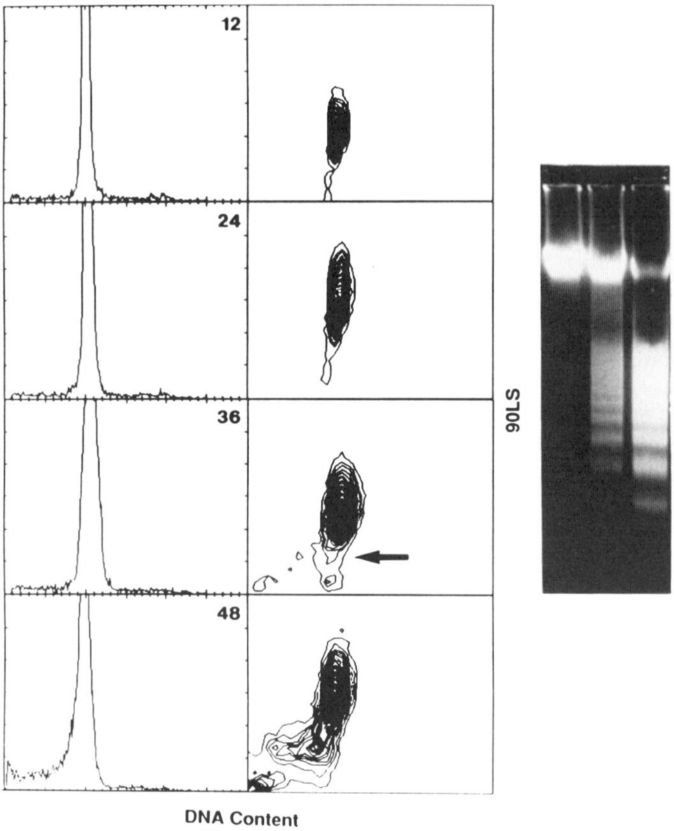

Figure 5. Induction of apoptosis in HeLa S3 cells blocked at the G1/S boundary with aphidicolin (2.0 ug/ml). Synchronized cells were exposed continuously to aphidicolin beginning 4 hrs after mitosis. Apoptotic cells are first apparent in DNA histograms beginning 36 hrs post-mitosis (indicated by arrows).

When the aphidicolin concentration is reduced so cells are slowed through S but not entirely blocked, the time to apoptosis (from G1) is increased in inverse proportion to the aphidicolin concentration and apoptosis occurs progressively later in S-phase (as determined by cellular DNA content). Figure 6 shows a series of DNA histograms obtained with cells treated with 0.6 ug/ml APC and the appearance of cells with reduced light scatter and DNA staining occurs significantly later in time than when cells are blocked at the G1/S boundary and after cells have proceeded partway through S-phase. Degradation of DNA into nucleosome-sized fragments occurs in this situation as well (data not shown). We have observed a similar temporal and cell-cycle pattern of apoptosis in cells treated with cis-platinum as well (J. Sheridan, in prep.)

At very low aphidicolin concentrations, eg. 0.2 ug/ml, cells will eventually progress to the next mitosis where they will undergo apoptosis in an arrested mitosis or undergo multipolar mitosis identical to that induced by low colcemid concentrations (Figure 7) .The microcell daughter cells undergo apoptosis in "pseudo-G1" with kinetics similar to colemid-induced microcells. A few cells are able to enter a tetraploid cell cycle. It is important to note that at these lower concentrations (<0.5 ug/ml), HeLa cells undergo unbalanced cell growth making definition of their cell cycle state problematic[13,14]. Thus, aphidicolin, a specific inhibitor of DNA synthesis, can induce apoptosis in essentially any cell cycle phase as a function of the drug concentration and exposure duration. The cell cycle pathways leading to aphidicolin-induced apoptosis are diagramed in Figure 8.

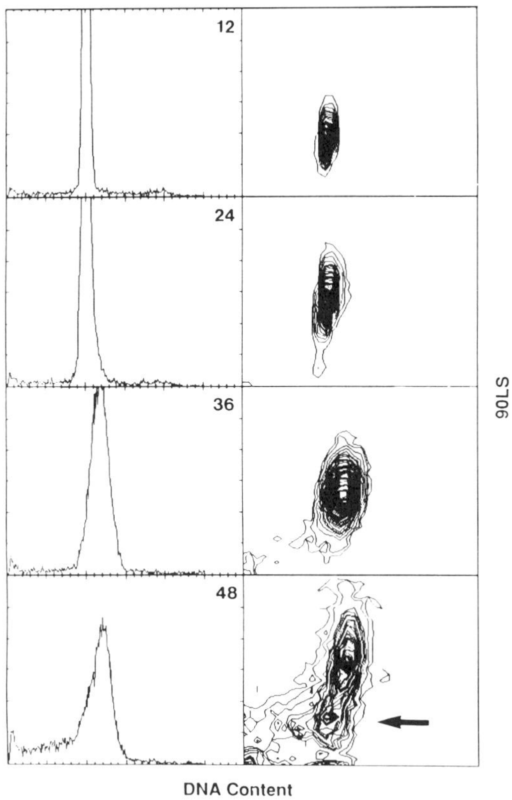

Figure 6. Induction of apoptosis in HeLa cells slowed in transit through S phase by low concentrations of aphidicolin. Synchronized cells were exposed to aphidicolin (0.3 ug/ml beginning 4 hrs after mitosis. DNA histograms indicate that apoptotic become not apparent at increasingly later times (post-mitosis) as they are slowed in progression through S phase.

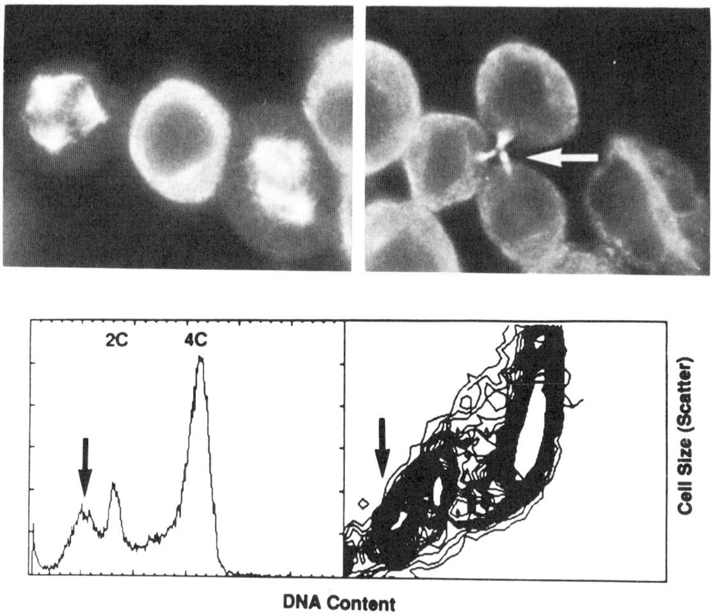

Figure 7. Induction of multipolar mitosis in HeLa cells by prolonged exposure to low concentrations of aphidicolin. Asynchronous cells were exposed to 0.2 ug/ml aphidicolin for 60 hrs. Top panel shows anti-tubulin immunofluorescence staining of multipolar spindles and the three daughter cells of a multipolar mitosis (midbodies of the tripolar anaphase spindle indicated by arrow). DNA histograms show that aphidicolin-induced microcells (arrows) are identical to colcemid-induced microcells in being of <G1 DNA content and smaller than normal G1 cells.

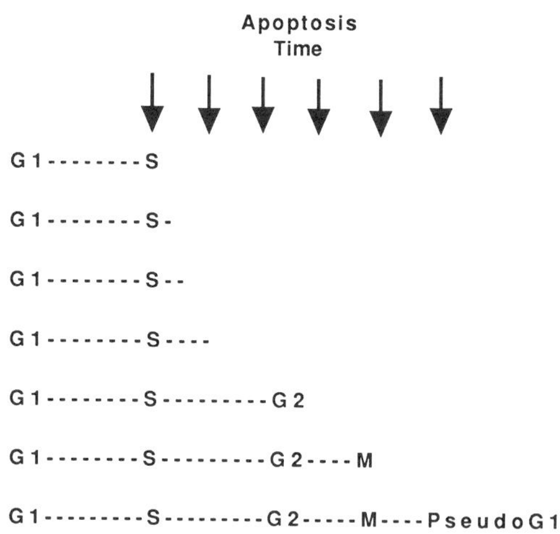

Figure 8. Summary of the temporal aspects of the induction of apoptosis in HeLa S3 cells by aphidicolin. As the aphidicolin concentration is reduced, cells undergo apoptosis at progressively later times (hrs post-synchronization) and after they have progressed further through S-phase. At very low concentrations, aberrant mitoses occur resulting in apoptosis in "pseudo-G1" microcells.

CONCLUDING REMARKS

We have attempted to define in cell cycle terms elements common to the response of cells to different cytotoxic agents which culminate in apoptosis. Our goal has been to determine how the signal(s) for the initiation of apoptosis and/or the progression of the process arise within the context of the mechanisms regulating cell cycle phase transitions. We find that the process of apoptosis can occur in any phase of the cell cycle, at least as defined by DNA content using flow cytometry and as deduced from kinetic data using synchronized cells. These results show that the the metabolic machinery required for apoptosis is present in the cell throughout the cell cycle, implying that if specific apoptosis effector molecules are induced, their induction is not regulated/limited in a cell-cycle related manner. Furthermore, the occurrence of apoptosis in microcells shows that cells "deleted" for a large proportion of the genome are capable of undergoing apoptosis, and strongly suggests that apoptosis effector molecules are constutively present in the cell. Finally, the fact that apoptosis occurs in cells arrested in mitosis with colcemid, conditions under which a nuclear envelope fails to reform, indicates that apoptosis effectors are present and functional under conditions in which the structural organization of the cell is abnormal.

While our data indicate that process of apoptosis can occur in any cell cycle phase, the the nature of the signal(s) initiating the process remains undefined. An element that seems to be common to all of the situations we've examined is that a period of cell cycle "stasis" preceeds the onset of apoptosis. We observe this as the inability of HeLa cells to move through (or exit from) mitotic arrest induced with high concentrations of colcemid, the failure to transit from G1 to S in cells treated with high concentrations of aphidicolin and in the failure of "pseudo-G1" cells to progress into S-phase, presumably because of severe genetic deficiencies (induced by low concentrations of colcemid, or APC). Mitosis and the G1/S boundary represent cell cycle phase transition points subject to checkpoint controls[15] and the onset of apoptosis can be mapped with respect to cell cycle position and these checkpoints, as well as in time. Many of the treatment protocols we have utilized result in the cells being held at a checkpoint prior to the onset of apoptosis. The duration of this "wait" is, in several situations, approximately one cell cycle before apoptosis is detectable. Thus, cells arrested in mitosis must remain blocked for approximately 18 hrs before they become committed to the apoptosis pathway and post-mitotic microcells wait in pseudo-G1 for a similar length of time (Figure 4).

Apoptosis in our experiments occurs after cells have been exposed to drugs for a relatively long period of time. As mentioned above, the cytotoxic events induced by colcemid, mitotic arrest or microcell arrest in "pseudo-G1", are equivalent to 1 cell cycle (G1 to G1). Such "clock-like' behavior in the induction of apoptosis is also evident in cells treated with aphidicolin, but the temporal pattern is more complex. Apoptosis induced in cells arrested at the G1/S boundary by aphidicolin occurs after 36hrs, i.e. the time of 2 cell cycles, while the time to apoptosis varies as a function of the rate at which cells progress through S-phase (i.e. aphidicolin concentration) at concentrations which do not completely block progression through S phase. We are intrigued with this apparent time-dependency of apoptosis in our system, and in particular, in the possibility of a "clock" influencing when apoptosis begins which is somehow reset during the cell cycle. such a clock could be modeled along the lines of a time-dependent decline in the amount of "anti-apoptosis" gene product(s), eg. bcl-2, which is made during any given cell cycle in an amount sufficient to delay the onset of apoptosis for one but not two cell cycles. Cell cycle stasis might put the cells into a state where the intracellular level of these products decline (due to turnover combined with the failure of new proteins to be synthesized) below some threshold level, resulting in apoptosis. The long delay in the onset of apoptosis under the conditions we have examined might then be explained by the turnover of postulated anti-apoptosis gene products relative to the time and/duration of cell cycle stasis induced by a given treatment.

Our results with HeLa cells indicate that apoptosis occurs after a period of suspended cell cycle progression equivalent in time to 1-2 cell cycles. It is well-known that some cell types undergo apoptosis much more rapidly when exposed to various treatments[16]. Other studies we have conducted indicate that the time from cell cycle arrest to the onset of apoptosis varies between cell lines and some cell lines (eg. Tera-2, SB and CEM cells) initiate apoptosis much sooner following colcemid-arrest than do HeLa (or CHO) cells (Kim and Yilmaz, unpubl. obs.). It is known that the time of apotosis can also be influenced by the expression of cellular proteins such as bcl-2. Clinically, this time-dependent aspect of apoptosis might be significant. If the time of drug treatment required for cells to undergo apoptosis varies between cell/tumor type, then a limited duration of drug exposure as occurs clinically may be within the "apoptotic time" of one tumor but not another. Thus, inherent differences in chemosensitivity of tumor-specific cell types (in short, the reason some tumors are treatable and others are not) may reflect differences in the rate at which the "apoptotic clock" ticks in different cell types. Succesful treatment requires that a therapeutic agent attain and sustain adequate in-vivo levels for a critical period of time. As we are beginning to define the critical duration of treatment required in cultured cells, it appears to be determined to some extent, at least, by the duration of cell cycle stasis induced and the postulated apoptotic clock of the cells targeted. The sensitivity and/or mechanisms of action of cell cycle checkpoints may be important in determining when the clock begins ticking after cells are exposed to cytotoxic agents. In many respects, these properties of cells represent mechanisms of drug resistance unrelated to drug-target interaction.

Our results with aphidicolin and colcemid imply that the specific target of drug action can be a factor of secondary importance in the induction of apoptosis and cytotoxicity. Thus, the cellular effects associated with the toxicity of both aphidicolin and colcemid, two agents with profoundly different targets, are identical under some circumstances. Microcell mitoses are generated by low doses of aphidicolin if the exposure is of sufficient duration, by low doses of colcemid and by pulse exposure to high doses of colcemid. Apoptotic cell death occurs in the genetically unbalanced daughter microcells after a period of arrest approximately 1 cell cycle in duration both in APC or colcemid induced microcells. Mechanistically, the element common to aberrant mitosis induced by both drugs is that cells transit mitosis at a reduced rate (relative to untreated cells). The multipolar spindles generated under these conditions may reflect disturbances in the centrosome cycle such that the event of centrosome repoduction and splitting which normally occurs in G1 occurs while the cells are proceeding through mitosis[17,18]. It is also possible that the spindle poles simply become disorganized. It is unclear whether all cell lines display this behavior; we have not observed microcell mitosis in VA-13 or human foreskin fibroblasts. However, we believe that aberrant mitosis of this kind and resulting apoptosis in genetically unbalanced microcells is a widespread effect of pharmacological agents, and insofar as multipolar mitoses represent a defining feature of many solid tumors, further understanding of the implications of aberrant mitosis is important.

Our studies demonstrate that the mechanisms of cytotoxicity and the cell cycle context of apoptosis induced by well-characterized agents in a "standard" cell line. can be complex. Taken together, the data show that in analysing apoptosis from the phenotype down, it is important to understand that seemingly different endpoints, eg. colcemid induced apoptosis in mitotic cells vs. colcemid induced apoptosis in interphase cells, can represent points of similarity (both involve cell cycle "stasis") as well as of dissimilarity (the cells are in different cell cycle phases). It is also important to note that defining the cell cycle state of cells treated with cell cycle-perturbing agents frequently results in dis-integration of cell cycle processes in such a way as to make it difficult to define the cell cycle phase in which cells reside at the time of apoptosis. Thus, for example, CHO cells treated with very low concentrations of APC undergo unbalanced cell growth resulting in a cell cycle state in which the cells have an S-phase DNA content but markedly elevated cyclin B levels indicating that the nuclear and cytoplasmic cell cycles have become de-coupled. While these observations add a level of complexity to interpreting apoptosis and the nature of cellular response eliciting apoptosis, we believe that they provide a framework to better identify and isolate the functionally important correlates of apoptosis.

ACKNOWLEDGEMENTS

We wish to thank Daphne Rush, David Vaux, Andrew Kung Jamie Sheridan, Louis Kim, Lal Yilmaz, Catherine Carswell-Crumption, Rakesh Sharma and other members of the lab for helpful discusions. This work was supported by grant NIH CA 16318 to RTS.

REFERENCES

1. Vaux, D. 1993. Toward an understanding of the mlecular mechanisms of physiological cell death. Proc. Nat. Acad. Sci. USA. 90:786-789.
2. Arends, M.J. and A.H. Wyllie. 1991. Apoptosis: Mechanisms and roles in pathology. Int. Rev. Expt. Path. 32: 223-241.
3. Williams, G.T. 1991. Programmed cell death: Apoptosis and oncogenesis. Cell 65:1097-1098.
4. Fawthrop, D.J., Boobis, A.R., and D.S. Davies. 1991.Mechanisms of cell death. Arch. Toxicol. 65:437-444.
5. Bursch, W., Oberhammer, F., and R. Schulte-Hermann. 1992. Cell death by apoptosis and its protective role against disease. TIPS 13:245-251.
6. Barry, M.A., Behnke, C.A., and A. Eastman. 1990. Activation of programmed cell death (apoptosis) by cisplatinum, other anticancer drugs, toxins and hyperthermia. Biocehem. Pharmacol. 40(10): 2353-2362.
7. Sen, S., and M. D'Incalci. 1992. Apoptosis: Biochemical events and relevance to cancer chemotherapy. Fed. European Biochem. Societies 307(1):122-127.
8. Schimke, R.T., Sherwood, S.W., Hill, A.B., and R.N. Johnson. 1986. Overreplication and recombination in higher eukaryotes: Potential consequences and biological implications. Proc. Nat. Acad. Sci. USA 83:2157-2161.
9. Kung, A.L., Zetterberg, A., Sherwood, S.W. and R.T. Schimke. 1990. Cytotoxic effects of cell cycle phase specific agents: Result of cell cycle perturbation. Cancer Res. 50:7307-7317.
10. Kung, A.L., Sherwood, S.W. and R.T. Schimke. 1990. Cell line-specific differences in the control of cell cycle progression in the absence of mitosis. Proc. Nat. Acad. Sci. USA 87:9553-9557.
11. Mariani, B.D., Slate, D.L. and R.T. Schimke. 1981. S-phase specific synthesis of dihydrofolate reductase in Chinese hamster ovary cells. Proc. Nat. Acad. Sci. USA 78:4985-4989.
12. Li, R., and A.W. Murray. 1991. Feedback control of mitosis in budding yeast. Cell 66: 519-531.
13. Hoyt, M.A., Totis, L. and B.T. Roberts. S. cerevesiae genes required for cell cycle arrest in response to loss of microtubule function. Cell 66: 507-517.
14. Kung, A.L., Sherwood, S.W. and R.T. Schimke. 1993. Differences in the regulation of protein synthesis, cyclin B accumulation and cellular growth in response to the inhibition of DNA synthesis in CHO and HeLa S3 cells. J. Biol. Chem. 268:23072-23080.
15. Sherwood, S.W., Ruch, D.F., Kung, A.L. and R.T. Schinke. Cyclin B1 expression in HeLa S3 cells studied by flow cytometry. Submitted.
16. Hartwell, L.H. and T.A. Weinert. 1989. Checkpoints: controls that ensure the order of cell cycle events. Science 246:629-634.
17. Bissonnette, R.P., Echeverri, F., Mahboubi, A., and D.R. Green. 1992. Apoptotic cell death induced by c-myc is inhibited by bcl-2. Nature. 359:552-556.
18. Kreyer, G., Ris, H., and H.H. Borisy. 1984. Centriole distribution during tripolar mitosis in Chinese hamster ovary cells. J. Cell Biol. 98: 2222-2229.
19. Sluder, G., and D.A. Begg. 1985. Experimental analysis of the reproduction of spindle poles. J. Cell Sci. 76:35-51.

Discussion

D. Livingston

So, at the very lowest concentration of the aphidicolin, where you can induce apoptotic death a number of hours later, is there or is there not DNA replication?

S. Sherwood

Yes, there is.

D. Livingston

Can you arrive at a set of circumstances where you set the process in motion, get no DNA replication and observe, for example, death in M or G2 or wherever? I am trying to get at the question as to whether there are two clocks. A traditional cell cycle clock and a death clock. You stop the traditional cell cycle clock, the death clock still runs. For example, the idea that you presented early on, that cells swell and they eventually die. Presumably there is no DNA replication going on during that period.

S. Sherwood

My feeling right now is that there are two clocks but they are linked. Ideally, what would be perfect, would be to have data showing that *bcl*-2 or something similar is cell cycle regulated and is made some time in the cell cycle particularly early. That would make the perfect model. And then if you hold the cells long enough, they are not able to get into the next cell cycle and make enough *bcl*-2 which drops below a threshold within about one cell cycle. That is where I am headed with looking for an answer. But I believe the two clocks are running independently but they might be tied together, particularly if these postulated anti-apoptosi, the proteins are made in a temporarily restricted time in the cell cycle.

D. Livingston

Is it absolutely inconceivable that there might be some concentration of any of the DNA synthesis inhibitors that one could use that would allow you to pass, without DNA replication, to a state of effectively G2-M at which death might occur?

S. Sherwood

I think actually some of the aphidicolin concentrations approach that.

D. Livingston

Or where, for example, you do not go from 2N to 4N, but to something less than that but nonetheless all the morphological data plus the cyclin-B activation of CDC2 and so forth, indicate that you are well beyond the end of DNA synthesis and then death occurs.

S. Sherwood

Well you can certainly, based on where they are by DNA content in the cell cycle, get them part way through S and have them apoptose. Under those conditions, cyclin-B1 levels are starting to go up but MPF activity is not activated.

D. Livingston

Yes, that is what I mean. I mean if you had MPF then you clearly would have passed, at least in the traditional sense.

S. Sherwood

You would have passed the point where cyclin goes up, which we mapped with our assays. Cells look like they reside in G2 for a little while before cyclin-B begins accumulating. So they will get past that point but they will not get to the full blown activation of the MPF activity.

P. Nicotera

Did you look at other cyclins in the cycle?

S. Sherwood

No, we have the antibodies and it is high on the list of things to do as soon as I get back.

P. Nicotera

Since you correlated all these things with your sorting of the cells, did you also think to use other markers like, for example, PCNA, the proliferating cell nuclear antigen that could tell you something specific about some other changes that have occurred?

S. Sherwood

No, we have not. One thing we have done is to try to label microcells with BRDU and they do not label, or very few label, so we are quite confident that they are not making it into S-phase, that they are really stuck in G1. But beyond that we have not characterized it.

E. Mihich

With reference to the length of time that in some cases it takes to get to the end point, and you indicated that you can start from any phase of the cell cycle, is there a time point beyond which you do not need protein synthesis? In other words, is cycloheximide blocking the apoptotic event in each case, at any time during the long period of incubation?

S. Sherwood

We do not know that. We do know that it does not matter whether they see cycloheximide in G1, or once they pass into early S. Beyond that, we do not know. Then you get the equivalent rescue of the cells from all the toxic effects of aphidicolin and hydroxyurea. We have not systematically pulsed them with cycloheximide through the cell cycle so I can only give you a partial answer.

D. Hockenbery

There have been a number of studies that show you can measure protein modifications and transglutamination and so on as early events in apoptosis. Can you exclude a model where there is a limited activation of apoptosis, as an early process within the cell and the apoptosis, in proportion to the degree of its activation, affects progression through the cell cycle, rather than progression through the cell cycle affecting apoptosis? There are a number of models in which things that activate apoptosis affect the cell cycle without an obvious reason why.

S. Sherwood

No, I cannot. One point I did not make, and I should have, is that the apoptotic process kicks on very late in the cell cycle arrest so with the high concentration of colcemid we know if we let them go up to 12, maybe 14 hours, they will undergo aberrant mitosis but they will recover. They will not be blocked in mitosis permanently and undergo apoptosis there. If we hold them for 18 hours, all of them will apoptose in mitosis, so sometime between 12 and 18 hours is when they are really committed to undergo apoptosis in mitosis. Therefore, I think that that activation step is coming quite late but they have to be held in the static state for some time for that to kick in.

M. D'Incalci

Did you have a look at other cell lines in which you have a similar inhibition of DNA synthesis by aphidicolin and that do not undergo apoptosis? Already many people have used aphidicolin to synchronize cells without causing toxicity. Can you comment on this point?

S. Sherwood

Certainly one of the lessons we would leave you with is: Do not use induction synchronies. There is an important time dependency because at least with the hamster cells the cell cycle clock stays on time. I am not sure what happens if you take them out of the aphidicolin before they get to that point. Clearly they are larger than normal, so a fair amount of disturbance has already occurred but that specific marker, I do not know what its state would be, but certainly we long ago concluded that induced synchronies is not a great thing to do. And that applies to thymidine too. You get the same effect with thymidine.

LYMPHOCYTE ACTIVATION INDUCED APOPTOSIS BY THE APO-1 CELL SURFACE RECEPTOR

Peter H. Krammer

Division of Immunogenetics
German Cancer Research Center
D-69120 Heidelberg
Germany

INTRODUCTION

Growth of lymphocytes is controlled by cell surface molecules and their ligands including cytokines. To characterize cell surface molecules involved in control of growth of malignant lymphocytes we raised monoclonal antibodies against the human B lymphoblast cell line SKW6.4. One antibody, anti-APO-1, reacted with a 48-kilodalton antigen (APO-1) on a set of activated human lymphocytes, on malignant human lymphocyte lines, and on a fraction of leukemic cells from patients. Nanogram quantities of anti-APO-1 completely blocked proliferation of APO-1 positive cells in vitro in a manner characteristic of a process called programmed cell death or apoptosis . Cell death was preceded by changes in cell morphology and fragmentation of DNA. This process was distinct from antibody and complement dependent cell lysis and mediated by the antibody alone[1]. In the following, results obtained in this experimental system are discussed in more detail.

RESULTS AND DISCUSSION

Morphology of anti-APO-1 induced Apoptosis

SKW6.4 cells undergoing apoptosis showed the following morphological changes: condensation of the cytoplasm and karyoplasm, formation of large electron-opaque aggregates of the chromatin lining the nuclear membrane, "blebbing" of the cell membrane at an early stage of apoptosis, and dilatation of the mitochondria. Two hours after adding anti-APO-1, the nuclear membrane was ruptured. Occasionally, large vesicular enlargement of the endoplasmic reticulum or the Golgi apparatus appeared in the cytoplasm. Finally, total breakdown of all cell membranes and cellular disintegration was observed[2]. Thus, anti-APO-1 induced cell death of SKW6.4 cells fulfilled all criteria normally seen in apoptosis induced by other experimental manipulations.

Purification and Molecular Cloning of the APO-1 Cell Surface Antigen

To characterize the APO-1 cell surface molecule and to better understand its role in induction of apoptosis, the APO-1 protein was purified to homogeneity from membranes of SKW6.4 B lymphoblastoid cells by solubilization with sodium deoxycholate, affinity chromatography with anti-APO-1 antibody, and reversed phase high performance liquid

Apoptosis, Edited by E. Mihich and R.T. Schimke
Plenum Press, New York, 1994

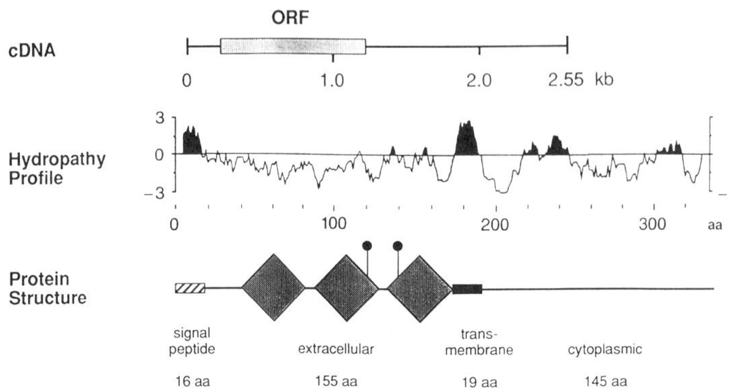

Figure 1. The molecular structure of APO-1. The open reading frame (ORF) of the APO-1 cDNA and the hydropathy profile and predicted protein structure of APO-1 are shown as detailed in Oehm et al.[4] aa = amino acids. ●— = glycosylation site.

chromatography. Each purification step was followed by an APO-1 specific solid phase enzyme-linked immunosorbent assay using the monoclonal antibody anti-APO-1. Endoproteinase-cleaved peptides of the APO-1 protein were subjected to amino acid sequencing and corresponding oligonucleotides were used to identify a full-length APO-1 cDNA clone from an SKW6.4 cDNA library (Figure 1). The deduced amino acid sequence of APO-1 showed sequence identity with the Fas antigen[3], a cysteine-rich transmembrane protein of 335 amino acids with significant similarity to the members of the tumor necrosis factor/nerve growth factor receptor superfamily. The APO-1 antigen was expressed upon transfection of APO-1 cDNA into BL60-P7 Burkitt's lymphoma cells and conferred sensitivity towards anti-APO-1 induced apoptosis to the transfectants[4].

The Human APO-1 Gene

Using cosmid cDNA containing the APO-1 gene as a probe for fluorescence *in situ* hybridization, we have mapped the gene to a subregion of chromosomal band 10q23. The human APO-1 locus lies within a conserved synteny segment present on mouse chromosome 19 consistent with the previous chromosomal assignment of the corresponding mouse antigen[5,6]. This synteny region is defined by the genes coding for interstitial/plasma retinol binding protein (RBP4) terminal deoxynucleotidyltransferase (DNTT), the α-2 and ß-1 adrenergic receptors (ADRA2 and ADRB1, respectively), soluble glutamic-oxaloacetic transaminase 1 (GOT1), the subfamily IIC cytochrome P450 (CYP2C), and phosphoglycerate mutase A (PGAM1).

APO-1 Expression in Normal Human Tissue

By immunohistochemistry APO-1 is a non-lineage antigen constitutively expressed in a variety of epithelial cells. This includes the immature layers of various squamous epithelia, transitional epithelium and columnar epithelium of the biliary tract and intestine. Among the epithelial cell types of the reproductive system of both genders, APO-1 expression is complex. Except the satellite cells of autonomic ganglia all cells of the nervous tissue are APO-1 negative. Among mesenchymal cells, constitutive APO-1 expression is rare but detectable in various kinds of activated cells, e.g. fibroblasts, osteoblasts, and subpopulations of endothelial cells. Within the immune system, APO-1 is broadly distributed among histiocytic cells but restricted to minor subpopulations of peripheral T and B cells. Immature T cells, i.e. thymocytes, are APO-1 negative by immunohistochemistry but show low positivity by immunofluorescence. Consistently, there was an *in situ* induction of APO-1 in several types of glandular epithelium in microtopographic association with lymphohistiocytic infiltration. This inflammation-associated APO-1 induction went along with increased expression of this molecule within the lymphocytic compartment of the lesion[7].

The Role of APO-1 in Activation induced Apoptosis in T and B Cells

As described above, the mouse APO-1/Fas antigen gene was localized to a region of chromosome 19 identical to the *lpr* locus[6]. The *lpr* defect in mutant mice is characterized by accumulation of $CD4^{-}CD8^{-}$, $B220^{+}$, TCR^{+} aberrant T cells in the peripheral lymphoid organs, particularly in the lymph nodes, and by autoimmunity e.g., autoantibodies, similar to those seen in systemic lupus erythematosus in man. A lack of Fas antigen expression was shown to be the prominent defect in the mutant animals. Watanabe-Fukunaga et al. hypothesized that this defect might lead to defective thymic selection and, thus, explain the *lpr* pathology[8]. Other experiments in TCR transgenic mice, however, indicated that selection of T cells in the thymus in *lpr* mice was perfectly normal[9]. In addition, we found that APO-1 was not detectable by immunohistology but showed only low staining in immunofluorescence on human thymocytes. These data suggested that Fas/APO-1 might be involved in peripheral T cell selection whereas its role in thymic selection is still unclear. The failure of apoptosis and peripheral selection might lead to the *lpr* syndrome[11,12].

To further characterize the APO-1 pathway we studied APO-1 mediated apoptosis of human peripheral blood T lymphocytes. APO-1 is not expressed on cord blood and the majority of resting T cells, but on activated T cells. One day activated T cells in culture showed activation induced resistance to apoptosis (ARA). However, after prolonged *in vitro* culture, 6 day activated T cells acquired sensitivity to activation induced sensitivity to apoptosis (ASA) (Figure 2). Restimulation of the ASA^{+} activated T cells by triggering TCR/CD3 or CD2 induced proliferation and apoptosis in a fraction of the cells. In the surviving fraction of ASA^{+} activated T cells, however, this treatment reinduced a transient ARA^{+} phenotype. Thus, activation of resting mature T cells or restimulation of activated T cells may induce a transient resistance to apoptotic signals. Activation signals may interfere with the APO-1 pathway and may prevent elimination of activated T cells in the periphery (peripheral selection)[12].

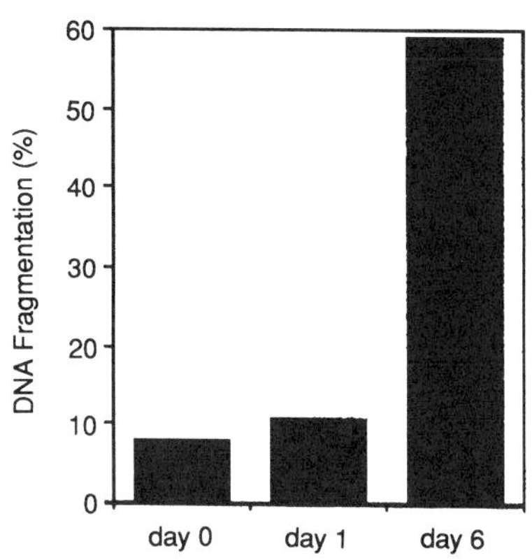

Figure 2. Differential sensitivity to anti-APO-1 induced apoptosis in human T cells. Human peripheral blood T cells were activated by PHA and cultured in IL-2 for 6 days. Apoptosis was determined on the days of culture indicated by *in situ* nick translation as described by Klas et al.12

On B cells APO-1 is an activation molecule. It was induced/enhanced on dense and buoyant tonsillar B cells, respectively, through surface Ig cross-linking in combination with interleukin-2 or by interferon-τ together with tumor necrosis factor-α. Immunohistologically, APO-1 was detectable at low levels in a subpopulation of follicle center B blasts and, at higher levels, in sinusoidal B cells. APO-1 was undetectable in plasma cells. In isolated tonsillar B cells, APO-1 was expressed in $CD10^{+}$ follicle center cells. Expression of APO-1 in B cells of different stages of differentiation might suggest a role of this molecule in the induction of B-cell apoptosis[13]. On the functional level sensitivity of B cells to APO-1 mediated apoptosis *in vitro* and *in vivo* in SCID mice reconstituted with HPBL is activation-dependent, antigen specific and restricted to a defined time after initial antigen triggering (P. Daniel and P.H. Krammer, to be published).

CONCLUSION

Taken together, our data show that APO-1 is a non-lineage restricted cell surface receptor that mediates apoptosis upon triggering by anti-APO-1. Potentially, the same effect might be induced by the as yet elusive natural ligand for APO-1. The pattern of expression of APO-1, its upregulation on activated T and B cells, the defect of Fas/APO-1 expression in mice with the *lpr* phenotype and its role in activation induced apoptosis of activated peripheral T and B lymphocytes suggest that APO-1 is a key molecule for peripheral deletion and tolerance. APO-1 may therefore be decisive in the regulation of the peripheral immune response and in its dysregulation in autoimmunity.

REFERENCES

1. B.C. Trauth, C. Klas, A.M.J. Peters, S. Matzku, P. Möller, W. Falk, K.-M. Debatin and P.H. Krammer. Monoclonal antibody-mediated tumor regression by induction of apoptosis. *Science* 245:301 (1989).
2. H.-R. Köhler, J. Dhein, G. Alberti and P.H. Krammer. Ultrastructural analysis of apoptosis by the monoclonal antibody anti-APO-1 on a lymphoblastoid B cell line (SKW6.4). *Ultrastructural Pathology* 14:513 (1990).
3. N. Itoh, S. Yonehara, A. Ishii, M. Yonehara, S-I. Mizushima, M. Sameshima, A. Hase, Y. Seto, and S. Nagata. The polypeptide encoded by the cDNA for human cell surface antigen Fas can mediate apoptosis. *Cell* 66:223 (1991).
4. A. Oehm, I. Behrmann, W. Falk, G. Maier, C. Klas, M. Li-Weber, M. Pawlita, S. Richards, J. Dhein, B.C. Trauth and P.H. Krammer. Purification and molecular cloning of the APO-1 antigen, a new member of the TNF/NGF receptor superfamily: sequence identity with the Fas antigen. *J Biol Chem* 267:10709 (1992).
5. P. Lichter, H. Walczak, S. Weitz, I. Behrmann, and P. H. Krammer. The human APO-1 antigen maps to 10q23, a region which is syntenic with mouse chromosome 19. *Genomics* 14:179 (1992).
6. R. Watanabe-Fukunaga, C.I. Brannan, N, Itoh, S, Yonehara, N.G. Copeland, N.A. Jenkins and S. Nagata. The cDNA structure, expression and chromosomal assignment of the mouse Fas antigen. *J Immunol* 148:1274 (1992).
7. F. Leithäuser, J. Dhein, G. Mechtersheimer, K. Koretz, S. Brüderlein, C. Henne, A. Schmidt, K.-M. Debatin, P.H. Krammer and P. Möller. Constitutive and induced expression of APO-1, new member of the NGF/TNF receptor superfamily, in normal and neoplastic cells. *Lab Invest*. In press.
8. R. Watanabe-Fukunaga, C.I. Brannan, N.G. Copeland, N.A. Jenkins, and S. Nagata, Lymphoproliferation disorder in mice explained by defects in Fas antigen that mediates apoptosis. *Nature* 356:314 (1992).
9. C.L. Sidmann; J.D. Marshall, and H. von Boehmer. Transgenic T cell receptor interactions in the lymphoproliferative and autoimmune syndromes of lpr and gld mutant mice. *Eur J Immunol* 22:499 (1992).
10. P.L. Cohen and R.A. Eisenberg. The lpr and gld genes in systemic autoimmunity: life and death in the Fas lane. *Immunol Today* 13:427 (1992).
11. P.H. Krammer and K.-M. Debatin. When apoptosis fails. *Curr Biol* 2:383 (1992).
12. C. Klas, K.-M. Debatin, R.R. Jonker and P.H. Krammer. Activation interferes with the APO-1 pathway in mature human T cells. *Internatl Immunol*. 5:625 (1993).
13. P. Möller, C. Henne, F. Leithäuser, A. Eichelmann, A. Schmidt, S. Brüderlein, J. Dhein and P.H. Krammer. Co-regulation of the APO-1 antigen with ICAM-1 (CD54) in tonsillar B cells and coordinate expression in follicular center B cells, in follicle center and mediastinal B cell lymphomas. *Blood* 81:2067 (1993).

Discussion

S. Schlossman

The other members of the APO-1 family, TNF, CD27, CD30, if you cross link them is apoptosis induced?

P. Krammer

Yes. TNF acts as a trimer and you need to trimerize the TNF receptor to induce TNF effects.

S. Schlossman

What about the others?

P. Krammer

I do not think it is known for NGF. I do not know it at least.

S. Schlossman

A second question relates to the controls that you use in the scid mice since cells coated with antibody are susceptible to ADDC or natural killing, which would also result in apoptosis induced or killer cells. Are the control antibodies that you use ones that coat the TALL's or are they simply an irrelevant immunoglobulin?

P. Krammer

No, they are the same isotype. We have shown in previous experiments that when you use an isotype which is not suitable to induce ADCC, you nevertheless can induce apoptosis like, for example, with IGA.

S. Schlossman

But it is the same isotype that coats the tumor cell so it would be a binding antibody of the same isotype as the anti APO-1, rather than just a nonbinding antibody of the same isotype?

P. Krammer

Yes, we did that.

T. McDonnell

You piqued our interest with your last statement there regarding the ligand. Can you give us any insight with respect to progress you are making in that regard? The second question is, can you tell us anything about the signal transduction mechanism that is linked to Fas/APO-1 and have you ever looked for expression in follicular lymphomas and, if it is there, is it functional?

P. Krammer

It is expressed in follicular lymphomas; in almost all of them it is highly expressed. Signal transduction is a very hot area studied in my laboratory and in various other laboratories, and I can describe for you spots and bands on gels but nothing cloned. The same applies to the ligand. There are several groups chasing it, including ours, but there is no molecular information.

P. Nicotera

We all generally assume that cyclohexamide will block protein synthesis and in your case I understand that the sensitization of the cells by cyclohexamide makes these cells go into apoptosis, similar to what we have observed with TNF-α in some other systems, which would be explained by the lack of a repressor or something which blocks apoptosis. Have you considered another possibility and that is that cyclohexamide provides you with the additional second signal for apoptosis by having, for example, super induction and bypassing the protein synthesis block?

P. Krammer

Yes, of course we have considered that. I do not like the idea because we would like to get at the repressor in molecular terms. If it is the second signal, we would be stuck.

E. White

What we did, this is just a comment, is that we took a cell that was sensitive to anti APO-1 and we put *bcl*-2 into it and it became resistant. I think the observation that you made looking at all the different leukemias, that some were sensitive, and some were resistant, and that did not correlate with *bcl*-2 expression, actually suggests a more interesting possibility; namely, that there are many ways to suppress apoptosis that may be independent of *bcl*-2.

P. Krammer

Yes, still it is unclear. I just quote Andreas Strasser's data. He has taken the cell line which we used, SKW6.4 - the B-lymphoblastoid cell line, which is exquisitely sensitive and undergoes apoptosis very fast after incubation with anti APO-1, and he has transfected this cell line with *bcl*-2 and the expression level on the protein level is about 20 times higher than in the nontransfected cells. And it does not change sensitivity when you titrate anti-APO-1 on this cell line. So, I still think it is unclear. It might be cell type specific.

T. Taniguchi

Two questions, one is: You showed that T-cells activated in culture for 6 days become more sensitive to anti-APO-1. Is this due to the increased expression of the APO-1 protein itself or do you think there are other mechanisms?

P. Krammer

No, the amount of APO-1 on the cell surfaces is approximately the same.

T. Taniguchi

I see, so it is a different mechanism?

P. Krammer

It is probably due to the development of a sensitive pathway, whatever that is.

T. Taniguchi

Second question is brief: Is there any evidence for homophillic interactions of the APO-1?

P. Krammer

We have thought about it. We have no evidence for that.

D. Hockenbery

What do you think APO-1 expression in the thymus means? Is it important for thymocyte selection?

P. Krammer

I think there is room for a role of APO-1/Fas in the thymus, but that is speculation. Some people tell me there may be something. It could be that negative selection involves a short activation step with expression of APO-1 and then the cells are knocked out and that you would not see easily.

D. Livingston

I assume from what you say the prediction would be that a resistant cell would be dominant to a sensitive. Have heterokarions been made and is that the phenotype?

P. Krammer

It has not been made yet. It is on the list.

D. Livingston

Second question is that when you induce sensitivity with cyclohexamide, the antibody reactivity of the sensitive and of the parent are identical? In other words, when you do a simple titration of antibody in a reactivity assay, does antibody react equally well with APO-1 on the surface of a resistant and a sensitive cell?

P. Krammer

There is no downregulation or downmodulation of the antigen.

D. Livingston

Thirdly, if you trigger APO-1 with an antibody, does its cytoplasmic domain get phosphorylated?

P. Krammer

That is not clear yet.

D. Livingston

And finally, if you put APO-1 into a tissue culture cell that never had it in it before, certainly was not expressed under a constitutive promoter, does it become sensitive now to the antibody and does it go through apoptosis?

P. Krammer

We have used the mouse fibroblasts AL929 and transfected them with APO-1, human APO-1, and they become sensitive towards anti APO-1 induced apoptosis. So they have the program probably.

D. Papermaster

Is APO-1 expressed in the nervous system?

P. Krammer

It is not expressed in the brain. I would have to look up the long list of all the various tissues, which I cannot keep in mind, but a paper with APO-1 expression from head to toe is coming out in *Laboratory Investigations*.

A. Eastman

Can you clarify something for me? Obviously, anti-Fas antibody is not what naturally kills the cell. When you are looking for the ligand, are you looking for a ligand that activates the receptor and therefore kills the cell, or is your model that the antibody is in fact blocking the ligand?

P. Krammer

I think it is a death ligand because $F(ab^1)2$ anti-APO-1 should block but it does not induce apoptosis. It does not cross link enough. Again the *lpr* defect mouse would suggest that in this animal apoptosis fails.

A. Eastman

Could not the *lpr* mice be constitutively active now that it has a point mutation in the internal domain. It could be active as a receptor and therefore keep the cell alive.

P. Krammer

It does not have many receptors. The molecule is practically absent. The *lpr* cg mouse has the molecule but has a mutation in the intracellular tail.

D. Delia

What cytokines do actually modulate, even upregulate the APO-1 antigen? Is upregulation of the molecule by any possible cytokine increase sensitivity to killing?

P. Krammer

The promoter is sensitive to interferon and TNF. It upregulates expression.

D. Delia

So would it be possible, for example, in experiments with TLL's where even in scid mice you get a low percentage of apoptotic response to APO-1, to increase the response by pre-treatment of the mice with interferon?

P. Krammer

It was not done yet but it is on the list also.

D. Papermaster

Can you give us some sense of the antibody concentration dependence of the induction of apoptosis? Do you need many of the molecules of the APO-1 antigen to be cross linked, or can just three or four of them be enough to trigger the death of the whole cell? That is one question. The other is: Can you induce the cell to cap and shed its APO-1? If so, does it then become resistant and does it regain sensitivity at a certain pace based on resynthesis?

P. Krammer

Well, we do not have exact data on how many molecules have to be covered, but it is not one, two, or three. It has to be more. Maybe there is a threshold but we have not decently investigated that yet. But from all the data I overview, I would think that there is a threshold. So it needs to be in the range of several thousands, which is surprising. Shedding? We have looked at that in the SKW6.4 cells, which is our pet animal, so to speak, and it is shed. It is seen in the supernatant.

G. Bellomo

It has been suggested that C2 ceramide could induce apoptosis. Do you have any evidence that ceramide concentration is changed or is increased in anti-APO-1 treated cells?

P. Krammer

We did not test that yet. We went through the papers in the journal club but we have not tested it yet.

E. Mihich

I have a somewhat related question. Are there any relationships in sensitivity and resistance between your anti APO-1 and TNF? In other words, in the cell which is at the stage of resistance to APO-1, is that cell sensitive to TNF and vice versa? Is there any synergism or competition between the two under conditions of sensitivity?

P. Krammer

Yes, we asked that question about two years ago and it is unclear. We did the following experiments. We took MCF7 cells which are sensitive to TNF and express APO-1 and are sensitive to anti APO-1 induced apoptosis. We used that cell line and subcloned it and looked whether sensitivity or resistance to TNF and to anti APO-1 co-segregate and I could never make sense of these data. In addition, we used MCF7 and transfected it with Mn SOD and, again, the clones we got were not really informative. So I think the issue is not closed yet. It is just that our data are not good enough.

C. Croce

What is the frequency of recombination between APO-1 and *lpr*, if any?

P. Krammer

The defect?

C. Croce

So, that is a defect and it is in the membrane domain of the protein?

P. Krammer

Well, the defect is nonexpression of Fas/APO-1. That is the *lpr* defect. It is probably a splicing defect in the gene. It is not entirely clear yet. Groups do not agree and we are awaiting Nagata's paper to come out on the structure of the gene.

P. Nicotera

I would just like to come after the question that Dr. Mihich had for you before. The MCF7 cells have the receptor for APO-1, but can you have apoptosis in the MCF7 cells treated with anti-APO-1, plus or minus cyclohexamide?

P. Krammer

We can do it without cyclohexamide.

D. Livingston

Sorry to go back to this, but the assumption is, and I assume there is proof of this, that the antibody is activating a pathway for which APO-1 is a receptor. Is there any possibility, whatsoever, that the exact opposite is happening? That the antibody is actually turning off a pathway for which APO-1 would normally be the receptor?

P. Krammer

That question was asked before and I think that if you turn something off, that is, for example, binding of a survival factor to this receptor, then it should be done also by the $F(ab^1)2$ anti-APO-1. And that is not the case. Well, it hits the same determinant.

D. Livingston

Are there multiple antibodies to this ligand, multiple monoclonals, for example, to this ligand that always give either the same or no phenotype? Do you ever get the opposite phenotype?

P. Krammer

We have only one. We are in the process of making more.

E. Mihich

Have you tried, or do you think it would be useful, to make the anti-idiotype to your anti-APO-1 and would you learn something by trying to immunize against your APO-1 positive cells?

P. Krammer

To find the ligand, you mean? Well, I think that has been an idea which was around for a long time and it has been tried but, as far as I know, it has not properly worked. The more fruitful approach is to use fusion proteins. APO-1 hooked to immunoglobulin Fc, so that it is secreted and then use those for staining for the ligand. And that has been done in various groups.

J. Cleveland

Your story of sensitivity and resistance sounds a lot like the interferon story where George Stark and Ian Kerr have isolated a series of mutants that can be complemented by elements in the signal transducing pathways of interferon. I was wondering if you were starting to test that out? It seems like a comparable system.

P. Krammer

We have not done that yet.

T. McDonnell

Speaking a little more directly to what you have been saying in the context of what Fas might be doing in malignant transformation or tumor progression, and on the basis of what you presented and what you think, you would speculate that this is mediating a death signal and, if so, have you found any indication in any tumors that you might have surveyed that there might be inactivating mutational events or gene deletion?

P. Krammer

Well, I was speaking to Carlo Croce about that. There is one B-CLL which proliferates upon incubation with anti-APO and co-incubation with SAC and IL-2. That is in fact published in the *European Journal of Immunology* which just recently came out. That is a leukemia that has a deletion of one chromosome 10 in the area where APO-1 is located. That is all we know so far. But I think it is worthwhile studying.

M. Kastan

You raised concerns about whether this would be a clinically useful antibody because of potential cross-reaction with other tissues in the body. Did you see any evidence of that when you injected it into the nude mice that had the tumors?

P. Krammer

The nude mouse loves it, but the nude mouse does not have the antigen.

M. Kastan

So it is not cross reactive with the mouse tissues?

P. Krammer

No, but data from primate studies would indicate that anti-APO-1 is very effective. In high quantities and unmodified it is lethal. But you can maybe use it *ex vivo*. That would have to be tried because bone marrow itself is not very sensitive to the antibody. It is less sensitive than to drugs. You could also envisage a strategy where you make hybrid chimeric antibodies where one part of the antibody is a Schlepper antibody which targets the hybrid molecule to a tumor cell specifically, while the other part, anti APO-1, would be the effector part to induce apoptosis. But the prerequisite of that would be that heterologous cross linking of antigen-x as the Schlepper target and APO-1 would induce apoptosis. That was not tested yet.

D. Delia

Since you mentioned the use of APO-1 in bone marrow purging, do you have any evidence to rule out that haemopoetic stem cells are APO-1 negative? Do you have colony assays to rule out this possibility?

P. Krammer

We did *in vitro* studies looking at colony formation and, as I said, when you use drugs, you normally use for tumor treatment, *in vitro*, on bone marrow, then bone marrow is more sensitive to these drugs than to anti APO-1. But there is a depression, interestingly, in the erythroid lineage.

M. D'Incalci

You have mentioned that anti-APO-1 is lethal. What are the most sensitive organs to anti-APO-1 toxicity?

P. Krammer

It seems that the liver is very sensitive, but you can also see apoptosis *in vivo* in peripheral lymphoid cells. Nevertheless, I think the liver is the most sensitive organ we can pick up. Whether it is due to a primary or a secondary effect, at the moment, is unclear. It could be that the lymphoid tissue in the periphery is sensitive, spits out cytokines after induction of apoptosis, which then induce a whole process of deleterious effects in other organs. But I do not think that is the case because the liver is the primary target and it shows low positivity.

ENDONUCLEASES ASSOCIATED WITH APOPTOSIS

Alan Eastman, Michael A. Barry, Catherine Demarcq, Jinfang Li and Jason E. Reynolds

Department of Pharmacology and Toxicology
Dartmouth Medical School
Hanover, NH 03755

INTRODUCTION

Apoptosis was originally defined by morphological criteria.[1,2] One of the earliest morphological events in apoptotic cells is chromatin condensation to the nuclear membrane. This occurs well in advance of loss of membrane integrity. Concurrent with chromatin condensation, the nuclear DNA is usually digested into oligonucleosomal fragments by an endogenous endonuclease. This chromatin condensation together with DNA digestion has been called one of the cardinal elements of apoptosis.[3] Furthermore, it has been shown that the chromatin condensation occurs as a consequence of DNA digestion.[4] Because DNA digestion occurs while other organelles remain intact, it is presumed that the endonuclease is of non-lysosomal origin. It is commonly believed that the endonuclease is constitutively present in the nucleus, and that it needs activation by some signal transduction pathway originating from the initial stimulus or insult. Confusion in studies of apoptosis arises from a misperception that there should be a single signal transduction pathway to activate the endonuclease. However, apoptosis is a product of many stimuli which act through many pathways culminating in the activation of an endonuclease. Perhaps the only common event in all these pathways is the final activation of the endonuclease, if indeed there is even one single endonuclease responsible for apoptotic DNA digestion. Therefore, an understanding of the events that trigger activation of the endonuclease is essential for elucidating the pathways upstream of the eventual apoptosis. This paper will review the available literature and propose a model for activation of an endonuclease involved in apoptosis.

Ca^{2+}/Mg^{2+}-DEPENDENT ENDONUCLEASES

Mammalian cells contain a variety of endonucleases which could be involved in apoptosis (reviewed in ref. 5). However, a Ca^{2+}/Mg^{2+}-dependent endonuclease has

Apoptosis, Edited by E. Mihich and R.T. Schimke
Plenum Press, New York, 1994

received almost universal acceptance as the endonuclease involved in apoptosis. The most cited evidence for the involvement of this endonuclease in apoptosis is a study of DNA digestion in thymocytes following glucocorticoid treatment.[6] However, this conclusion was based solely on the detection of this endonuclease in thymocytes; no consideration was given to the possible presence of other endonucleases. An earlier paper had also suggested that the Ca^{2+}/Mg^{2+}-dependent endonuclease was involved in apoptosis, in this case in thymocytes following irradiation.[7] It was concluded that this endonuclease was involved because its initial rate of activity was increased in nuclei with increasing irradiation, although its absolute level decreased slightly. This latter paper was aware of the presence of other endonucleases, including a Mg^{2+}-dependent and an acidic endonuclease. Indeed, it was reported that the acidic endonuclease was four-times as active as the Ca^{2+}/Mg^{2+}-dependent endonuclease, but that it did not change following irradiation. Accordingly, it was considered an unlikely candidate to be involved in apoptosis.

Since these initial reports, supporting evidence for the role of the Ca^{2+}/Mg^{2+}-dependent endonuclease has come from observations that glucocorticoids and other toxic agents induce increased intracellular Ca^{2+} levels during apoptosis of thymocytes, and furthermore that cells can be protected from both DNA digestion and death by co-treatment with intracellular Ca^{2+} chelators.[8,9] Also, incubation of thymocytes with Ca^{2+} ionophores induces DNA digestion and death.[10,11] There are a few problems with this model since the observed intracellular Ca^{2+} levels were approximately 800 nM while the Ca^{2+}/Mg^{2+}-dependent endonuclease requires mM concentrations of Ca^{2+} *in vitro*. It remains to be determined whether this 1000-fold difference in Ca^{2+} requirement reflects an artifact of removing the endonuclease from its normal environment.

Demonstration of the involvement of Ca^{2+}/Mg^{2+}-dependent endonuclease is further confounded by numerous reports of such endonucleases possessing a range of relative molecular masses from 18-65 kilodaltons.[5] These enzymes are all reported to require mM concentrations of Ca^{2+} and Mg^{2+}, but some prefer Mn^{2+} to Mg^{2+}, and Zn^{2+} can variously inhibit or activate them. The inhibition by Zn^{2+} of both apoptotic DNA digestion in cells and Ca^{2+}/Mg^{2+}-dependent endonuclease in cell extracts has often been suggested as evidence for the involvement of this endonuclease.[6] However, Zn^{2+} also influences many other biological reactions, not least of which is its ability to inhibit calcium channels and hence deregulate intracellular Ca^{2+} homeostasis.[12] Aurintricarboxylic acid is another inhibitor of both apoptosis and Ca^{2+}/Mg^{2+}-dependent endonuclease and accordingly has been cited as evidence for the involvement of this endonuclease. Aurintricarboxylic acid is a very non-specific inhibitor that functions by inhibiting protein-protein interactions including those between virus and protein, which has led to a proposal for its use as an anti-AIDS drug.[13] More definitive methods will need to be used to adequately test the hypothesis that Ca^{2+}/Mg^{2+}-dependent endonuclease is involved in apoptosis.

One Ca^{2+}/Mg^{2+}-dependent endonuclease purified from thymocytes has been shown to have a mass of 18 kilodaltons by denaturing electrophoresis and hence has been named NUC18. The original report of this endonuclease[14] was disputed because of the possibility that it was histone H_2B which created an artifact in the assay.[15] However, subsequent work has confirmed that it is not a histone.[16] In control thymocytes, this protein is associated with a high molecular weight complex whereas during apoptosis, some of the endonuclease was detected as a 25 kilodalton form under non-denaturing conditions, which is presumed to be the same as the 18 kilodalton form measured under denaturing conditions. It was concluded that the glucocorticoid treatment released the active form from a high molecular weight repressor complex.

Another recent report has suggested that the Ca^{2+}/Mg^{2+}-dependent endonuclease is deoxyribonuclease I (DNase I).[17] The authors extracted an endonuclease from cells and used antibodies to DNase I to identify it. They also immuno-depleted cell extracts with antibodies to DNase I and prevented DNA digestion in an *in vitro* assay. Transfection of a cDNA for

DNase I into cells having little endogenous endonuclease activity led to the internucleosomal digestion in isolated nuclei. These results confirm the presence of DNase I in nuclear extracts, but they do not confirm that it is the mediator of apoptotic DNA digestion. Of particular concern is the fact that they only performed incubations under conditions optimum for Ca^{2+}/Mg^{2+}-dependent endonuclease. These cells could contain other endonucleases that are either unstable in their extracts or that require different incubation conditions.

Many of these observations have been made in thymocytes, whereas in many other models of apoptosis, there is no apparent requirement for Ca^{2+}, nor do all cells appear to contain a Ca^{2+}/Mg^{2+}-dependent endonuclease. This raises the possibility that different endonucleases, and therefore different signals exist in different models, and that the only common event in apoptosis is the morphologic change that is a consequence of DNA digestion. Alternately, there is the possibility that another endonuclease is the real contributor to all cases of apoptosis.

Mg^{2+}-DEPENDENT ENDONUCLEASE

It is worth noting another endonuclease that occurs in many tissues and as such has the potential to contribute to apoptosis. This Mg^{2+}-dependent but Ca^{2+}-independent endonuclease was recently isolated from myeloid leukemia cell nuclei and shown to produce internucleosomal DNA digestion.[18] This endonuclease has been known for sometime and its tissue distribution has been determined in rats.[19] Activity was highest in kidney with significant levels in heart, spleen and liver. In contrast, the Ca^{2+}/Mg^{2+}-dependent endonuclease had highest levels in spleen, with appreciable levels also in kidney, liver and thymus, but appeared to be absent in lung and heart. It was not considered whether this was the same Ca^{2+}/Mg^{2+} dependent endonuclease in each tissue. Several Mg^{2+}-dependent endonucleases have been purified with fairly similar characteristics and may therefore represent the same or closely related enzymes. One endonuclease purified from rat liver nuclei appeared to be a homodimer of 36 kilodaltons,[20] whereas a homodimeric protein of 29 kilodaltons was identified as the endonuclease at high level in bovine heart mitochondria.[21] This latter observation is consistent with the high levels of Mg^{2+}-dependent endonuclease identified in heart, but this location would not seem to support its involvement in apoptosis. Currently, there is no proposed mechanism for the activation of this endonuclease if it were to be involved in DNA digestion during apoptosis. It should be emphasized that any endonuclease present in a cell has the potential to produce apoptotic DNA digestion as the internucleosome spacer region is the most exposed region and hence most susceptible to digestion.

DEOXYRIBONUCLEASE II

Identification of Deoxyribonuclease II

Although the Ca^{2+}/Mg^{2+}-dependent endonuclease is reported to be highest in spleen, the endonuclease classically purified from spleen is deoxyribonuclease II (DNase II). This is the endonuclease rediscovered in our studies and which we have now implicated in at least some cases of apoptosis.

Studies from this laboratory have been aimed at understanding how cytotoxic anti-cancer agents kill cells. Our particular focus has been on cisplatin which produces interstrand and intrastrand cross-links in DNA. We established that Chinese hamster ovary (CHO) cells underwent apoptosis following incubation with cisplatin, and subsequently reported that this occurs following many cytotoxic insults.[22] In a few cases such as following hyperthermia,

DNA digestion was observed within a few hours of a lethal treatment. However, in most cases including cisplatin, DNA digestion did not occur until 48 - 72 hours after damage. This delayed apoptosis resulted from progression of cells to the G_2 phase of the cell cycle where they arrested before eventually undergoing apoptosis.[23,24] These studies were performed at equitoxic conditions for each agent (IC_{90}); apoptosis could be induced sooner if cells were incubated under conditions of greater lethality and this could occur at any stage of the cell cycle. To understand how these agents all induced DNA digestion, we set out to purify the endonuclease that was responsible.

Our initial efforts to extract a Ca^{2+}/Mg^{2+}-dependent endonuclease from CHO cells were unsuccessful. However, we detected an alternate endonuclease that was active below pH 7, and had no apparent requirement for Ca^{2+} or Mg^{2+}.[25] This acidic endonuclease could be activated in intact cells by reducing intracellular pH below 7 with a proton ionophore. This activity generated the oligonucleosome-length DNA fragments characteristic of apoptosis. Nuclear extracts contained a cation-independent endonuclease with identical pH-dependent activity. The endonuclease was purified and compared to commercially available bovine DNase II. The two endonucleases were found to be practically identical by all tests including sensitivity to various inhibitors, purification by the same chromatographic steps, a similar molecular mass, and recognition by an antibody raised against the bovine enzyme. Finally, readdition of either the CHO or bovine endonuclease to isolated nuclei produced internucleosomal DNA digestion up through pH 6.5. A partial amino acid sequence obtained from several peptides has not identified any homologous proteins in the Genbank database.

Activation of Deoxyribonuclease II

The above results demonstrated that DNase II is an endonuclease able to induce apoptosis in cells. However, it did not establish whether the necessary conditions existed in cells following stimuli that normally induce apoptosis. It was therefore necessary to establish whether cells undergoing apoptosis also undergo intracellular acidification consistent with the activation of DNase II. CHO cells were not a good model for these experiments because they did not load well with the pH-sensitive fluorescent dye carboxy-SNARF, and furthermore, the cells died heterogeneously over a long period of time which would make it difficult to observe acidification if it also occurred heterogeneously. Accordingly, we selected an alternate system, human promyelocytic leukemia HL-60 cells, which had previously been shown to undergo rapid apoptosis following a variety of insults.[26]

HL-60 cells rapidly digested their DNA following incubation with the calcium ionophore ionomycin.[27] However, when we titrated extracellular Ca^{2+}, it was evident that the DNA digestion correlated only with the presence of ionomycin and not with the Ca^{2+} concentration. DNA digestion even occurred after chelation of the Ca^{2+} with EGTA. We measured both intracellular Ca^{2+} and pH using the fluorescent dyes INDO-1 and carboxy-SNARF respectively. DNA digestion was observed under conditions of both high and undetectable levels of intracellular Ca^{2+}. In contrast, DNA digestion correlated with intracellular acidification. These results were consistent with the hypothesis that DNase II rather than a Ca^{2+}/Mg^{2+}-dependent endonuclease was responsible for the DNA digestion.

The acidification observed following incubation of cells with ionomycin was only about 0.3 pH units, which is probably insufficient to activate DNase II. However, these studies had been performed in whole populations of cells. The subsequent experiments were designed to measure the pH in individual cells using flow cytometry. HL-60 cells were again used but the cells were killed with the topoisomerase inhibitor etoposide.[28] Cells were incubated with etoposide for only 30 min so that all subsequent measurements were a consequence of this initial damage rather than a result of continuing damage. Apoptotic DNA digestion appeared after an additional 2.5 h and was maximum by 4-5 h (Figure 1). This DNA digestion arose from a sub-population of cells that were in S phase and particularly sensitive to etoposide. This constituted about 20% of the cell population. A second episode of apoptosis occurred after the remainder of the cells had progressed through the cell cycle and arrested at G_2 for more than 48 hours. The following experiments were therefore performed at short times after incubation with etoposide to assess the intracellular ionic changes associated with the rapid phase of DNA digestion.

Following incubation with etoposide, HL-60 cells were loaded with the pH sensitive fluorescent dye carboxy-SNARF and analyzed on the flow cytometer. The results were displayed in a two-dimensional dot plot according to the fluorescent intensities at two different wavelengths (Figure 2). When displayed in this format, cells with identical intracellular pH lie on the same line extending outward from the origin; the distance from the origin reflects the amount of dye loaded into each cell. More acidic cells lie to the right while more alkaline cells lie to the left. Following etoposide, an increasing number of cells appeared to the right of the control population. Four hours after etoposide treatment, about 20% of the cells underwent acidification. The dose response and kinetics of acidification paralleled the appearance of DNA digestion. The undamaged cells showed an intracellular pH of 7.2. Etoposide did not cause a continuum of cells with decreasing pH, rather a discrete population with an intracellular pH of about 6.4. In parallel experiments with the Ca^{2+}-sensitive dye INDO-1, the intracellular Ca^{2+} remained at 30 nM. The flow cytometer was used to sort the cells on the basis of intracellular pH, and only the acidic cells showed the morphology and DNA digestion characteristic of apoptosis. These results are consistent with DNase II as the endonuclease responsible for the apoptotic DNA digestion.

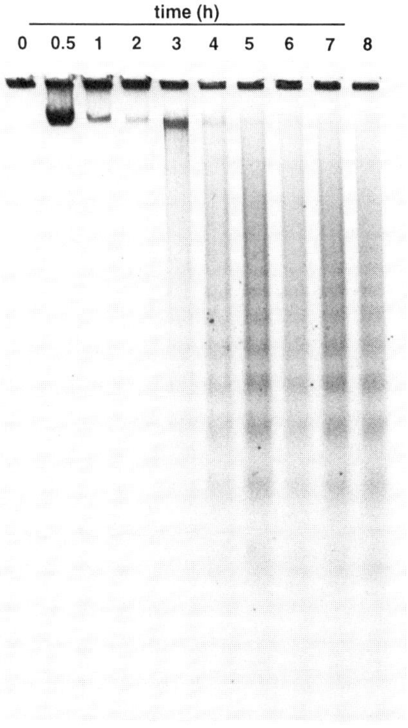

Figure 1. DNA digestion induced in HL-60 cells following incubation with etoposide. Cells were incubated with 80 μg/ml etoposide for 30 min (0.5 h time point), the drug was removed, and cells were incubated for additional times up to 7.5 h (8 h time point). The 0 point represents undamaged cells. Cells were harvested and the DNA was separated by gel electrophoresis and visualized with ethidium bromide. Reproduced with permission from ref. 28.

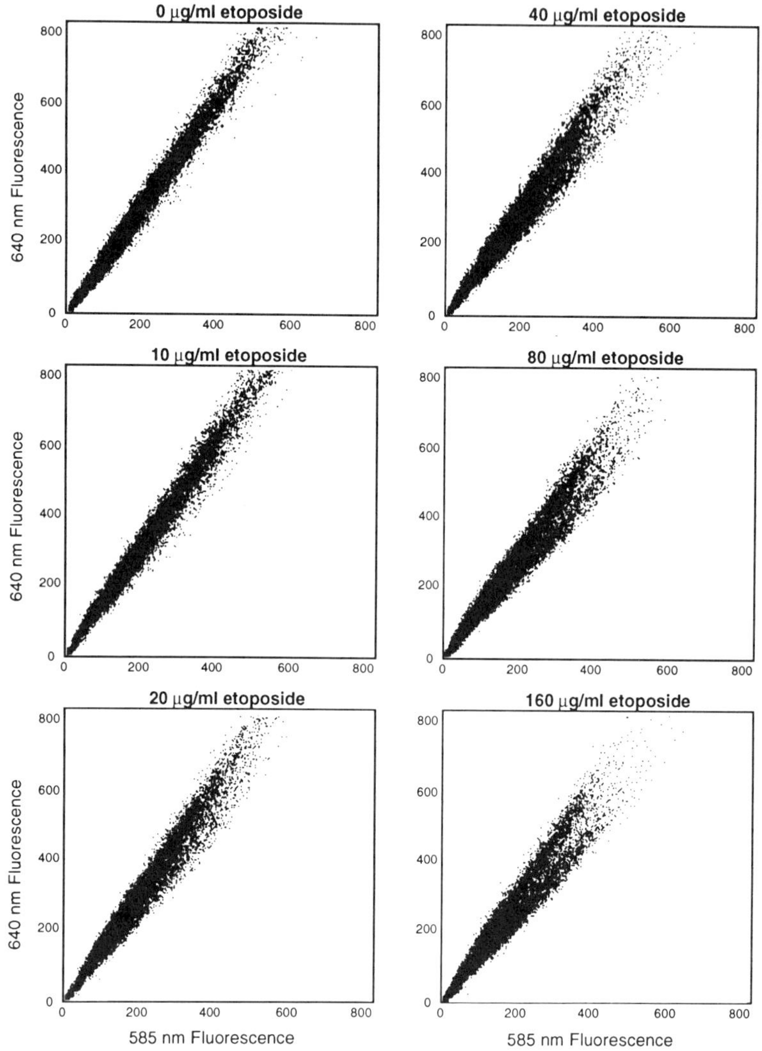

Figure 2. Flow cytometry analysis of intracellular pH changes in individual HL-60 cells damaged with etoposide. Cells were incubated with 0-160 μg/ml etoposide for 30 min, the drug was removed, and the cells were incubated for an additional 3.5 h prior to analysis. The cells were loaded with carboxy-SNARF during the final hour, analyzed by flow cytometry, and displayed as a two-dimensional dot plot according to the fluorescence intensities at emission wavelenghts of 585nm and 640 nm. Reproduced with permission from ref. 28.

Continuing experiments are addressing whether acidification occurs in other pathways of apoptosis. We have preliminary results from studies on an interleukin-2-dependent cytotoxic T lymphocyte cell line CTLL-2. Upon withdrawal of IL-2, cells first accumulate in the G_1 phase of the cell cycle and begin to digest their DNA after about 16 h. Concurrent with this DNA digestion we observed small increases in Ca^{2+} as well as decreased pH. Cells were incubated with EGTA to chelate extracellular Ca^{2+}, and this was shown to prevent the increase in intracellular Ca^{2+} following removal of IL-2. However, chelation of Ca^{2+} did not protect the cells, rather acidification was still observed and the cells still digested their DNA. These results are again consistent with DNase II being responsible for the DNA digestion associated with apoptosis.

Regulation of Intracellular pH and Apoptosis

The ability to measure intracellular pH and Ca^{2+} is dependent upon membrane integrity as the cells must retain the fluorescent dye for the measurements to be made. Furthermore, the results from assay of intracellular Ca^{2+} demonstrate that the cells retain an electrochemical gradient across their cytoplasmic membrane. To maintain low intracellular Ca^{2+} against a Ca^{2+} gradient requires metabolic activity. A metabolically inactive or necrotic cell would have intracellular pH and Ca^{2+} equal to the extracellular environment. Under conditions of a normal electrochemical gradient, selective inhibition of pH regulation would cause the intracellular pH to drop 1 pH unit.[29] This is consistent with the level of intracellular acidification observed in these studies, and supports the hypothesis that apoptotic DNA digestion results from selective inhibition of pH homeostasis, without interference in the homeostasis of other ions.

Cells regulate their pH through a variety of ion transport mechanisms including Na^+/H^+ antiporters, ATP-driven H^+-pumps, and several bicarbonate exchangers; the latter influence intracellular H^+ concentration by altering the intracellular concentration of HCO_3^- (Figure 3). The Na^+/H^+ antiporter is a major H^+-extruding mechanism which is driven by the inward-directed Na^+ gradient. This antiporter has a high affinity for H^+ at pH 6, but is inoperative at neutral pH. However, a wide variety of external signals including growth factors enhance the affinity of the antiporter for H^+ at neutral pH leading to alkalinization of the cells. This modulation of intracellular pH is brought about by phosphorylation of the antiporter.[30,31] In the presence of bicarbonate, cells can use alternate mechanisms to regulate their intracellular pH. However, the Na^+/H^+ antiporter has been shown to be important for tumor growth,[32] perhaps because bicarbonate transporters alone are unable to regulate the degree of acidification that may occur during growth of a tumor *in vivo*. Further evidence for the importance of pH regulation for tumor growth comes from experiments in which a yeast H^+/ATPase was expressed in NIH/3T3 cells. This led to increased intracellular pH and transformation of the cells.[33] Another oncogene Ha-*ras*, has also been shown to activate the Na^+/H^+ antiporter,[34] which would suggest that it may also be able to suppress cell death.

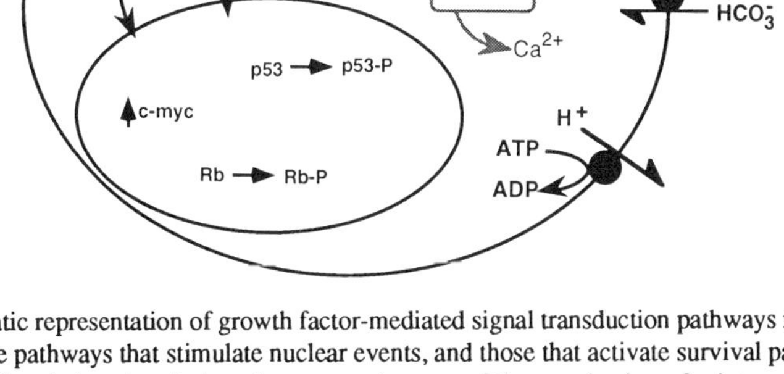

Figure 3. Schematic representation of growth factor-mediated signal transduction pathways in a cell, emphasizing those pathways that stimulate nuclear events, and those that activate survival pathways such as through PKC-mediated phosphorylation of proton antiporters. Other mechanisms for intracellular pH regulation are also shown. The abbreviations for the components in the pathways are described in the text.

The interaction of growth factors with their receptors leads to activation of a number of different signal transduction pathways (Figure 3). For example, phosphatidylinositol-3-kinase (PI3K) and *ras*-GTPase activating protein (GAP) activate pathways that lead to changes in nuclear transcription events such as upregulation of c-*fos*, c-*jun* and c-*myc*, as well as phosphorylation of the tumor suppressor proteins Rb and p53.[35] A third signal transduction pathway involves the activation of phospholipases such as phospholipase C (PLC) which produces diacylglycerol, an activator of protein kinase C (PKC) and inositol-1,4,5-triphosphate (IP_3), which mediates release of Ca^{2+} from intracellular stores. This release of Ca^{2+} probably does not contribute to apoptosis as it is induced by stimuli that protect cells. In contrast, activation of PKC appears to be directly involved in the survival of cells. Activation of PKC can cause induction of nuclear events, but more importantly for the current discussion, is its mediation of the phosphorylation of several ion transport proteins including the Na^+/H^+ antiporter.[31] Direct activation of PKC with phorbol esters, which also causes phosphorylation of the Na^+/H^+ antiporter, can also overcome apoptosis induced by removal of growth factors.[36] Accordingly, the PKC pathway, and its mediation of phosphorylation of the Na^+/H^+ antiporter, as well as of other ion transporters, appears to be critical for cell survival. Hence the available information is consistent with apoptosis being induced by intracellular acidification that can be prevented by alteration in the capacity of cells to regulate pH changes. One caveat is the possibility that the Na^+/H^+ antiporter may be phosphorylated by PKC-independent pathways[34] suggesting multiple survival strategies for a cell.

This model also addresses the susceptibility of certain cells to undergo apoptosis upon over-expression of c-*myc* or wild-type p53. These are nuclear oncogenic events which lead to apoptosis in the absence of activation of a survival pathway such as mediated by PKC. For example, the induction of apoptosis by both c-*myc* and wild-type p53 can be suppressed by growth factors.[37,38] Furthermore, tumor cells over-expressing c-*myc* frequently appear to have mutated or deleted p53, suggesting that wild-type p53 may be a mediator of the c-*myc*-induced apoptosis. Therefore, deletion of p53 may be one way to prevent apoptosis induced by over-expression of c-*myc*.

Activation of Deoxyribonuclease II by Cisplatin

The above discussion has suggested the importance of phosphorylation of membrane ion transporters in protecting cells from apoptosis. Induction of apoptosis would require dephosphorylation of these transporters. This poses the question as to whether there is any relationship between cytotoxic agents such as cisplatin and inhibition of ion homeostasis. It was mentioned earlier that cisplatin causes a G_2 arrest prior to inducing apoptosis. We have investigated the sequence of events in considerable detail and can now suggest a mechanism whereby cisplatin can cause dephosphorylation of these transporters.

These experiments used a DNA repair-deficient CHO cell line so that the level of DNA damage would remain constant following a 1 h incubation with cisplatin.[39] Cells were synchronized at the G_1/S border by mitotic shake followed by aphidicolin, then released from the block and damaged with cisplatin. These cells progressed to the G_2 phase where they arrested for up to 16 h. Entry into mitosis is regulated by the $p34^{cdc2}$ kinase.[40] Inactive kinase is hyperphosphorylated, while activation is associated with dephosphorylation. The G_2 arrested cells contained the hyperphosphorylated form of $p34^{cdc2}$ demonstrating that the arrest was caused by lack of dephosphorylation. At later times (24 h) cells eventually dephosphorylated $p34^{cdc2}$ and entered mitosis. However, during mitosis the chromosomes segregated abnormally leading to G_1 progeny with unequal numbers of chromosomes. This abnormal chromosome segregation appeared to arise from aberrant mitotic spindle formation. Apoptosis occurred at even later times and apparently as a consequence of lethal mitosis.

Cells arrested in G_2 following cisplatin could be forced to enter mitosis by addition of

5 mM caffeine; within 2 h, $p34^{cdc2}$ was dephosphorylated and mitosis was completed by 4 h. Under these conditions, the chromosomes also segregated unequally and apoptosis occurred more rapidly. In contrast, apoptosis could be delayed by inhibiting protein synthesis with cycloheximide. This occurs because protein synthesis is required for normal passage through mitosis, and without the appropriate proteins, cells can not undergo a lethal mitosis. This is an example of the proteins required for apoptosis being normal cell cycle regulatory proteins.

Following a lethal mitosis, cells began to float off the culture dish. This is a common observation when cells die but its relevance to apoptosis has been generally over-looked, in part because the majority of investigations on apoptosis have been performed in suspension cultures. Only the detached cells contained the DNA digestion characteristic of apoptosis. We surmise that the extracellular matrix, in this case the culture dish, acts like a growth factor. Extracellular matrix proteins such as fibronectin, which communicate through integrin-type receptors have been shown to activate a signal transduction pathway through a focal adhesion kinase.[41] The extracellular matrix can also activate the Na^+/H^+ antiporter leading to intracellular alkalinization.[42] Therefore, loss of attachment is likely to inactivate the Na^+/H^+ antiporter leading to intracellular acidification and activation of DNase II. This is parralleled by observations *in vivo* where one of the earliest events in apoptosis is the shrinking of cells away from their neighbors, and this loss of contact could lead in a similar manner to apoptotic DNA digestion.

In summary, these experiments have established a sequence of events that occur during the induction of apoptosis by cisplatin. Damaged cells progress to and arrest in G_2 due to suppression of $p34^{cdc2}$ kinase activity. Upon subsequent dephosphorylation of $p34^{cdc2}$, cells undergo an aberrant mitosis with unequal chromosome segregation. This is probably the first irreversible step in the induction of apoptosis. In the subsequent G_1 phase, cells lacking critical chromosomes detach from the extracellular matrix and undergo apoptotic DNA digestion. It is hypothesized that DNA digestion results from loss of a signal transduction pathway originating from the extracellular matrix which leads to intracellular acidification and subsequent activation of DNase II.

SUMMARY

Internucleosomal DNA digestion is a common occurrence in cells undergoing apoptosis. Identifying the endonucleases involved will facilitate an understanding of their regulation and hence the signal transduction events required for their activation. A number of different endonucleases have been suggested to be involved in this process. Most frequently cited is the Ca^{2+}/Mg^{2+}-dependent endonuclease which is activated by increases in intracellular Ca^{2+}. However, this endonuclease can not explain many cases of apoptosis, and we suggest that DNase II may be a likely alternative. Following a variety of inducing stimuli, we have observed intracellular acidification that is sufficient for activity of DNase II. The acidification appears to result from selective inhibition of proton regulatory mechanisms as intracellular Ca^{2+} remains normal. The Na^+/H^+ exchanger, and probably other ion regulators, are activated by phosphorylation and this occurs when cells are incubated with growth factors. We suggest that apoptosis resulting from removal of a growth factor occurs because the cells dephosphorylate and inactivate these ion regulators. During oncogenesis, cells frequently become independent of growth factors due to mutations in signal transduction pathways. Such cells frequently have increased intracellular pH. However, some oncogenic changes when expressed alone, particularly changes in nuclear proteins such as c-*myc* and p53, can be lethal to a cell, presumably because there is no activation of a survival pathway. Such cells can be rescued by growth factors or other genetic changes such as *bcl*-2 that activate a survival pathway. The observation that DNase II is involved in apoptosis suggests that this survival pathway involves regulation of intracellular pH. This hypothesis is consistent with the current knowledge of intracellular signalling mechanisms, as well as with the current understanding of the functions of oncogenes and tumor suppressor genes.

ACKNOWLEDGMENTS

This study was supported by NIH Grant CA50224 and Cancer Center Support Grant CA23108. Michael Barry and Jason Reynolds were supported by a National Cancer Institute Training Grant CA09658.

REFERENCES

1. J.F.R. Kerr, A.H. Wyllie and A.R. Currie, Apoptosis: a basic biological phenomenon with wide-ranging implications in tissue kinetics, *Br. J. Cancer* 26:239 (1972).
2. A.H. Wyllie, J.F.R. Kerr and A.R. Currie, Cell death: the significance of apoptosis, *Int. Rev. Cytol.* 68:251 (1980).
3. A.H. Wyllie, Apoptosis: cell death in tissue regulation, *J. Pathol.* 153:313 (1987).
4. M.J. Arrends, R.G. Morris and A.H. Wyllie, Apoptosis: the role of the endonuclease, *Am. J. Path.* 136:593 (1990).
5. A. Eastman and M.A. Barry, The origins of DNA breaks: a consequence of DNA damage, DNA repair or apoptosis, *Cancer Invest.* 10:229 (1992).
6. J.J. Cohen and R.C. Duke, Glucorticoid activation of a calcium-dependent endonuclease in thymocyte nuclei leads to cell death, *J. Immunol.* 132:38 (1984).
7. L.V. Nikonova, P.A. Nelipovich and S.R. Umansky, The involvement of nuclear nucleases in rat thymocyte DNA degradation after γ-irradiation, *Biochim. Biophys. Acta* 699:281 (1982).
8. D.J. McConkey, P. Hartzell, S.K. Duddy, H. Hakansson and S. Orrenius, 2,3,7,8-Tetrachlorodibenzo-p-dioxin kills immature thymocytes by Ca^{2+}-mediated endonuclease activation, *Science* 242:256 (1988).
9. D.J. McConkey, P. Nicotera, P. Hartzell, G. Bellomo, A.H. Wyllie and S. Orrenius, Glucocorticoids activate a suicide process in thymocytes through an elevation of cytosolic Ca^{2+} concentration, *Arch. Biochem. Biophys.* 269:365 (1989).
10. A.H. Wyllie, R.G. Morris, A.L. Smith and D. Dunlop, Chromatin cleavage in apoptosis: association with condensed chromatin morphology and dependence on macromolecular synthesis, *J. Pathol.* 142:67 (1984).
11. D.J. McConkey, P. Hartzell, P. Nicotera and S. Orrenius, Calcium-activated DNA fragmentation kills immature thymocytes, *FASEB J.* 3:1843 (1989).
12. D. Busselberg, M.L. Evans, H. Rahmann and D.O. Carpenter, Zn^{2+} blocks the voltage activated calcium current of Aplysia neurons, *Neurosci. Lett.* 117:117 (1990).
13. Y.X. Gan, J.L. Weaver, P.S. Pine, K.C. Zoon and A. Aszalos, Aurintricarboxylic acid, the anti-AIDS compound, prevents the binding of interferon-alpha to its receptor, *Biochem. Biophys. Res. Commun.* 172:1298 (1990).
14. M.M. Compton and J.A. Cidlowski, Identification of a glucocorticoid-induced nuclease in thymocytes. A potential "lysis gene" product, *J. Biol. Chem.* 262:8288 (1987).
15. E.S. Alnemri and G. Litwack, Glucocorticoid-induced lymphocytolysis is not mediated by an induced endonuclease, *J. Biol. Chem.* 264:4104 (1989).
16. M.L. Gaido and J.A. Cidlowski, Identification, purification and characterization of a calcium dependent endonuclease (NUC18) from apoptotic rat thymocytes: NUC18 is not Histone H_2B, *J. Biol. Chem.* 28:18580 (1991).
17. M.C. Peitsch, B. Polzar, H. Stephan, T. Crompton, H.R. MacDonald, H.G. Mannerherz and J. Tschopp, Characterization of the endogenous deoxyribonuclease involved in nuclear DNA degradation during apoptosis (programmed cell death), *EMBO J.* 12:371 (1993).
18. H. Kawabata, N. Anzai, H. Masutani, T. Hirama, Y. Yoshida and M. Okuma, Detection of Mg^{2+}-dependent endonuclease activity in myeloid leukemia cell nuclei capable of producing internucleosomal DNA cleavage, *Biochem. Biophys. Res. Commun.* 191:247 (1993).
19. C. Giannakis, I.J. Forbes and P.D. Zalewski, Ca^{2+}/Mg^{2+}-dependent nuclease: tissue distribution, relationship to inter-nucleosomal DNA fragmentation and inhibition by Zn^{2+}, *Biochem. Biophys. Res. Commun.* 181:915 (1991).

20. Y. Hibino, T. Yoneda and N. Sugano, Purification and properties of a magnesium-dependent endodeoxyribonuclease endogenous to rat-liver nuclei, *Biochim. Biophys. Acta* 950:313 (1988).
21. O.W. Cummings, T.C. King, J.A. Holden and R.L. Low, Purification and characterization of the potent endonuclease in extracts of bovine heart mitochondria, *J. Biol. Chem.* 262:2005 (1987).
22. M.A. Barry, C.A. Behnke and A. Eastman, Activation of programmed cell death (apoptosis) by cisplatin, other anticancer drugs, toxins and hyperthermia, *Biochem. Pharmacol.* 40:2353 (1990).
23. C.M. Sorenson and A. Eastman, Influence of cis-diamminedichloroplatinum(II) on DNA synthesis and cell cycle progression in excision repair proficient and deficient Chinese hamster ovary cells, *Cancer Res.* 48:6703 (1988).
24. C.M. Sorenson, M.A. Barry and A. Eastman, Analysis of events associated with cell cycle arrest at G_2 and cell death induced by cisplatin, *J. Nat. Cancer Inst.* 82:749 (1990).
25. M.A. Barry and A. Eastman, Identification of deoxyribonuclease II as an endonuclease involved in apoptosis, *Arch. Biochem. Biophys.* 300:440 (1993).
26. S.H. Kaufmann, Induction of endonucleolytic DNA cleavage in human acute myelogenous leukemia cells by etoposide, camptothecin, and other cytotoxic anticancer drugs: a cautionary note, *Cancer Res.* 49:5870 (1989).
27. M.A. Barry and A. Eastman, Endonuclease activation during apoptosis: the role of cytosolic Ca^{2+} and pH, *Biochem. Biophys. Res. Commun.* 186:782 (1992).
28. M.A. Barry, J.E. Reynolds and A. Eastman, Etoposide-induced apoptosis in human HL-60 cells is associated with intracellular acidification, *Cancer Res.* 53:2349 (1993).
29. I.H. Madshus, Regulation of intracellular pH in eukaryotic cells, *Biochem. J.* 250:1 (1988).
30. C. Sardet, A. Franchi and J. Pouyssegur, Molecular cloning, primary structure, and expression of the human growth factor-activatable Na^+/H^+ antiporter, *Cell* 56:271 (1989).
31. C. Sardet, L. Counillon, A. Franchi and J. Pouyssegur, Growth factors induce phosphorylation of the Na^+/H^+ antiporter, a glycoprotein of 110 kD, *Science* 247:723 (1990).
32. D. Rotin, D. Steele-Norwood, S. Grinstein and I. Tannock, Requirement of the Na^+/H^+ exchanger for tumor growth. , *Cancer Res.* 49:205 (1989).
33. R.J. Gillies, R. Martinez-Zaguilan, G.M. Martinez, R. Serrano and R. Perona, Tumorigenic 3T3 cells maintain an alkaline intracellular pH under physiologic conditions, *Proc. Natl. Acad. Sci. USA.* 87:7414 (1990).
34. K. Maly, F. Uberall, H. Loferer, W. Doppler, H. Oberhuber, B. Groner and H.H. Grunicke, Ha-*ras* activates the Na^+/H^+ antiporter by a protein kinase C-independent mechanism, *J. Biol. Chem.* 264:11839 (1989).
35. S.A. Aaronson, Growth factors and cancer, *Science* 254:1146 (1991).
36. K. Swann and M. Whitaker, Stimulation of the Na/H exchanger of sea urchin eggs by phorbol ester, *Nature* 314:274 (1985).
37. G.I. Evan, A.H. Wyllie, C.S. Gilbert, T.D. Littlewood, H. Land, M. Brooks, C.M. Waters, L.Z. Penn and D.C. Hancock, Induction of apoptosis in fibroblasts by c-myc protein, *Cell* 69:119 (1992).
38. E. Yonish-Rouach, D. Resnitzky, J. Lotem, L. Sachs, A. Kimchi and M. Oren, Wild-type p53 induces apoptosis of myeloid leukaemic cells that is inhibited by interleuken-6, *Nature* 352:345 (1991).
39. C. Demarcq, D. Creswell and A. Eastman,Involvement of $p34^{cdc2}$ kinase in cisplatin-induced G_2 arrest and apoptosis in Chinese hamster ovary cells, submitted.
40. P. Nurse, Universal control mechanism regulating onset of M-phase, *Nature* 344:503 (1990).
41. I. Zachary and E. Roxengurt, Focal adhesion kinase ($p125^{FAK}$): a point of convergence in the action of neuropeptides, integrins and oncogenes, *Cell* 71:891 (1992).
42. M.A. Schwartz, C. Lechene and D.E. Ingber, Insoluble fibronectin activates the Na/H antiporter by clustering and immobilizing integrin $\alpha_5\beta_1$, independent of cell shape, *Proc. Natl. Acad. Sci. USA.* 88:7849 (1991).

Discussion

J. Isaacs

DNA. The A in DNA stands for acid. And yet, when you polymerize DNA, you essentially neutralize the acidity of the nucleotides. How do you know that when you are starting to get such massive fragmentation that this is not what actually leads to the change in the pH in the cell? So in other words, acidification is occurring in the process but not due to the fact that it is activating DNA fragmentation but as a consequence of it.

A. Eastman

I would think that is very unlikely. If the acid came from any metabolic process, we would expect to see a continuum of decreasing pH. However, we always see this unique population of acidic cells whose pH shift is consistent with the Nernst equation. This suggests that the acidification is really a consequence of the cell selectivity turning off pH regulation.

J. Isaacs

I do not mean this as a criticism, but we have made those calculations and there is enough acid in the DNA to make those type of changes. If one could show that you had cells that had a change in pH down, let us say to 6.4, and you sorted them and their DNA was all high molecular weight and you could show that you could then see subsequent degradation in those populations, that would be excellent. But I think that with simultaneous appearance, which is what you have demonstrated, it is going to be difficult to dissociate that possibility.

A. Eastman

I would agree with that. These events were occurring concurrently. It is very hard to discriminate really which came first.

P. Nicotera

I have a couple of questions and one comment. The first question is: In one of your early slides, you showed the effect of EGTA but I did not see the concentrations used in that experiment. I think it was the experiment where you added ionomycin to the cells. I believe you had EGTA present on one lane and you could still see fragmentation. Which were the conditions for that experiment?

A. Eastman

The conditions were 20 μM ionomycin and 5 mM EGTA.

P. Nicotera

And did you include it from the beginning? It was a pre-incubation with EGTA?

A. Eastman

The EGTA and ionomycin were added concurrently, 4h before assessing DNA digestion and intracellular pH.

P. Nicotera

In that condition usually you deplete completely all intracellular Ca^{2+} pools. EGTA works as a sink for calcium. So, in other words, you do not only buffer your extracellular calcium but you lose your calcium from the intracellular stores. You may expect, in fact, to have very little calcium left in the cell. Which ion does exchange with protons at that point?

A. Eastman

Ionomycin, in the absence of calcium, actually acts as a proton ionophore.

P. Nicotera

Yes it does, as showed by the pH measurements. However, it does so by exchanging with a counter ion. It does not specifically open channels for protons, it promotes exchange with other cations. Should it work as a simple proton transporter, it would follow a pH gradient. You should have acidic pH in the incubation medium to be able to drive protons inside.

A. Eastman

I see what you are saying. It could exchange with magnesium.

P. Nicotera

In that case you would not expect to have endonuclease activity because you would have very little magnesium in the cell and these enzymes usually require magnesium or calcium.

A. Eastman

No, our endonuclease does not require magnesium at all.

P. Nicotera

The other point was: Have you tried amiloride to be able to block the sodium proton exchange and rescue cells from apoptosis?

A. Eastman

We have been trying that sort of experiment. We have put cells on EIPA (an amiloride analog). This can drive the pH down and thereby drive apoptosis. It would simply mimic the effect because it would effectively just turn off the Na/H exchanger.

P. Nicotera

Did you try other topoisomerase inhibitors? Did you check whether etoposide works as an ATPase inhibitor blocking proton pumps, thereby accumulating protons?

A. Eastman

No, I have not checked whether etoposide is an ATPase inhibitor but, unlike novobiocin, it does not interact with the ATP binding site so probably is not an ATPase inhibitor.

G. Bellomo

Is the decrease in intracellular pH high enough to trigger DNA fragmentation?

A. Eastman

In the HL60 cells, clearly yes.

G. Bellomo

There are some similarities of this model to other models of cell killing by necrosis, not by apoptosis, in which the cell death is preceded by intracellular acidification and in this condition the best way to protect against cell killing is to decrease the extracellular pH, so to acidify also the extracellular medium. Is it possible to protect against DNA fragmentation in your conditions by changing the pH in the extracellular medium?

A. Eastman

I would hope it is. Such experiments are currently underway. If we were able to maintain a cell at an extracellular pH of 8.5, then selected loss of H exchange should, according to the Nernst equation, only facilitate an intracellular shift to pH 7.5, and therefore DNA digestion should not occur. We are currently testing this hypothesis.

G. Bellomo

I think it would be interesting because it could dissociate apoptosis from what is happening with necrosis in which, if you increase the extracellular pH, you actually enhance cell killing.

P. Nicotera

In our experiments we have used isolated nuclei and incubated them at different pH levels in the presence or absence of calcium. In permeabilized nuclei at pH 6.4, even if we have one millimolar calcium, we do not see any DNA fragmentation. We have attributed that to the state of chromatin condensation. In fact, if we pre-treat the nuclei at the same pH with heparin to unfold the chromatin, we recover back the activity. Now under normal conditions, measurements performed by Lechene have shown that the nucleus is fairly acidic. This is actually common knowledge. Nuclei take up acridine orange. So would you assume that under those conditions your endonuclease would be active, inactive, or would you suggest that there is a repressor, or is it the chromatin structure which plays a role in that case?

A. Eastman

Our current methods only allow us to measure the cytosolic pH. We hope to be able to measure nuclear pH soon. It has been reported that there are proton pumps which drive nuclear pH down, so we are guessing that the nucleus will be more acidic.

P. Nicotera

So would you assume that under resting conditions your enzyme is normally active and therefore you need some repressor or some conditions in the chromatin to prevent cleavage?

A. Eastman

We are getting some very preliminary data which suggests that the drop in pH may not be sufficient. In this regard, the CHO cells may be unusual in that they have an active DNase II all the time. We started with CHO cells because our initial screen showed they had lots of this endonuclease. We purified it from nuclei originally. We found over 70% of the cellular enzyme in the nucleus. But in other cells, we have not seen as much DNase II. We have some suggestion from one model, the CTLL-2 cells, that another step may be required. When we incubated with a proton ionophore at pH 5, we saw little DNA digestion. However, on removal of the necessary growth factor, the intracellular pH still went acidic and the cells then digested their DNA. This suggests there may be an activation step as well as an acidification step that is required.

R. Schimke

You have used CHO cells and HL60 cells and find that, at reasonable cisplatinum concentrations, these aberrant mitotic events are seen. We have looked at a number of cell types, again not with cisplatinum, but predominantly with colcemid and find, under the conditions that Steve Sherwood talked about, at low concentrations of colcemid that both CHO and HeLa will undergo these aberrant polar mitotic events but surprisingly enough there are some cell lines that do not undergo that process. We think it may well be that from the standpoint of cytotoxicity, a mechanism of self killing, that is more frequent than you would think, is responsible for these aberrant mitoses. There are a number of instances described in the literature where cells will die but take a long time to do so, as if they were going through several or at least one extra round of mitosis, and I think you have written about that. So if this is a more general mechanism of cell killing than had been thought, namely, that the initial event is not a prolonging inhibition to the point of apoptosis but essentially a partial inhibition, either in time or in concentration to produce these effects, then one possible mechanism of resistance would be the generation of cells that will not undergo these multipolar mitoses. They just sit there until they are going to undergo a mitosis and then they are back into the normal state. And so I wonder how many cell lines have you looked at? I realize, just as you do, that nobody wants to look at every cell line under the sun, but I wonder if you have had a chance to look at this and see whether there are some cell lines that are either inherently more resistant to these low levels of cisplatinum, or otherwise do not undergo these multipolar mitosis events?

A. Eastman

We feel like we have looked at 100 steps in one cell line rather than looking at a whole series of different cell lines. But you bring up one very important point. CHO cells suffer from aneuploidy and may be much more prone to die in the first mitosis. A lot of other cells tend to take a couple of rounds before they die, perhaps because they segregate an odd chromosome and they can survive without it until they segregate another one wrongly the next time, which I think is what you're alluding to. No, we have not looked at any other cell lines.

R. Schimke

This is really one of the questions that I think was asked five years ago, and triggered, in part, this particular meeting because I thought it was very important to know the intimate details of how cells were dying. If you do not understand the intimate details you really do not understand necessarily what sort of response the cells have that are preventing this. And these are obviously under conditions where the drug is functioning.

D. Delia

Along these lines I would like to report that a group working in our Institute who has been looking at the effects of cisplatinum on a variety of human tumor cell lines by looking at DNA fragmentation and ultrastructural analysis, have been unable to demonstrate apoptosis in a wide number of cell lines. My question is not this one though. I want to get a correct idea of the protective effects of TPA on apoptosis. Nicotera yesterday reported that the protective effects may be ascribed to upregulated levels of spermine. Today you are talking about the unchanged levels, or possibly that TPA may affect potassium levels or pH levels. So I would like to understand more clearly which of these phenomena is more relevant to apoptosis. Do you have an idea? We do observe in a variety of cell lines similar protective phenomena by TPA.

A. Eastman

This is why I raised the same question yesterday. TPA does many, many things. It activates PKC and there are a hundred or more events that are reported to occur downstream, phosphorylations, inductions, etc. We have no idea which events are most important. We have our biases. I obviously like the phosphorylation of the Na/H antiporter which is mediated by some other kinase activated by PKC. There are also non-PKC mediated ways to phosphorylate the antiporter. Anywhere, there are many things that PKC can do.

D. Delia

Can you overcome the TPA protective effects by drugs which lower pH?

A. Eastman

Yes.

D. Delia

You do? After TPA treatment? If you pretreat cells with TPA and then put drugs which lower pH, do you actually get apoptosis?

A. Eastman

Yes, my experiment would be rather crude because I would incubate cells with a proton ionophore and a pH of 5.5. The pH load would be far too great for TPA to protect the cells.

R. Schimke

Are you going to bite the bullet and look in individual cells where you have introduced a DNase-II antibody? Or otherwise make an antisense construct to wipe out the DNase-II and find out what happens?

A. Eastman

We are proposing definitive experiments to address these issues. We have already spent a couple of years trying to clone the gene. We obtained a lot of amino acid sequence and we have been using PCR to try to isolate DNA sequences. We have not yet obtained the appropriate bands, so we are making more primers. Clearly the way to attack these questions is to knock out the endonuclease and see what happens. The antibody we raised to DNase II was very poor. Although it did recognize the hamster proteins, it was far too weak for any microinjection experiments. So we are considering genetic approaches to delete the gene. I think this will give us the definitive answers we need.

INDEX